2007
PHYSICS EDUCATION
RESEARCH
CONFERENCE

D1445077

Proceedings in the Series of
Physics Education Research Conferences

Year	Title	Publisher	ISBN
2007	2007 Physics Education Research Conference	AIP Conf. Proceedings Vol. 951	978-0-7354-0465-6
2006	2006 Physics Education Research Conference	AIP Conf. Proceedings Vol. 883	978-0-7354-0383-3
2005	2005 Physics Education Research Conference	AIP Conf. Proceedings Vol. 818	0-7354-0311-2
2004	2004 Physics Education Research Conference	AIP Conf. Proceedings Vol. 790	0-7354-0281-7
2003	2003 Physics Education Research Conference	AIP Conf. Proceedings Vol. 720	0-7354-0200-0

To learn more about this title, or the AIP Conference Proceedings Series,
please visit the webpage **http://proceedings.aip.org**

2007
PHYSICS EDUCATION RESEARCH CONFERENCE

Greensboro, North Carolina   1 – 2 August 2007

EDITORS

Leon Hsu
University of Minnesota, Minneapolis, Minnesota

Charles Henderson
Western Michigan University
Kalamazoo, Michigan

Laura McCullough
University of Wisconsin-Stout, Menomonie, Wisconsin

All papers have been peer-reviewed

SPONSORING ORGANIZATION
American Association of Physics Teachers (AAPT)

Melville, New York, 2007
AIP CONFERENCE PROCEEDINGS ■ VOLUME 951

Editors:

Leon Hsu
Department of Postsecondary Teaching and Learning
University of Minnesota
128 Pleasant Street, S.E.
Minneapolis, MN 55455
USA

E-mail: lhsu@umn.edu

Charles Henderson
Physics Department
Western Michigan University
1903 West Michigan Avenue
Kalamazoo, MI 49008
USA

E-mail: Charles.Henderson@wmich.edu

Laura McCullough
Physics Department
University of Wisconsin-Stout
P.O. Box 790
Menomonie, WI 54751
USA

E-mail: mcculloughl@uwstout.edu

Authorization to photocopy items for internal or personal use, beyond the free copying permitted under the 1978 U.S. Copyright Law (see statement below), is granted by the American Institute of Physics for users registered with the Copyright Clearance Center (CCC) Transactional Reporting Service, provided that the base fee of $23.00 per copy is paid directly to CCC, 222 Rosewood Drive, Danvers, MA 01923, USA. For those organizations that have been granted a photocopy license by CCC, a separate system of payment has been arranged. The fee code for users of the Transactional Reporting Services is: ISBN/978-0-7354-0465-6/07/$23.00.

© 2007 American Institute of Physics

Permission is granted to quote from the AIP Conference Proceedings with the customary acknowledgment of the source. Republication of an article or portions thereof (e.g., extensive excerpts, figures, tables, etc.) in original form or in translation, as well as other types of reuse (e.g., in course packs) require formal permission from AIP and may be subject to fees. As a courtesy, the author of the original proceedings article should be informed of any request for republication/reuse. Permission may be obtained online using Rightslink. Locate the article online at http://proceedings.aip.org, then simply click on the Rightslink icon/"Permission for Reuse" link found in the article abstract. You may also address requests to: AIP Office of Rights and Permissions, Suite 1NO1, 2 Huntington Quadrangle, Melville, NY 11747-4502, USA; Fax: 516-576-2450; Tel.: 516-576-2268; E-mail: rights@aip.org.

L.C. Catalog Card No. 2007938312
ISBN 978-0-7354-0465-6
ISSN 0094-243X
Printed in the United States of America

CONTENTS

INVITED PAPERS

PEER-REVIEWED PAPERS

PREFACE

The theme of the Physics Education Research Conference in August 2007 was "Cognitive Science and Physics Education Research." The invited papers in this volume and several of the contributed papers represent the intersections and interstices of these fields. The Editors thank the organizers of the conference, Steve Kanim, Michael Loverude, and Chandralekha Singh, for their work in producing a successful and exciting meeting.

The PERC Proceedings continue to grow and the process continues to improve and become more efficient. The work of the referees in reviewing papers in a short time frame makes the Proceedings possible. As always, we thank the referees for their efforts: Dodie Alber, Brad Ambrose, Bijaya Aryal, Gordon J. Aubrecht II, Jonte Bernhard, Tom Bing, Katrina Black, Scott Bonham, Jennifer Blue, Eric Brewe, David Brookes, Alice Churukian, Luke Conlin, Edgar Corpuz, Karen Cummings, Melissa Dancy, Cynthia D'Angelo, Chuck De Leone, Dedra Demaree, Karim Diff, Paula Engelhardt, Eugenia Etkina, Adam Feil, Noah Finkelstein, Tom Foster, Gary Gladding, Ayush Gupta, Richard Hake, Kastro Hamed, Danielle Harlow, Kathleen Harper, Andrew Heckler, Charles Henderson, Paula Heron, Leon Hsu, Andy Johnson, Kyesam Jung, Spartak Kalita, Shulamit Kapon, Anna Karelina, Christopher Keller, Eunsun Kim, Patrick Kohl, Lauren Kost, Gyoungho Lee, Yu-fen Lee, Laura Lising, David Maloney, Jeff Marx, Fran Mateycik, Dyan McBride, Kate McCann, Laura McCullough, Jose Mestre, Donald Mountcastle, Sahana Murthy, Valerie Otero, Stephen Pellathy, Noah Podolefsky, Evan Pollock, Steve Pollock, Ed Price, Susan Ramlo, N. Sanjay Rebello, Joe Redish, David Rosengrant, Maria Ruibal-Villasenor, Mel Sabella, Jeff Saul, Antti Savinainen, Thomas Scaife, David Schuster, Michael Scott, Kim Shaw, Vazgen Shekoyan, Chandralekha Singh, John Thompson, Eugene Torigoe, Chandra Turpen, Joel Van Deventer, Alan van Heuvelen, Michael Wittmann, Dean Zollman, and Xueli Zou.

Over its seven years in print, the Proceedings has gone from a young publication working through its growing pains to a mature venue for publishing PER results and I am proud of the part I have played in that ongoing process. But there is always room for improvement and every year the editors work to streamline the submission and review process. I hope and expect the PERC Proceedings will continue to serve the PER community for many years, and I know that I leave the process in good hands as I end my time as editor.

Laura McCullough
Outgoing Editor

PROGRAM

2007 PHYSICS EDUCATION RESEARCH CONFERENCE
GREENSBORO, NORTH CAROLINA

WEDNESDAY, August 1, 2007

When	What
	Bridging Session: Invited Talks & Panel Discussion
3:30 - 5:30	**Making Physics Learning Inviting - A View from Cognitive Science** *Janet Kolodner,* Georgia Tech
	Problem Solving and Learning for Physics Education *Brian Ross,* University of Illinois at Urbana-Champaign
	Naive Physics/Savvy Science: Causal Learning in Very Young Children ... and the Rest of Us *Laura Schulz,* MIT
6:00 - 8:00	**Dinner Banquet** *Presider: Steve Kanim* *Banquet Speaker: Art Kramer,* University of Illinois at Urbana Champaign **Cognitive Neuroscience Explorations of Cognitive Plasticity & Human Performance**
8:00 -10:00	**Contributed Poster Session**

THURSDAY, August 2, 2007

When	What
	Workshops , Targeted Poster Sessions & Roundtable Discussions- I
8:15 - 9:45 (Parallel Sessions)	**Workshop A:** **Cognitive Analysis of Student Learning Using LearnLab** *Brett van de Sande and Kurt VanLehn,* University of Pittsburgh
	Workshop B: **Physics Learning in the Context of Scaffolded Diagnostic Tasks** *Edit Yerushalmi & Bat Sheva Eylon,* Weizmann Institute of Science *Chandralekha Singh,* University of Pittsburgh
	Targeted Poster Session TP-A: **Experimental Paradigms from Cognitive Science to Learn About Learning** *Jose Mestre,* University of Illinois, Urbana-Champaign
	Roundtable Discussion A: **What We Can Learn from Neuroscience from Encoding to n-Coding** *Nathaniel Lasry,* Harvard University *Mark Aulls,* McGill University
9:45 - 10:00	**Break**
10:00 - 12:00	**Invited Talks & Panel Discussion (Session II)**
	Cognitive Science: The Science of the (Nearly) Obvious *Bruce Sherin,* Northwestern University
	Facilitating Conceptual Learning Through Analogy and Explanation *Timothy Nokes,* University of Pittsburgh
	Socializing Learning and Transfer *Dan Schwartz,* Stanford University

12:00 - 1:30	**Luncheon Banquet** Presider: *Steve Kanim*
	Workshops , Targeted Poster Sessions & Roundtable Discussions- II
1:45 - 3:15 (Parallel Sessions)	**Workshop B:** **Physics Learning in the Context of Scaffolded Diagnostic Tasks** *Edit Yerushalmi & Bat Sheva Eylon*, Weizmann Institute of Science *Chandralekha Singh*, University of Pittsburgh
	Workshop C: **Publishing & Refereeing Papers in PER** *Leon Hsu*, University of Minnesota *Robert Beichner*, North Carolina State University *Karen Cummings*, Southern Connecticut State University *Janet Kolodner*, Georgia Tech *Laura McCullough*, University of Wisconsin, Stout
	Targeted Poster Session TP-A: **Experimental Paradigms from Cognitive Science to Learn About Learning** *Jose Mestre*, University of Illinois, Urbana-Champaign
	Targeted Poster Session TP-B: **A Conversation About Models, Modeling, Representations and Cognitive Science** *Brant Hinrichs*, Drury University *Eric Brewe*, Florida International University
	Roundtable Discussion C: **Experiential Learning in Physics Courses** *Tetyana Antimirova*, Ryerson University, Toronto, Ontario
3:15 - 3:30	**Break**
	Workshops , Targeted Poster Sessions & Roundtable Discussions- III
3:45 - 5:15 (Parallel Sessions)	**Workshop A:** **Cognitive Analysis of Student Learning Using LearnLab** *Brett van de Sande and Kurt VanLehn*, University of Pittsburgh
	Workshop C: **Publishing & Refereeing Papers in PER** *Leon Hsu*, University of Minnesota *Robert Beichner*, North Carolina State University *Karen Cummings*, Southern Connecticut State University *Janet Kolodner*, Georgia Tech *Laura McCullough*, University of Wisconsin, Stout
	Roundtable Discussion B: **Student Views of Learning in a First Semester College Physics Course: A Study** **Using Q Methodology** *Susan Ramlo*, The University of Akron
	Targeted Poster Session TP-B: **A Conversation About Models, Modeling, Representations and Cognitive Science** *Brant Hinrichs*, Drury University *Eric Brewe*, Florida International University

POSTER TITLES AND AUTHORS

Physics Problem Solving Component Skills and Evaluation
Adams, Wendy (wendy.adams@colorado.edu) University of Colorado at Boulder
Wieman, Carl (cwieman@exchnage.ubc.ca) University of British Columbia

College Students' Responses to Inquiry-based Group Work in a Reformed Pedagogy Classroom
Alber, Delores (dodiealber@yahoo.com) Southern Illinois University Edwardsville
Puchner, Laurel (lpuchne@siue.edu) Southern Illinois University Edwardsville
Lindell, Rebecca (rlindel@siue.edu) Southern Illinois University Edwardsville

New Dimensions to Probing Student Thinking about Oscillations in Two Dimensions
Ambrose, Bradley (ambroseb@gvsu.edu) Grand Valley State University

Investigating Peer Scaffolding in Learning and Transfer of Learning Using Teaching Interviews*
Aryal, Bijaya (bijaya@phys.ksu.edu) Kansas State University
Zollman, Dean (dzollman@phys.ksu.edu) Kansas State University

Comparison of Student Perceptions of Three Different Physics by Inquiry Classes
Aubrecht, Gordon (aubrecht@mps.ohio-state.edu) Dept. of Physics, Ohio State University

Humans, Intentionality, Experience and Tools for Learning: Some Contributions from Post-Cognitive Theories to the Use of Technology in Physics Education.
Bernhard, Jonte (jonbe@itn.liu.se) Engineering Ed. Res. Group, ITN, Campus Norrköping, Linköping University

Scaffolded Reflection: A Chemist's View of Conceptual Change
Bhattacharyya, Gautam (gautamb@clemson.edu) Department of Chemistry, Clemson University

Students' Difficulties with Concepts Related to Conductors and Insulators
Bilak, Joshua (jdbilak@gmail.com) University of Pittsburgh
Singh, Chandralekha (clsingh@pitt.edu) University of Pittsburgh

Different Types of Mathematical Justification in Upper Level Physics
Bing, Thomas J. (tbing@physics.umd.edu) University of Maryland
Gupta, Ayush (ayush@glue.umd.edu) University of Maryland
Redish, Edward F. (redish@umd.edu) University of Maryland

Mapping Student Reasoning about Math- and Physics-oriented Differential Equation Solutions
Black, Katrina E. (katrina.black@umit.maine.edu) University of Maine Department of Physics and Astronomy
Wittmann, Michael C. (wittmann@umit.maine.edu) University of Maine Department of Physics and Astronomy

Measuring Student Effort and Engagement in an Introductory Physics Course
Bonham, Scott (scott.bonham@wku.edu) Western Kentucky University

Voltage is the Most Difficult Subject for Students in Physics by Inquiry's Electric Circuits Module
Bowman, Carol (bowman.79@osu.edu) The Ohio State University at Marion
Aubrecht, Gordon (aubrecht@mps.ohio-state.edu) The Ohio State University at Marion

Do Introductory Astronomy Texts Promote Higher Order Thinking?
Bracey, Georgia (georgia_bracey@hotmail.com) Southern Illinois University Edwardsville
Lindell, Rebecca (rlindel@siue.edu) Southern Illinois University Edwardsville

Improving the Spread of PER-based Instructional Approaches: A Case Study of Dissemination within the Modeling Physics Project
Brewe, Eric (eric.brewe@gmail.com) Hawaii Pacific University
Dancy, Melissa (mhdancy@uncc.edu) University of North Carolina at Charlotte
Henderson, Charles (Charles.Henderson@wmich.edu) Western Michigan University

The Dynamics of Students' Behaviors and Reasoning during Collaborative Physics Tutorial Sessions

Conlin, Luke D. (luke.conlin@gmail.com) University of Maryland, College Park
Gupta, Ayush (ayush_umd@yahoo.com) University of Maryland, College Park
Scherr, Rachel E. (rescherr@gmail.com) University of Maryland, College Park
Hammer, David (davidham@umd.edu) University of Maryland, College Park

The Use of Hands-On and Minds-On Modeling Activities in Improving Students' Understanding of Microscopic Friction

Corpuz, Edgar (ecorpuz@utpa.edu) Department of Geology & Physics, University of Texas -- Pan American
Rebello, N. Sanjay (srebello@phys.ksu.edu) Physics Department, Kansas State University

Problem Solving Behaviors of Math and Science Teachers: Striving for an Answer or for Understanding

D'Angelo, Cynthia (cynthia.dangelo@asu.edu) Arizona State University

From FCI to CSEM to Lawson Test: A Report on Data Collected at a Community College.

Diff, Karim (karim.diff@sfcc.edu) Santa Fe Community College, Gainesville FL
Tache, Nacira (nacira.tache@sfcc.edu) Santa Fe Community College, Gainesville FL

The Effectiveness of a Physical Science by Inquiry Program for K-12 Teachers*

Endorf, Robert (Robert.Endorf@UC.edu) University of Cincinnati
Koenig, Kathleen (kathy.koenig@wright.edu) Wright State University

Spending Time on Design: Does it Hurt Physics Learning?

Etkina, Eugenia (etkina@rci.rutgers.edu) Rutgers University, Graduate School of Education
Van Heuvelen, Alan (alanvan@physics.rutgers.edu) Rutgers, University, Department of Physics and Astronomy
Karelina, Anna (anna.karelina@gmail.com) Rutgers University, Graduate School of Education
Ruibal-Villasenor, Maria (ruibal_villasenhor@yahoo.com) Rutgers University, Graduate School of Education

Design and Non-design Labs: Does Transfer Occur?

Karelina, Anna (anna.karelina@gmail.com) Rutgers University, Graduate School of Education
Etkina, Eugenia (etkina@rci.rutgers.edu) Rutgers University, Graduate School of Education
Ruibal-Villasenor, Maria (ruibal_villasenhor@yahoo.com) Rutgers University, Graduate School of Education
Rosengrant, David (rosengra@eden.rutgers.edu) Rutgers University, Graduate School of Education

From Physics to Biology: Helping Students Attain All-terrain Knowledge

Ruibal-Villasenor, Maria (ruibal_vilasenhor@yahoo.com) Rutgers University, Graduate School of Education
Etkina, Eugenia (etkina@rci.rutgers.edu) Rutgers University, Graduate School of Education
Karelina, Anna (anna.karelina@gmail.com) Rutgers University, Graduate School of Education
Rosengrant, David (rosengra@eden.rutgers.edu) Rutgers University, Graduate school of Education

Visual Physics: A Case Correlation for Introductory Calculus-based Physics Re-Design

Ezrailson, Cathy (cezrailson@tamu.edu) Texas A&M University
McIntyre, Peter (p-mcintyre@physics.tamu.edu) Texas A&M University
Kamon, Teruki (kamon@physics.tamu.edu) Texas A&M University
Loving, Cathleen C (cloving@tamu.edu) Texas A&M University

Using a Recognition Memory Test to Measure Expert-Novice Differences in the Encoding of Physics Diagrams.

Feil, Adam (adamfeil@uiuc.edu) University of Illinois
Mestre, Jose (mestre@uiuc.edu) University of Illinois

Using Models of Student Thinking to Predict Variability in Responses to Motion Questions

Frank, Brian (bwfrank@physics.umd.edu) University of Maryland
Kanim, Stephen New Mexico State University
Ortiz, Luanna Arizona State University

SECTION I

Invited Papers

Publishing And Refereeing Papers In Physics Education Research

Leon Hsu[1], Robert Beichner[2], Karen Cummings[3], Janet L. Kolodner[4], and Laura McCullough[5]

[1]Department of Postsecondary Teaching and Learning, University of Minnesota, Minneapolis, MN 55414, USA
[2]Department of Physics, North Carolina State University, Raleigh, NC 27695, USA
[3]Department of Physics, Southern Connecticut State University, New Haven, CT 06515, USA
[4]College of Computing, Georgia Institute of Technology, Atlanta, GA 30332, USA
[5]Department of Physics, University of Wisconsin–Stout, Menomonie, WI 54751, USA

Abstract. At the 2007 Physics Education Research Conference, a workshop on publishing and refereeing was held with a panel of editors from four different publishing venues: the physics education research section of the American Journal of Physics, the Journal of the Learning Sciences, Physical Review Special Topics–Physics Education Research, and the Physics Education Research Conference Proceedings. These editors answered questions from participants regarding publishing in their respective venues, as well as writing referee reports that would be useful to both journal editors and authors. This paper summarizes the discussion.

Keywords: Publishing, Referee reports, Journals, Conference proceedings
PACS: 01.30.-y, 01.30.Cc, 01.30.Rr

INTRODUCTION

The purpose of the workshop on publishing and refereeing at the 2007 Physics Education Research Conference (PERC) was to help participants learn more about four different publishing venues for physics education research, including what types of papers are appropriate for each, what happens in the editorial office once a paper is received, the role of referees in the decision to publish or reject, and how to write a referee report that will be most useful to the editors and authors of a paper.

Table 1 gives a quick comparison of some of the parameters for each publication. For purposes of this paper, the PERC Proceedings will be referred to as a "journal." Similarities and differences between each will be elaborated below.

CHOOSING A JOURNAL

The biggest factor in deciding to which journal an author will submit a paper should be the audience for which the publication is aimed. The descriptions given here should be taken only as general guidelines. There is overlap between the readers of each publication and one should not base a decision on which journal to submit to solely on the description given here.

Papers submitted to the American Journal of Physics (AJP)–Physics Education Research Section (PERS) should be addressed primarily to an audience consisting of consumers of physics education research (PER), *i.e.*, those who are interested in reading about it and using it, rather than those who are conducting the research. Thus, the papers should have some relevance, whether direct or indirect, to classroom practice. One should also keep in mind that certain types of PER articles can be published in the non-PER section of AJP. However, those articles should be aimed at an even wider audience, as AJP is received by many physicists, only a fraction of whom are interested in teaching and an even smaller fraction of whom are interested in research on learning and teaching. If you wish the referees of your paper to be PER-savvy, then you should indicate specifically when submitting that you wish your paper to be considered for the PER section of the journal.

The Journal of the Learning Sciences (JLS) is primarily interested in articles that address learning in real-life (non-laboratory) situations. Articles describing a research study should discuss what learning happened, what was done to make it happen,

CP951, *2007 Physics Education Research Conference*, edited by L. Hsu, C. Henderson, and L. McCullough
© 2007 American Institute of Physics 978-0-7354-0465-6/07/$23.00

TABLE 1. A comparison of some PER publishing venues.

	American Journal of Physics–Physics Education Research Section	Journal of the Learning Sciences	Physical Review Special Topics–Physics Education Research	Physics Education Research Conference Proceedings
Primary audience	Consumers of PER	Researchers interested in learning in non-laboratory environments	PER researchers	A combination of PER consumers and researchers
Acceptance rate	21%	15-20%	33%	70-80%
Time frame from submission to acceptance	~9 months	~10 months	~4-5 months	~2 months
Length guidelines (journal pages)	Roughly 7 pages	Up to 50 or 60 pages	No set limit	4 pages

and what is the evidence that it happened. JLS is also interested in the impact of environmental factors on learning, especially in the context of today's technological society. As the title of the journal implies, the focus of the research should be on learning, rather than on teaching.

Manuscripts sent to Physical Review Special Topics (PRST) should be aimed primarily at those who conduct research in PER, rather than consumers. Presentations of curriculum development, new teaching techniques, etc. are not appropriate unless there is a strong, explicitly discussed basis in PER. Theoretical papers may be more appropriate for PRST than for AJP. The research described should advance the field in some way. Because PRST is a purely electronic journal, it is particularly suited to papers accompanied by large amounts of data, such as interview transcripts or even video clips, which can be archived on the journal's website. Another advantage of electronic publishing is the rapid turnaround time once a manuscript is accepted, so PRST is also open to short papers describing notable results that are in need of quick publication.

The audience for the PERC Proceedings is designed to be a record of the PERC given to all attendees, who are a mix of PER consumers and researchers. The Proceedings provides a snapshot of the field and as such is open to preliminary results and research in progress, as well as papers that would simply be thought-provoking to the PER community.

In addition to the audience, other resources an author can consult to help decide on a journal include the mission statements and guidelines for referees often found on a journal's website. The references cited in the paper to be submitted, in addition to the literature and conceptual foundations that the research draws upon can also be a useful guide.

One concern that many authors have is the page charges for publishing in PRST. Beichner, the current editor, points out that both the American Association of Physics Teachers and the American Physical Society have established a fund to help defray these charges for authors in need and this fund has not yet been depleted. Thus, page charges should not discourage authors from submitting to PRST.

Finally, authors should remember that they must, in the end, choose only one journal. It is unethical to have the same manuscript be considered for publication by more than one journal at a time. If preliminary results have already been published in the PERC proceedings, more research or analysis must be added in order for the same study to be submitted to a journal such as PRST, *i.e.*, there must be some "value added" by the additional publication.

THE EDITORIAL PROCESS

Once a manuscript is received by the editor of a journal, the process for each of the four publications described here is similar.

At PRST, Beichner will skim the submitted paper and decide on two referees for the paper based on expertise and previous referee assignments. A graduate student may occasionally be asked to serve as a third referee. The referees are asked to complete their review within three weeks. Depending on their reports, the paper may be accepted, rejected, or returned to the author for revision and resubmission. Thus far, the acceptance rate for PRST has been 33%, though this number is skewed downward by the submission of inappropriate papers (such as those discussing parity violation or special relativity). Of the papers that have been accepted, the median time between submission and acceptance has been 133 days.

At the AJP-PERS, the acceptance rate has been 21%, although this number too, is skewed downward by the submission of inappropriate papers. A paper submitted to AJP-PERS will sometimes be returned to the author because of length concerns. In the past, the typical AJP article has been approximately seven journal pages. If a submission is significantly longer than this, *e.g.*, more than 10 pages, then its likelihood of being returned without review increases. One quick method suggested by Cummings, the current editor, to estimate the journal page length of an article is to take the number of manuscript pages (11-point font, double spaced) including figures and references, and divide by three. Although there is no length limitation for papers published in PRST, long papers are less likely to be read and also will take more time to be reviewed by referees. The most frequent complaint by PRST referees is excessive length.

At JLS, Kolodner, the current editor will typically send a submission to an "action editor" in the appropriate field, who will skim the paper and choose referees. At this time, the likely recipient of PER-focused papers is David Hammer. Each submission is sent to three referees, two in the paper's specific area and one that is somewhat outside to insure that the paper is accessible to non-specialists. Each referee is given six weeks to complete a review before he or she is sent a reminder. When the referee reports are returned, the action editor will read them carefully in addition to the paper itself, then write a detailed decision letter. This letter is sent to the main editor (Kolodner) and then to the author. The most common outcome of the first round of review is a recommendation of "revise and resubmit." Each round of review is about four months long and a paper that is published typically goes through three rounds. The last round is usually quite short, consisting only of a reading by the editor.

Papers published in JLS have been as long as 60 journal pages, allowing the author to include enough data to provide evidence for the conclusions drawn. However, there is a limit beyond which submissions will be returned without review and the usual caveats about fewer people reading very long articles apply.

Although historically, most PER papers have been published in either in AJP or PRST, Kolodner encourages authors to send appropriate manuscripts to JLS not only to help researchers outside of PER become aware of PER work and the PER community, but also to spur PER workers to begin reading journals outside of just AJP and PRST in order to become aware of other research in learning being conducted outside of the context of physics.

Because of the short time required to process papers and to send them to AIP for publication, the editorial process for the PERC proceedings differs significantly from the other three journals. Each submitted paper is sent to three referees, chosen from the pool of authors and other members of the PER community who volunteer to serve as reviewers. In choosing referees, the editors aim for a balance between researchers in a paper's field and those slightly outside of that field in addition to a balance of more and less experienced PER researchers.

The majority of papers receive recommendations of "publish" or "publish with minor modifications" from all three referees. If one or more referees recommends that a paper not be published, then at least one (and in most cases, all three) of the editors will read the paper carefully to judge (1) whether the objections raised by the referee are sufficient to prevent the paper from being published and (2) whether the objections can be overcome with minor modifications. Papers that would require major changes to be made publishable cannot be accepted because there is no time for a second round of reviews. Authors of accepted papers then have a chance to make modifications suggested by the referees before publication.

Although preliminary research results or even descriptions of experiments that have not yet produced results are welcome in the PERC Proceedings, authors should take care to state the limitations of their conclusions. The four-page limit means that authors must often restrict themselves to describing only a few parts of their work. Also, research in which claims are made based on student quotes can be difficult to publish because of the very limited amount of evidence that can be presented.

THE ROLE OF REFEREES

One important fact to keep in mind is that where AJP, PRST, and JLS are concerned, referee reports are not votes on whether or not a paper should be published. It is possible for a paper to receive two favorable referee reports, yet for the paper not to be accepted, or at least not accepted right away. Conversely, it possible for a paper to receive two negative referee reports, yet not be rejected. The editor uses the referee reports, along with her own reading of the paper and knowledge of the journal's mission and audience to make a decision.

For the PERC Proceedings, the referee reports take on a more important role because quick deadlines prevent the editors from reading every paper. Because of the wide range of experience and knowledge of the referees, the editors use the referee reports as votes in deciding whether to accept or reject a paper, weighted by the expertise and experience of the referees. The particular comments, criticisms, and suggestions made by the referees also play an important role in the decision. Each year, 15-20% of the papers (approximately 10) have mixed referee reports that require the editors to read the papers themselves and come to a consensus on the decision.

There are two primary audiences for which a referee report should be written: the author(s) of the paper and the editor. For the editor, the referee's job is to help her make an informed decision about the paper by looking at the science and how it fits into the big picture. Are the research questions interesting? Do they move the work of the journal forward? Are the claims stated clearly and backed by evidence? Is the methodology appropriate to the research questions? Is additional data or analysis needed and if that additional data cannot be gathered, is the paper still publishable if the authors soften the claims?

For the author(s), the function of the referee report is to help them make the best possible presentation of their work and also perhaps to help guide their research. The referee should think about whether the paper is written in such a way as to allow the readers of the journal to understand and appreciate the significance of the work. Both praise and suggestions for improvement should be included. One good way to decide what should be included in the report is for the referee to think about what kind of help or suggestions she would want if she were the author of the paper.

On the other hand, the referee should not spend much, if any, time performing the functions of a copy editor. Grammatical or typographical mistakes can be ignored. Also, the referee should bear in mind that writing styles are personal and should not try to enforce her writing style on the paper's author(s). It is not the referee's paper! Along these lines, the referee report should not be a wish list of items that the referee would have included in the paper unless those items are crucial to supporting the message of the work. Referees for the PERC Proceedings, however, should pay attention to the formatting of submitted manuscripts because it is possible that they are the only ones who will check to make sure that the paper follows all the stylistic elements required for AIP publication. The formatting of the references is a place where errors are commonly found.

WRITING A GOOD REFEREE REPORT

One of the most important qualities of a good referee report is that it be completed on time. Nothing is more frustrating for a journal editor than for a referee to sit on a manuscript for weeks or months, and then to say that he or she cannot do it. Not only must the editor then find an alternate referee, the alternate referee must also be given some reasonable amount of time to review the paper, lengthening the total time in which the paper is in process. If a referee is really swamped with work, it is much better to let the editor know right away that the review cannot be completed.

There are many guides to writing good referee reports that can be found on the web (see the following section on additional resources). However, two general guidelines are the following. (1) Summarize the paper briefly. This lets the author and editor know that the

referee actually read the paper and whether or not there are any major misunderstandings in the reading. (2) To paraphrase a quote by Einstein, a report should be as short as possible to get the message across, but no shorter. Use bullet points to make the report easy to read and to respond to.

A referee who does not think that she is sufficiently competent in a particular topic to be a reviewer should keep in mind that editors will often send papers to people who are somewhat outside of the field discussed in the paper to insure that the paper is accessible to a broad audience. In many cases, it is allowable for a referee to show a manuscript to a knowledge colleague for help with writing a report. However, papers under review should never be disseminated widely. If in doubt, a referee should always feel free to contact the journal editor for clarification.

Finally, referees should remember that authors are human beings and so they should write reports that they themselves would be happy to receive. Regardless of how bad a paper is, one can always find something good to say about it. Editors will never reveal the identities of the referees to authors without their permission. However, it is sometimes possible to guess who a referees is. A referee should never write a report that she would be embarrassed by if it became known that she was the author.

ADDITIONAL RESOURCES

Websites for the publications discussed here are:
AJP: http://scitation.aip.org/ajp/
JLS: http://www.cc.gatech.edu/lst/jls/
PRST: http://prst-per.aps.org/
PERC Proceedings: http://web.phys.ksu.edu/perc2007

A web article with good general advice on writing referee reports can be found at http://www.psychologicalscience.org/observer/getArticle.cfm?id=2157. An article with advice on the actual writing of a report is located at http://people.bu.edu/rking/JME_files/guide_for_referees.htm. Although these articles are not specific to the journals discussed here, much of their advice can be applied directly.

A useful article for authors on dealing with rejection and revising papers is http://www.roie.org/howj.htm.

ACKNOWLEDGMENTS

The authors would like to thank the organizers of the 2007 PERC for helping to make the workshop on which this paper is based a success.

Facilitating Conceptual Learning Through Analogy And Explanation

Timothy J. Nokes* and Brian H. Ross[†]

*Learning Research and Development Center and Department of Psychology
University of Pittsburgh, Pittsburgh, PA 15260, USA

[†]Beckman Institute and Department of Psychology, University of Illinois, Urbana, IL 61801, USA

Abstract. Research in cognitive science has shown that students typically have a difficult time acquiring deep conceptual understanding in domains like mathematics and physics and often rely on textbook examples to solve new problems. The use of prior examples facilitates learning, but the advantage is often limited to very similar problems. One reason students rely so heavily on using prior examples is that they lack a deep understanding for how the principles are instantiated in the examples. We review and present research aimed at helping students learn the relations between principles and examples through generating explanations and making analogies.

Keywords: Analogy, cognitive science, conceptual learning, explanation, problem solving
PACS: 01.40-d; 01.40Fk; 01.40gb; 01.40Ha

INTRODUCTION

How can we facilitate students' deep learning of new concepts? One approach to this problem is to examine what knowledge components comprise 'expert understanding' and then design learning environments to help novices construct that knowledge. Research on expertise has shown that experts 'perceive' the deep structure and principles of domain relevant problems, use that knowledge to identify and execute a set of strategies and procedures appropriate for the task, and flexibly transfer their knowledge to new contexts [1-4]. This research suggests that a key component of expert knowledge is their understanding of the relations between the principles and the problem features.

Unlike experts, novices do not work forwards from the domain principles but instead use backwards-working strategies [5] and rely on prior examples to solve new problems [6-8]. Although using examples enables novices to make progress when solving new problems, they are often only able to apply such knowledge to near transfer problems with similar surface features [9]. Furthermore, students prefer to use examples even when they have access to written instructions or principles [10, 11].

One reason that students may rely so heavily on prior examples is that they lack a deep understanding for how the principles are instantiated in the examples.

That is, they lack the knowledge and skills required for relating the principle components to the problem features. Learning activities to help students acquire these relations should improve their conceptual understanding and future problem solving.

Research in cognitive science has shown that two powerful learning activities are generating explanations and making analogical comparisons. In the current work we explore how these two activities can facilitate students learning of the conceptual relations between principles and examples.

EXPLANATION

Much work in cognitive science has shown that explanation can facilitate learning and transfer [12-14]. Self-explanation, or the process of explaining to oneself with the goal of making sense of new information, is positively correlated with learning gains in a variety of domains including biology [15], computer programming [16], mathematics [14, 17], and physics [18] among others. Self-explanation is not merely correlated with greater learning, but laboratory experiments have also documented a causal connection. For example, Chi et al. [15] have shown that students who were prompted to self-explain while reading a biology text generated more inferences and acquired a deeper conceptual understanding than an

CP951, *2007 Physics Education Research Conference*, edited by L. Hsu, C. Henderson, and L. McCullough
© 2007 American Institute of Physics 978-0-7354-0465-6/07/$23.00

unprompted group as measured by pre- and post-test questions.

The two cognitive mechanisms hypothesized to underlie the self-explanation effect are generating inferences and repairing mental models [12]. Generating inferences helps the learner construct causal connections between concepts and develop a more accurate mental model of the principle or problem. Self-explanation also provides an opportunity to notice gaps, misunderstandings, or faulty assumptions in a learners reasoning, giving them an opportunity to revise and correct it.

In the next section we report a study that examined how particular types of explanations can help students learn the relations between principles and examples.

Explanation Experiment: Explaining to Generalize versus Specialize

Math and science instruction often consists of both principles and examples. Since a learner could explain either type of information, we investigated how the *content of the explanation* affects what is learned and to what situations the resulting knowledge transfers. We examined two types of explanation: 1) using principles to explain examples and 2) using examples to explain principles.

We hypothesized that using a principle to explain an example (*specialization explanation*) would result in a specific understanding of the principle embedded in the context of the example being explained. This type of explanation should promote the construction of a mental model of the example and facilitate subsequent performance on problems that use the same principle with similar content. In contrast, using examples to explain a principle (*generalization explanation*) was hypothesized to result in an abstract understanding of the principle. This type of explanation should promote schema construction and facilitate subsequent performance on problems that use the same principle but with different content.

We tested these hypotheses in a laboratory experiment in which students gave either specialization or generalization explanations when learning about two elementary probability principles (permutations and combinations).

Methodology

Forty UIUC students were paid $8 for their participation.

The experiment had a between-subjects design and consisted of a learning and test phase. In the learning phase, participants were randomly assigned to either a specialization or generalization condition. Specialization participants first read a probability principle and a worked example, explained the solution to a second example, and then read the solution to that example. This procedure was then repeated for the second principle. Generalization participants first read the two worked examples, explained the principle, and then read the principle. This procedure was then repeated for the second principle.

In the test phase all participants solved ten probability word problems (5 for each principle). Four of the problems had the same surface content (i.e., same story line and objects but different numbers) and six had different content (i.e., different story line and objects). The participant's task was to first decide which formula applied and then solve the problem.

We predicted that the specialization group would perform better on the same content problems whereas the generalization group would perform better on the different content problems.

Results and Discussion

Table 1 shows the data for learners who successfully completed the learning task (2/3 of the learners). Both groups showed high accuracy for the same content problems. This suggests that both types of explanations facilitated an understanding of the principle that could be applied to problems with the same contents as the learning examples. However, the specialization group had faster decision times than the generalization group (Cohen's *d* effect size of .55) suggesting that their understanding of the principle was more closely tied to the contents of the example and facilitated fast access to the principle for problems with the same content.

TABLE 1. Mean (and se) accuracy and decision times (seconds) for the Specialization and Generalization groups on the same and different content problems.

	Accuracy	Decision-Time
Same Content		
Specialization	.79 (.07)	23.7 (2.2)
Generalization	.79 (.05)	31.5 (2.7)
Different Content		
Specialization	.39 (.05)	40.0 (4.7)
Generalization	.51 (.04)	41.7 (4.9)

For the different content problems the generalization group showed better accuracy than the specialization group (*d* = .93). This result suggests that using the examples to explain the principle resulted in a more abstract understanding of the concept that facilitated performance on problems with contents different from those of the learning problems.

In sum, the *type of explanation* is critical to what is learned and to where that knowledge transfers. Using principles to explain examples resulted in a specific understanding of the principle in the context of the example whereas using examples to explain the principle resulted in a more abstract understanding of the principle. Next we examine a second path to conceptual knowledge by analogical comparisons.

ANALOGY

Analogical comparisons have been hypothesized to facilitate the acquisition of a *problem schema* [19], a knowledge representation of a particular problem category. It includes declarative knowledge of the principle, concepts, and formulae, as well as the procedural knowledge for how to apply that knowledge to solve a problem. Schemas have been hypothesized as the underlying knowledge structures supporting expert performance [1-4].

Analogical comparison operates through aligning and mapping two example problem representations to one another and then extracting their commonalities [19-21]. This process discards the elements of the knowledge representation that do not overlap between two examples but preserves the common elements. The resulting knowledge structure typically consists of fewer superficial aspects than the examples, but retains the deep causal structure of the problems.

Several factors are known to improve schema acquisition including: more examples [19], greater variability of the examples [22, 23], instructions that focus the learner on structural commonalities [24, 25], or subgoals of the problems [26, 27], and examples that minimize students cognitive load [28].

In the current work we were interested in how different types of problem comparisons affect learning and transfer [29].

Analogy Experiment: Near-miss versus Surface-different Comparisons

The purpose of the experiment was to examine the effect of near-miss versus surface-different problem comparisons on learning and later problem solving. Near-miss comparisons involve two problems that have the same content and structure but with one critical surface change that highlights some aspect of the principle structure (e.g., switching the variable-object assignments across the problems). Surface-different comparisons consist of problems in which the same principle applies but each uses different contents.

We hypothesized that near-miss comparisons would focus the learner on how the variables were instantiated in the problem and would benefit performance on tests of principle use (which emphasize assigning objects to variables). In contrast, surface-different comparisons were hypothesized to focus the learner on the fact that multiple contents can be associated with a principle and would benefit performance on tests of principle access (that emphasize choosing the relevant principle). We were also interested in whether poor learners, who generally rely more on content, might show stronger effects.

Methodology

Thirty UIUC students were paid $8 for their participation.

The experiment had a within-subjects design and consisted of a learning and a test phase. In the learning phase participants learned about four probability principles by reading a worked example and then solving either a near-miss or surface-different practice problem. For near-miss comparisons participants solved problems that had the same content as the worked example but with reversed object correspondences (the same objects assigned to different variables). For surface-different comparisons participants solved problems that had different (but analogous) content as the worked example but with reversed object correspondences. Participants learned two principles with near-miss comparisons and two principles with surface-different comparisons.

In the test phase participants solved four use and four access test problems (one for each principle). All of the test problems had new contents and non-obvious object correspondences to that of the learning problems. For use problems participants were asked to assign the values from the problem statement to the correct variables in the given formula. For the access problems the participants were given all four formulae and their task was to choose the correct equation.

Results

Participants were split into two learning groups (good and poor learners) based on a median split of their performance on the practice problems (M = .84, SD = .09 and M = .55; SD = .15 respectively). Table 2 shows the mean performance for each learning group on the use and access tests.

TABLE 2. Mean use and access scores (and se) as a function of learning condition for good and poor learners. Adapted from Nokes & Ross [29].

	Use	Access
Good Learners		
Near-miss	.96 (.02)	.91 (.05)
Surface-different	.93 (.04)	.79 (.06)
Poor Learners		
Near-miss	.90 (.03)	.46 (.09)
Surface-different	.79 (.08)	.58 (.10)

The good learners showed high performance across all of the tests with unexpected lower performance in the surface-different condition on the access test. In contrast, the poor learners showed the predicted effect: the near-miss comparisons improved principle use ($d = .51$) whereas the surface-different comparisons improved principle access ($d = .32$).

These results suggest that near-miss comparisons help poor learners understand how the structure of the problem relates to the variables in the formula whereas surface-different comparisons help them learn how to tell which principles are relevant. This interaction suggests that different types of comparisons may be useful for teaching (poor) students different aspects of the principle knowledge.

CONCLUSIONS

Research in cognitive science shows that there are multiple paths to facilitate conceptual learning. One is through generating explanations and a second is through making analogical comparisons. The research presented in this paper shows that the type of explanation (explaining the principle or example) and the kind of problem comparison (near-miss or surface-different) is critical to what is learned and to where that knowledge transfers. In ongoing work we examine how these activities can improve students' learning of physics in the college classroom (see the Pittsburgh Science of Learning Center for current projects - http://www.learnlab.org).

ACKNOWLEDGMENTS

Preparation for this paper was supported by Grant SBE0354420 from the Pittsburgh Science of Learning Center and Grant R305B070085 from the Institute of Education Sciences. The empirical work was supported by a Beckman Fellowship from the University of Illinois to the first author. We thank David Brookes, Robert Hausmann, Jose Mestre, Kurt VanLehn, and Anders Weinstein for discussions.

REFERENCES

1. W. G. Chase, and Simon, *Cognitive Psychology* 4, 55-81 (1973).
2. M. T. H. Chi, P. Feltovitch, and R. Glaser, *Cognitive Science* 5, 121-152 (1981).
3. J. Larkin, J. McDermott, D. P. Simon, and H. A. Simon, *Science* 208, 1335-1342 (1980).
4. P. J. Feltovich, M. J. Prietula, and K. A. Ericsson, "Studies of expertise from psychological perspectives", in *The Cambridge handbook of expertise and expert performance*, edited by K. A, Ericsson, N. Charness, P. J. Feltovich, and R. R. Hoffman, Cambridge University Press, Cambridge, MA, 2006.
5. D. P. Simon, and H. A. Simon, "Individual differences in physics problem solving", in *Thinking: What develops?*, edited by R. Siegler, NJ, Erlbaum, Hillside, 1978.
6. J. R. Anderson, J. G. Greeno, P. J. Kline, and D. M. Neves, "Acquisition of problem-solving skill", in *Cognitive skills and their acquisition*, edited by J. R. Anderson, Erlbaum, Hillside, NJ, 1981.
7. B. H. Ross, *Cognitive Psychology* 16, 371-416 (1984).
8. K. VanLehn, *Cognitive Science* 22, 347-388 (1998).
9. L. M. Reeves, and W. R. Wiessberg, *Psychological Bulletin* 115, 381-400 (1994).
10. J. LeFerve, and P. Dixon, *Cognition and Instruction* 3, 1-30 (1986).
11. B. H. Ross, *Journal of Experimental Psychology: Learning, Memory, and Cognition* 13, 629-639 (1987).
12. M. T. H. Chi, "Self-explaining expository texts: The dual processes of generating inferences and repairing mental models", in *Advances in instructional psychology*, edited by R. Glaser, Erlbaum, Mahwah, NJ, 2000.
13. M. Roy, and M. T. H. Chi, "The self-explanation principle", in *The Cambridge handbook of multimedia learning*, edited by R. E. Mayer, Cambridge University Press, Cambridge, MA, 2005.
14. R. S. Siegler, "Microgenetic studies of self-explanation", in *Microdevelopment: Transition processes in development and learning*, edited by N. Garnott and J. Parziale, Cambridge University Press, Cambridge, MA, 2002.
15. M. T. H. Chi, N. de Leeuw, M. Chiu, and C. LaVancher, C. *Cognitive Science* 18, 439-477 (1994).
16. P. Pirolli, and M. Recker, *Cognition and Instruction* 12, 235-275 (1994).
17. B. Rittle-Johnson, *Child Development* 77, 1-15 (2006).
18. M. T. H. Chi, M. Bassok, M. W. Lewis, P. Reimann, and R. Glaser, *Cognitive Science* 13, 145-182 (1989).
19. M. L. Gick, and K. J. Holyoak, *Cognitive Psychology* 15, 1-38 (1983).
20. D. Gentner, *Cognitive Science* 7, 155-170 (1983).
21. J. E. Hummel, and K. J. Holyoak, *Psychological Review* 110, 220-264 (2003).
22. Chen, Z. *Journal of Educational Psychology* 91, 703-715 (1999).
23. F. G. W. C. Paas, and J. J. G. Van Merrienboer, *Journal of Educational Psychology* 86, 122-133 (1994).
24. D. D. Cummins, *Journal of Experimental Psychology: Learning, Memory, and Cognition* 18, 1103-1124 (1992).
25. D. Gentner, J. Loewenstein, and L. Thompson, *Journal of Educational Psychology* 95, 393-408 (2003).
26. R. Catrambone, *Journal of Experimental Psychology: Learning, Memory, and Cognition* 22, 1020-1031 (1996).
27. R. Catrambone, *Journal of Experimental Psychology: General* 127, 355-376 (1998).
28. M. Ward, and J. Sweller, *Cognition and Instruction* 7, 1-39 (1990).
29. T. J. Nokes, and B. H. Ross, "Near-miss versus surface-different comparisons in analogical learning and generalization", in *Proceedings of the 29th Annual Cognitive Science Society*, edited by D. S. McNamara and J. G. Trafton, Cognitive Science Society, Austin, TX, 2007.

Cognitive Science: Problem Solving And Learning For Physics Education

Brian H. Ross

Beckman Institute and Department of Psychology, University of Illinois, Urbana, 61801, USA

Abstract. Cognitive Science has focused on general principles of problem solving and learning that might be relevant for physics education research. This paper examines three selected issues that have relevance for the difficulty of transfer in problem solving domains: specialized systems of memory and reasoning, the importance of content in thinking, and a characterization of memory retrieval in problem solving. In addition, references to these issues are provided to allow the interested researcher entries to the literatures.

Keywords: Cognitive science; memory; problem solving; learning
PACS: 01.40Fk; 01.40gb; 01.40Ha

INTRODUCTION

To better conduct research on how people learn physics, we need a systematic approach to the principles of learning and a wide range of methods for examining this learning. Over the last thirty plus years, much research in cognitive science has examined these issues. The goal of this paper is to very selectively present a small portion of this work that may be of particular relevance to transfer in physics education and to provide entry references for both general approaches and specific issues. To begin the last aim, there are a large number of introductory textbooks that examine cognitive psychology and cognitive science [1,2], including from a more cognitive neuroscience perspective [3].

Although many cognitive science topics might be of interest to this community, I chose three that I thought might be less obvious and have implications for transfer: specialized systems, the importance of content in thinking, and the way in which memory retrieval during problem solving may lead to analogies, categories, and other problem-solving behaviors.

SPECIALIZED SYSTEMS

It is common in education to mark out different types of knowledge and reasoning skills one wants the students to learn. The work in cognitive science has also considered a variety of knowledge and thinking types, so I present here some distinctions and how they might relate to transfer in physics learning. I focus on one distinction each in memory and reasoning that has important implications for this community. Different ways of representing and manipulating knowledge (i.e., memory and thinking) may have different operating principles that influence how people problem solve and learn.

Memory and Reasoning Systems

Much research in cognitive science, neuroscience, and psychology has found that memory is not a monolithic entity but instead consists of a number (we do not know how many) of specialized systems that have different purposes and at least somewhat different means of representing and retrieving knowledge. A critical distinction for education is between procedural and declarative memory [4].

Declarative memory consists of factual information, "knowing that" information that relates single concepts. Examples include the memory of a worked-out problem, a formula, or a definition. These are quickly (although sometimes imperfectly) learned, can be flexibly accessed by different elements of the memory, and flexibly applied. For example, one can learn $f = ma$, access it given the word "acceleration" in a simple problem, and solve for f or m or a given the other variables.

Procedural memory is "knowing how" to do something, a procedure. However, it is not the factual information of how the procedure is done (that would be declarative), but the actual processes to accomplish

CP951, *2007 Physics Education Research Conference*, edited by L. Hsu, C. Henderson, and L. McCullough
© 2007 American Institute of Physics 978-0-7354-0465-6/07/$23.00

the procedure. For example, although we can tie a shoelace, most of us could not tell someone how we do it without watching (or imagining) ourselves tying the lace (parents who have taught this recently are the exception). The declarative knowledge we began with has been turned into a representation that can guide the tying but is no longer accessible in a verbalizable format.

The evidence for this distinction includes a large number of behavioral studies showing these lead to very different patterns of performance, of neuroscience studies showing differences in brain regions and amnesias, and computational models that are able to account for many different findings by using both types of representations [5-8].

Although reasoning theorists argue about whether the differences in reasoning require positing different systems, I think the evidence tends to favor those who see at least two different systems [9,10]. According to these views, System 1 is a heuristic processing system that works unconsciously in an associative manner, pulling to mind some use of knowledge that relates to the current information being processed. This processing is similar to some of the claims about how experts process situations, but also captures more simple associative reasoning in everyday thinking and in novices. System 2 is a more analytical, deliberative processing, such as one might see in a novice figuring out how to fill numbers of a problem into an equation. System 1 is usually thought of as a faster initial processing that provides an answer or passes information to System 2.

Implications for Physics Learning

Why do these distinctions matter to you? As a simple motivation, I remind you of two observations that I am sure you have made. First, people often have the relevant knowledge but do not retrieve it at the time that it is needed [11]. Second, people who seem to be able to do one thing often cannot do something else for which the same knowledge is relevant. Although both of these observations can have multiple causes (e.g., age, intoxication, sleepiness), a common problem has to do with a mismatch of knowledge or reasoning processes. Thirty years ago, a principle of *transfer-appropriate processing* [12] proposed that performance depends not just on what you know and how you learned it, but on the similarity between the original processing and the way in which you are processing the information now. Because of the different memory and reasoning systems, the processing may differ enough from prior learning to lead to lack of transfer. Experimental studies show the importance of procedural overlap in predicting transfer

[13,14] and the lack of full transfer when the same knowledge is learned and tested in different ways [15]. A common situation I had when teaching statistics to Psychology majors is that a failing student would express dismay since s/he had understood the way to solve the problem perfectly when I explained it in class. I had to point out that the exam did not test their understanding of my solution, but their ability to generate their own solution. This mismatch in processing is a common and underappreciated influence on transfer.

CONTENT IN THINKING

Abstract thinking is hard. People usually think about specific people, things, situations, problems, etc. As educators we want them to learn the underlying abstract principles, but we underestimate the difficulty of abstract thinking probably because of our expertise in the domain and our interest in abstract principles. However, the extent and influence of specific (content-based) thinking may be much greater than we appreciate [16], both in novices and experts.

Novice physics students rely upon objects in problems when asked to sort problems by types [17]. However, the influence of superficial content information goes much further. A common practice in textbooks and classrooms is to present an abstract principle and then an example to illustrate the principle. We assume students use this example to refine exactly how the principle works, to clear up any ambiguities in what the words mean or how the formula might be applied. A more common result, unfortunately, is that students' understanding of the principle becomes intimately intertwined with the example used to illustrate it, so that details of that example are encoded as aspects of the principle [18-20]. In addition, although we might give multiple examples to avoid this overspecification of content and to allow the students to learn the intersection among the examples (i.e., the principle), they often still end up with knowledge that includes content [21,22]. Generalizations are often very conservative, partly because the novices do not know exactly which aspects are relevant and which are not. However, even when they do know that an aspect is irrelevant (such as the particular object), their declarative representation includes this information as they reason about the problem.

Why do they do this? One might argue that evolution has prepared us for dealing with the physical world and that very formal abstract principles are a much newer useful type of knowledge. Although I think this is true, it is also important to realize that even in formal domains, content is often correlated with the underlying structure, the principles. Given a problem is about an inclined plane, it is more likely to involve some principles than others.

Experts are also influenced by the content, although they may sort problems by underlying principles [23]. Given the correlation of content and structure in most domains and the relative ease of encoding content mentioned in the problem, the experts can at least use this predictive content to get a head start in accessing knowledge that is likely to be relevant. In algebra, the content and structure are so correlated that the names of many of the problem types are in terms of the content. For example, a usual "age problem" might be:

> Michelle is 4 times older than her niece. In 5 years, Michelle will be 3 times older than her niece. How old will they each be in 5 years?

However, one could also write the same underlying problem very differently:

> A mason mixed 4 times as much cement in one mixer as another. He added 5 liters of cement to each mixer. The first mixer now has 3 times as much cement. How much does each mixer have?

Although the problems have the same underlying solution, the content of the second is that typically found with "mixture" problems, so is unusual for this type of problem. Highly experienced algebra problem solvers were often misled by this unusual content when categorizing the problems and even had reduced solution accuracy for difficult problems. In addition, one algebra teacher showed the extent to which content is involved even with highly-learned problem schemas. When solving a problem from the river-current category (e.g., a boat goes downriver in t1 time and back upriver in t2 time, how fast is the current?) that substituted going back up a down escalator in a mall, she referred to the current of the escalator and going upstream. The schema variables include content because it makes it faster to map the problem variables to the schema variables. For domains in which content and structure are correlated (which is most domains), even experts will (and should) make use of it.

CONSIDERING MEMORY RETRIEVAL

What people know and retrieve has a huge influence on their problem solving. I give a perspective on this issue that I hope will be helpful to this community in understanding the role of memory in problem solving and its relation to different problem

solving such as equation-based, analogy, and the use of problem categories. A very simple view is that:

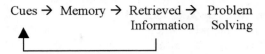

That is, cues are used to probe memory, and the retrieved Information may either be used for problem solving or to reformulate the cues (for further memory retrievals). Even this simple view, supplemented with the principle that memory retrieval includes some similarity matching between the Cues and Memory, provides enough to understand some of the observed problem solving for novices and experts.

Novices begin with cues that consist of easily available aspects of the problem, such as objects and variables (e.g., if acceleration is mentioned they might use "variable a" as a cue) [17]. Their memory is likely to consist of some problems, such as worked-out examples, and many formulae. Given these cues and that memory, two results are most likely. One, they might retrieve an earlier example that matches the current problem in the objects involved. Two, they might retrieve a formula (or formulae) that contain one or more of the variables. The problem solving is thus likely to be by analogy to an earlier example or more equation-based, as they keep probing with known and goal variables for formula that might allow them to make progress towards the solution.

Experts' cues from reading the problem also include the objects, but some structure of the problem as well. Their memory consists of a much larger range of earlier problems, many formula, but also a variety of problem categories, with means for identifying the problem type and associated procedures for solving them (i.e., problem schemas) [17]. Given these cues and memory, they might retrieve analogies and formulae too, but the most probable outcome is to retrieve a problem category (both because the cue containing structure provides a strong match to the category in memory, plus the category has become a very available memory from much earlier use, as compared to any of the examples). Given this retrieval, one is likely to see a category-based (or principle-based) problem solution, perhaps falling back to analogies or equations for difficult or unusual problems.

I am not trying to present a full theory of how memory is used in problem solving, but just enough to point out what I hope will be some useful points. First, the type of problem solving observed can often be understood in terms of the memory and cues. Second, analogy and category-based problem solving are similar in many ways. Early abstractions often arise from using analogies [18,24,25] and problem categories may arise with much more practice with these generalizations [26].

This is a very simplified picture, but I do want to mention two important other influences. First, the cues are not the problem elements, but the problem as interpreted by the problem solver [27]. This representation of the problem is critical and has huge influences on problem solving [28] at least partly due to what knowledge is retrieved. Second, novices might be able to improve their problem solving by learning to do some more structural analysis and storing the results in memory (to be accessed by cues for later problems), such as with self-explanations [29,30] and strategy-writing [31].

CONCLUSIONS

Cognitive science may have some useful principles and methods for researchers interested in physics education. In this paper, three issues are discussed (specialized systems, content in thinking, and memory retrieval in problem solving) and references are provided to permit entry to the relevant literatures.

ACKNOWLEDGMENTS

Preparation of this paper was supported by Grant R305B070085 from the Institute of Education Sciences. I thank David Brookes, Jose Mestre, Timothy Nokes, and Eric Taylor for discussions.

REFERENCES

1. J. R. Anderson, *Cognitive Psychology and Its Implications*, New York: Worth Publishing, 2005, Sixth Edition.
2. D. L. Medin, B. H. Ross, and A. B. Markman, *Cognitive Psychology*, Hoboken, NJ: John Wiley & Sons, Inc., 2004, Fourth Edition.
3. S. M. Kosslyn and E.E. Smith, *Cognitive Psychology: Mind and Brain*, Upper Saddle River, NJ: Prentice-Hall Inc., 2006.
4. N. J. Cohen and L. R. Squire, *Science* **210**, 207–210 (1980).
5. J. R. Anderson, *Language, memory, and thought*, Hillsdale, NJ: Erlbaum, 1976.
6. J. R. Anderson and C. Lebiere, *The atomic components of thought*, Mahwah, NJ: Lawrence Erlbaum Associates, 1998.
7. N. J. Cohen and H. Eichenbaum, *Memory, amnesia, and the hippocampal system*, Cambridge, MA: MIT Press, 1993.
8. H. Eichenbaum and N. J. Cohen, *From conditioning to conscious recollection: Memory systems of the brain*, New York: Oxford University Press, 2004.
9. S.A. Sloman, *Psychological Bulletin* **119**, 3-22 (1996).
10. K. E. Stanovich and R. F. West, *Behavioral and Brain Sciences* **23**, 645-726 (2000).
11. J. D. Bransford, J. J. Franks, N. J. Vye, and R. D. Sherwood, "New approaches to instruction: Because wisdom can't be told", in *Similarity and analogical reasoning*, edited by S. Vosniadou and A. Ortony, Cambridge: Cambridge University Press, 1989.
12. C. D. Morris, J. D. Bransford, and J. J. Franks, *Journal of Verbal Learning & Verbal Behavior* **16**, 519-533 (1977).
13. Z. Chen, *Journal of Experimental Psychology: Learning, Memory, and Cognition* **28**, 81-98 (2001).
14. M. K. Singley and J. R. Anderson, *The transfer of cognitive skill*, Cambridge, MA: Harvard University Press, 1989.
15. N. Pennington, R. Nicolich, and J. Rahm, *Cognitive Psychology* **28**, 175-224 (1995).
16. D. L. Medin, and B. H. Ross, "The specific character of abstract thought: Categorization, problem-solving, and induction" in *Advances in the psychology of human intelligence, Vol. 5*, edited by R. Sternberg, Hillsdale: Erlbaum, 1989.
17. M. T. H. Chi, P. Feltovitch, and R. Glaser, *Cognitive Science* **5**, 121-152 (1981).
18. B. H. Ross, *Cognitive Psychology* **16**, 371-416 (1984).
19. B. H. Ross, *Journal of Experimental Psychology: Learning, Memory, and Cognition* **13**, 629-639. (1987).
20. B. H. Ross, *Journal of Experimental Psychology: Learning, Memory, and Cognition* **15**, 456-468 (1989).
21. M. L. Gick, and K. J. Holyoak, *Cognitive Psychology* **15**, 1-38 (1983).
22. B. H. Ross and M. C. Kilbane, *Journal of Experimental Psychology: Learning, Memory, and Cognition* **23**, 427-440 (1997).
23. S. B. Blessing and B. H. Ross, *Journal of Experimental Psychology: Learning, Memory, and Cognition* **22**, 792-810 (1996).
24. D. Gentner, "The mechanisms of analogical reasoning" in *Similarity and analogical reasoning*, edited by S. Vosniadou and A. Ortony, Cambridge: Cambridge University Press, 1989.
25. D. Gentner, K. J. Holyoak, and B. N. Kokinov, *The analogical mind: Perspectives from cognitive science*, Cambridge, MA: MIT Press, 2001.
26. B. H. Ross, and P, T. Kennedy, *Journal of Experimental Psychology: Learning, Memory, and Cognition* **16**, 42-55 (1990).
27. A. Newell, and H. A. Simon, *Human problem solving*. Englewood Cliffs, NJ: Prentice-Hall, 1972.
28. S. Martin and M. Bassok, *Memory & Cognition* **33**, 471-478 (2005).
29. M. T. H. Chi, M. Bassok, M. W. Lewis, P. Reimann, and R. Glaser, *Cognitive Science* **13**, 145-182 (1989).
30. M. T. H. Chi, N. de Leeuw, M. Chiu, and C. LaVancher, C. *Cognitive Science* **18**, 439-477 (1994).
31. W. J. Leonard, R. J. Dufresne, and J. P. Mestre, J.P., *American Journal of Physics* **64**, 1495-1503 (1996).

Instrumentation In Learning Research

David A. Sears[*] and Daniel L. Schwartz[†]

[*]*Purdue University, BRNG 5130, 100 N. University Ave., West Lafayette, IN 47907*
[†]*School of Education, Stanford University, 485 Lasuen Mall, Stanford, CA 94305-3096*

Abstract. In physics experiments, a great deal of effort is spent calibrating instruments. These include instruments that precipitate some event, and instruments that measure the effects of those events. Design research in the learning sciences often focuses on precipitating learning events, but it does not pay equal attention to designing effective measures. We present the results of a study that compared two types of instruction on students working alone or in pairs. We show how one measure, common to most studies of learning, failed to detect any effects. Then we show how a second measure, called a Preparation for Future Learning measure, detected important differences. Specifically, pairs working to invent solutions to problems in statistics were more prepared to learn about new, related types of statistics than pairs who were shown how to solve the original problems, as well as individuals who invented or were shown how to solve the original problems.

Keywords: Preparation for Future Learning (PFL), transfer, collaborative learning, Innovation and Efficiency
PACS: 01.40.Ha, 01.40.G-, 01.40.gf, 01.50.Kw

INTRODUCTION

One of the major goals of education is preparation for future learning—the ability to use prior knowledge to learn new skills and understand new concepts. Despite this being a primary goal, the tests we give our students often focus exclusively on retrieving and applying prior knowledge rather than learning. For example, a typical test environment requires students to clear their desks and not talk to their peers (or other sources, like the internet). Bransford and Schwartz [1] called these Sequestered Problem Solving (SPS) environments. They attempt to measure mature knowledge, or the skills that students can apply without help. Having mature knowledge may be essential to developing deeper understanding, but, as Vygotsky [2] was careful to note, it is not the full measure of an individual's understanding. Schwartz and his colleagues developed a different type of assessment to measure this less mature form of knowledge, and they called it Preparation for Future Learning (PFL) assessments [1].

A couple of examples can help illustrate how a PFL assessment works. In an early use of a PFL test, Schwartz and Bransford [3] compared the learning effects of graphing data sets from psychological experiments versus writing summaries of a chapter about the same studies. Given an SPS test in the form of a multiple-choice test, the college students who had

done the graphing activity did not look better than the students who had summarized the chapter. For the PFL assessment, students in both conditions then heard a lecture, and afterwards, they had to make predictions about a novel experiment. Students who had completed the graphing activity learned much more from the lecture as indicated by their superior abilities to predict the outcomes of the novel experiment.

In a later study, Schwartz and Martin [4] examined the benefits of asking high-school students to invent their own solutions to statistics problems. The authors did not think the students would invent correct statistics. Instead, the invention activity was designed to prepare them to learn the conventional solutions more effectively. In the invention condition, the students invented mathematical procedures for how to compare scores from different distributions (i.e., normalize data). In the direct instruction condition, students received the same data set, and they were taught and then practiced the graphical equivalent of finding Z-scores (a way to normalize data).

The authors measured student learning by creating a posttest that covered several weeks of instruction. To experimentally evaluate which condition better prepared students for future learning, the authors made two posttest forms. Both forms of posttest included a difficult transfer problem near the end of the test relevant to computing normalized scores. The difference was that the PFL form included an

CP951, *2007 Physics Education Research Conference*, edited by L. Hsu, C. Henderson, and L. McCullough
© 2007 American Institute of Physics 978-0-7354-0465-6/07/$23.00

embedded worked example in the middle of the test. The SPS form of the test did not include the worked example. The worked example held the keys to solving the difficult transfer question. The goal of this target transfer question was to see which condition would show a greater benefit from the worked example. The direct instruction led to the same performance on the transfer problem regardless of whether the students received the SPS or PFL version of the test. These students had not been prepared to learn, and they did not learn from the worked example. In contrast, the invention students who received the PFL version of the test did roughly twice as well as the invention students who received the SPS version, and all the students from the direct instruction condition.

Thus, both studies showed that the PFL measure could detect effects of instruction missed by standard SPS assessments. The field of the learning sciences has spent a good deal of time creating instructional treatments that cause learning, but less time developing measures of those treatments. Given the effectiveness of PFL measures, we decided to design a study that put a PFL measure to work to identify which aspects of those treatments cause learning. In particular, we wanted to determine whether collaborative learning, which is common to many reform-minded pedagogies, improves preparation for future learning, and if so, under what conditions.

COLLABORATIVE LEARNING

Collaborative learning has been successful at promoting student achievement, as measured by SPS outcomes, especially when certain scaffolds are in place such as scripts and roles, individual accountability and interdependence, and training [5,6,7,8,9,10]. When collaborative learning is not scaffolded, it tends not to promote educational gains beyond what individuals accomplish studying alone [10]. This suggests at least two possibilities. One is to create collaborative arrangements that do not depend on instructional designs that are difficult for teachers to implement. The second is to determine whether there are specific situations where collaborative benefits are especially effective and merit the effort needed for implementation. We examine these two possibilities.

One way to state our question is to ask which types of tasks naturally benefit from group work, and for what types of outcomes. A broad framework for characterizing different types of tasks comes from recent developments in the field of transfer. Transfer is the generalization of knowledge to new contexts. A longstanding debate in the field has involved whether tasks featuring direct instruction, incremental mastery,

and immediate feedback were better for transfer than those that featured discovery, multiple-solutions, and delayed feedback. Schwartz, Bransford, and Sears [11] suggested a resolution to this debate by proposing that rather than seeing these tasks as being on opposite ends of a single dimension, they could be seen as on two complementary dimensions. They called these dimensions Efficiency and Innovation. Efficiency tasks often involve direct instruction followed by repeated practice. The goal is speed and accuracy, or routine expertise [12]. Examples include times-tables recitations and end-of-chapter review problems. Innovation activities involve opportunities for students to explore, invent, or discover solutions to problems rather than being told what to do from the outset. The outcomes of the innovation dimension of instruction have not been as clearly defined as the outcomes of instruction that emphasizes efficiency.

Schwartz et al. [11] hypothesized that one benefit of tasks aligned to the Innovation dimension is that they would prepare students for future learning. For example, by allowing students to attempt to find a solution before receiving the expert approach, they will be more likely to connect their prior knowledge to the new material and to notice key features that characterize the problem. When they receive the expert solution, they should be more prepared to appreciate how it works and how they can apply it flexibly in the future. These authors proposed a need to balance innovation and efficiency experiences in instruction, because each experience when appropriately combined contributes to a trajectory of adaptive expertise [12].

In the current study, we compared Efficiency only instruction with Innovation plus Efficiency instruction. We assessed the outcomes using both SPS measures of efficiency and PFL measures of students' abilities to learn given new resources during the test.

Of particular interest was whether initially working alone or collaboratively would affect performance on the SPS and PFL measures. Under what conditions are two heads better than one, and for what outcomes? For Efficiency tasks, one can imagine that knowing the procedure to use would allow groups to monitor and correct mistakes. If so, we should expect collaborators on efficiency tasks to do better on SPS assessments than those working alone or those collaborating on innovation tasks. Alternatively, one might imagine that innovation tasks would permit students to gain access to the benefits of different points of view so they can cull out the deep structure common to both their perspectives [13]. If true, then collaborating over innovation tasks should better prepare students to learn in the future, because they would develop a more structured understanding of the domain.

In the experiment, the Innovation and Efficiency framework was tested for its ability to characterize naturally productive tasks for collaborative learning. University students, mostly undergraduates, worked alone or in same-gender pairs to learn about the chi-square formula. On the posttest, all participants worked alone and answered three types of questions: 1) SPS calculations of the chi-square, 2) SPS comprehension measures that tapped the depth of understanding about how the formula worked, and 3) PFL measures of students' abilities to transfer their lessons to learn a related, but complicated statistical concept (i.e., computing inter-rater reliability).

METHODS

Participants: Seventy-six university students were randomly assigned to one of two conditions: Innovation or Efficiency. Participants either worked in same-sex pairs (36) or alone (40). Forty-eight women (24 in dyads) and 28 men (12 in dyads) with little or no background in statistics participated.

Materials—A nine-page learning packet about the chi-square formula and a seven-item posttest comprised the materials for this experiment. The learning packet consisted of three units about different aspects of the chi-square formula. Each unit contained three pages: Lesson, Problems, and Final Practice Example with solution provided. (The order of these pages was shuffled to implement the Efficiency v. Innovation plus Efficiency contrast, as described below.)

Time taken for the learning-packet and the posttest was recorded. The seven-item posttest included two problems requiring chi-square calculations, three comprehension questions about where and how the formula worked, and a difficult PFL assessment with two parts related to the statistics topic of inter-rater reliability. For the PFL assessment, a resource question introduced the topic of inter-rater reliability, and the target question built upon those principles and required participants to transform their chi-square formula into a measure analogous to Cohen's Kappa. This is a very far transfer task, because students had not been taught Cohen's Kappa during the initial instruction. The goal was to see which version of the task, Innovation or Efficiency, would better prepare students for learning how to solve the PFL target problem.

Procedures—The key difference between conditions was that participants in the Innovation condition had to invent solutions to the Problems *before* seeing the Lesson while those in Efficiency saw the Lesson and then applied what they learned to the Problems. Table 1 summarizes the procedures.

TABLE 1. Procedures

Step	Context	Innovation	Efficiency	Time
1	Alone / Dyads	9-page Learning Packet on the Chi-Square Formula		35 to 65min
		1) Problems (<u>INVENT a formula</u>) 2) Lesson (instruction) 3) Final Example (reinforce)	1) Lesson (instruction) 2) Problems (<u>APPLY the formula</u>) 3) Final Example (reinforce)	
		Short Break		5min
2	Alone	Posttest (7 problems: 2 calculations, 3 comprehension, PFL—resource & target)		25min

RESULTS

The posttest was consistent internally (alpha = .81) and across different raters (all Cohen's Kappas > .81). Time-on-task showed no significant differences between conditions as shown in Figure 1.

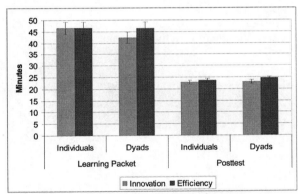

FIGURE 1. Time-On-Task. (SEM Bars in All Graphs).

As shown in Figure 2, studying in groups showed no benefits on the SPS calculation and comprehension measures for either condition. Thus, working in dyads did not improve efficiency outcomes relative to working alone, and working on innovation tasks (with subsequent efficient instruction) did not harm either individuals or dyads.[1] Thus innovation-based instruction can lead to good SPS outcomes.

_____

[1] Dyads scored lower on the comprehension questions than individuals. This was unexpected and likely due to them having to share the Lessons page rather than each having a copy. Post-hoc analyses indicated that dyads, particularly those in the Innovation condition, spent less time on the Lessons pages than individuals (Means: 9.2, 6.3, 9.9, and 9.2 minutes for Innovation individuals and dyads, and Efficiency individuals and dyads, respectively; SEMs: ± 0.7).

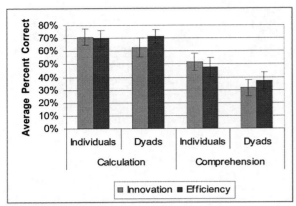

FIGURE 2. Innovation Outperformed Efficiency on the Transfer Measures.

Participants in the Innovation condition surpassed their Efficiency counterparts on the PFL measure. On the PFL target question, Innovation dyads exceeded all others (see Figure 3). Thus, they showed the greatest preparation for future learning, and we can tentatively conclude that innovation tasks are naturally suited to the benefits of working in groups. For SPS outcomes, it does not matter much how students are taught if it is a well-designed curriculum. However, for PFL outcomes, innovation activities using dyads provide a special benefit.

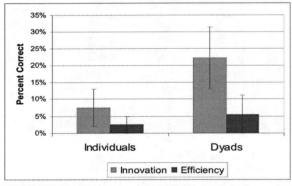

FIGURE 3. Performance on the PFL Target Question.

CONCLUSION

By using PFL assessments, we found that opportunities to invent solutions to problems followed by direct instruction on the canonical solution can facilitate transfer and collaborative learning. In so doing, we found that the Innovation and Efficiency framework provides a means of characterizing naturally productive tasks for group learning. If only traditional measures of performance were used, we would not have seen any differences between Innovation and Efficiency individuals or dyads. This

would have led to the erroneous conclusion that the novel Innovation method of instruction produced no benefits compared to the more traditional Efficiency instruction, and that working in groups has no natural benefits relative to working alone.

ACKNOWLEDGMENTS

This material is based upon work supported by the National Science Foundation under Grant No. SLC-0354453. Any opinions, findings, and conclusions or recommendations expressed in this material are those of the authors and do not necessarily reflect the views of the National Science Foundation.

REFERENCES

1. J. D. Bransford, and D. L. Schwartz, "Rethinking Transfer: A Simple Proposal with Multiple Implications," in *Review of Research in Education*, edited by A. Iran-Nejad and P. D. Pearson, Washington, D.C.: American Educational Research Association, 1999, pp. 61-100.
2. L. Vygotsky, *Mind in Society: The Development of Higher Psychological Processes*, edited by M. Cole, V. John-Steiner, S. Scribner, and E. Souberman, Cambridge, MA: Harvard University Press, 1978/1932.
3. D. L. Schwartz and J. D. Bransford, *Cognition and Instruction*, **16**, 475-522, (1998).
4. D. Schwartz and T. Martin, *Cognition & Instruction*, **22**, 129-184, (2004).
5. E. Coleman, *The Journal of the Learning Sciences*, **7**, 387-427, (1998).
6. M. R. Gillies and A. F. Ashman, *Journal of Educational Psychology*, **90**, 746-757, (1998).
7. D. Johnson and R. Johnson, *Theory into Practice*, **38**, 67-73, (1999).
8. A. King, "Discourse Patterns for Mediating Peer Learning," in *Cognitive Perspectives on Peer Learning*, edited by A. O'Donnell and A. King, Mahwah, NJ: Lawrence Erlbaum Associates, Inc, 1999, pp. 87-115.
9. A. O'Donnell, "Structuring Dyadic Interaction Through Scripted Cooperation," in *Cognitive Perspectives on Peer Learning*, edited by A. O'Donnell and A. King, Mahwah, NJ: Lawrence Erlbaum Associates, Inc, 1999, pp. 179-196.
10. R. Slavin, *Contemporary Educational Psychology*, **21**, 43-69, (1996).
11. D. Schwartz, J. Bransford and D. Sears, "Efficiency and Innovation in Transfer," in *Transfer of Learning from a Modern Multidisciplinary Perspective*, edited by J. Mestre, CT: Information Age Publishing, 2005, pp. 1-51.
12. G. Hatano, and K. Inagaki, "Two Courses of Expertise," in *Child Development and Education in Japan*, edited by H. Stevenson, H. Azuma, and K. Hakuta, New York, NY: Freeman, 1986, pp. 262-272.
13. M. L. Gick and K. J. Holyoak, *Cognitive Psychology*, **15**, 1-38, (1983).

Cognitive Science: The Science Of The (Nearly) Obvious

Bruce Sherin

School of Education and Social Policy, Northwestern University, Evanston, IL 60208 USA

Abstract. This article discusses the need for a "grand theory" of physics cognition and learning. The idea of the grand theory is that it would be a complete description of the knowledge students possess and how that changes over time, from before instruction, all the way through expertise. The first part of the paper discusses our current state of knowledge and possible strategies for making progress on the grand theory. The second part of the paper illustrates, with an example, how the program might work.

Keywords: cognitive science, learning theory, conceptual change
PACS: 01.40.-d, 01.40.Ha

TOWARD A "GRAND THEORY"

The theme for this-year's Physics Education Research Conference is "cognitive science and physics education research." I have decided to use this invited address as an opportunity to talk, somewhat generally, about how I believe cognitive science should be applied in the context of physics education research. What I am going to talk about is what might be worth calling a "grand theory" of physics and learning. The idea of the grand theory is that it would be a complete description of the knowledge students possess and how that changes over time, from birth through physics expertise and beyond.

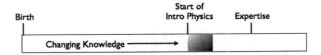

FIGURE 1. The big picture of physics learning.

So, as a field, how close are we to having the grand theory? The answer, of course, is that we are not very close. As physics education researchers, we've focused a lot of attention in the highlighted region of Figure 1, which includes the time during introductory physics, and extends a bit beyond. Within this range, we know quite a lot about student performance on short, qualitative questions of the sort that show up on the FCI and in misconceptions research. These instruments tell us something about conceptions prior to instruction, and how those conceptions change during introductory physics instruction, at least as they are revealed by short qualitative questions.

We also know a fair amount about how students solve textbook problems, from early introductory physics and somewhat into more advanced undergraduate study. Sometimes the same introductory textbook problems are given to experts for comparison. There has also been an emphasis on student epistemologies – their beliefs about physics knowledge and learning – and how those epistemologies affect what they learn.

However, outside of the central highlighted region of Figure 1, and the particular foci just described, physics education researchers have done much less. For the later part of the life cycle, post-undergraduate, there is a limited amount of real research. Where experts are studied, they are often studied solving problems that are trivial for them. There are certainly notable exceptions [1, 2]. But, if we want to study the rarified regions out at the right of the life cycle in Figure 1, then we should be studying physics experts doing the range of cognitive tasks that truly characterize physics expertise, such as reading challenging journal articles, designing an experimental apparatus, and deriving new theoretical results.

Down the far left end of the life cycle in Figure 1, physics education researchers have also not done much. Fortunately, there are some other fields, especially developmental psychology, that are filling in some relevant parts of the story.

What can we reasonably do?

Really constructing the grand theory is going to be extremely hard. Part of the problem is that physics is difficult and subtle, and expertise is built over years. If

CP951, *2007 Physics Education Research Conference*, edited by L. Hsu, C. Henderson, and L. McCullough
© 2007 American Institute of Physics 978-0-7354-0465-6/07/$23.00

we were to try to study experts at work, in their chosen specialties, simply understanding the physics involved might be non-trivial. But an even bigger problem, I believe, is that, if we are really going to do this right, the amount of knowledge we have to consider is potentially enormous, and the learning really does start at birth. For example, our knowledge of physics is built, in a very fundamental way, on what we know about the physical world from being a person in the world. That knowledge is enormous, and we learn it starting at birth. Similarly, part of our grand theory must take into account our mathematical knowledge, and that learning also starts from birth.

Given these difficulties, what, as a field, can we reasonably do? One mitigating factor is that there are other fields that are doing some of the work for us, especially in the early part of the life cycle. Psychology and mathematics education research are obvious places to look. We can make it our business to draw on that work and to make our own work consistent with it. Furthermore, I do not think it would be unreasonable for us to attempt to study physics expertise a bit more systematically. Finally, I believe that we can tweak our efforts in the highlighted part of Figure 1 – the time period that we already study – so that our efforts can fit more neatly in the grand program. I elaborate on this in what follows.

AN EXAMPLE FROM PSYCHOLOGY

I want to present a brief example from psychology, to illustrate the type of research that is being conducted down at the very earliest parts of the life cycle. I selected one article by Needham and Baillargeon [3], but there are many similar examples. This article investigates what 4.5-month-old babies think will happen when support of an object is removed. The way that these experiments with babies work is that babies are shown phenomena, and then the researchers measure how long the babies look at it. The longer they look, the more surprising or interesting the phenomenon is to the babies.

Needham and Baillargeon showed babies possible and impossible events relating to support, as shown in Figure 2. The top row in this figure shows the possible event; hands appear and place an object on top of another object. The bottom row shows an impossible event. The hands appear, the object moves across and then beyond the supporting object. But then the object just appears to hang, unsupported, in space.

What happens is that babies look longer at the impossible than the possible event, thus suggesting they know something about support. It turns out the story is a bit more complicated – and a bit more interesting. But the details are not too relevant here.

The point is just that this learning starts very early, and that psychologists are mapping out some parts of this early learning about the physical world.

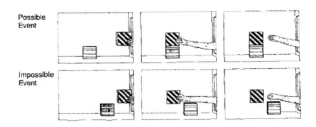

FIGURE 2. Possible and impossible events. From Needham and Baillargeon [3].

TWEAKING OUR EFFORTS

I stated that we can tweak our efforts in the highlighted region of Figure 1 so that they fit more neatly into the grand program. Here's where cognitive science as "the science of the nearly obvious finally comes in." The idea is that we can refocus our attention on what is nearly obvious to us, what is right before our eyes and that we do every day.

As I assert in my title, I believe that this is the primary business of cognitive science, to extract basic structure that we see in the world, the structure that is so basic to how we understand things that it can be all but invisible. The point is that changing our stance in this way will help us do work that fits more neatly within the grand program.

An example: The Alphabet

To illustrate what I mean by this change in stance, I am going to discuss an exercise that I use in introductory courses that I teach on cognitive theories of learning. What I have students do is form into pairs. Then one person in each pair interviews the other, and they ask some questions that have to do with the alphabet. They ask:

1. What letter is the 12th letter in the alphabet?
2. What's the 10th letter after C?
3. Say the alphabet backwards starting at Q.
4. Is G before or after J?
5. Is M before or after W?

Not surprisingly, the way that the subjects answer the first two questions is that they say the alphabet from the beginning, counting on their fingers as they go. Answering the third question proves to be a bit more challenging. The main way that the subjects solve this is that they will jump backward a few letters, say the alphabet forwards from that point, then say that chunk of the alphabet backwards. Then they'll jump back a bit more and iterate. Question 5, in contrast, turns out to be trivial; the subjects just answer it instantly. However, to answer question 4, and determine if G is before or after J, the subjects sometimes have to say a part of the alphabet to themselves.

ABCD EFG HIJK LMNOP QRSTUVW XYZ

FIGURE 3. A model of alphabet knowledge.

A shown in Figure 3 one way to understand these results is to hypothesize that our knowledge of the alphabet is in the form of a clumpy, ordered list. In this simple model, we can move forward in the list from where we are. We can enter into the list at points other than A, but only at the start of a clump. So when people jump backward, they don't have to go all the way back, but they have to jump back to the start of a clump. Also, in this model, given a letter, we can say what clump it's in. That's why comparison across clumps is easy and within clumps can be hard.

Now I want to make some points about this alphabet example. First, in retrospect, the clumpy structure might seem to be somewhat obvious. In fact, when we discuss this example in my courses, students are very comfortable talking about these clumps, as if they have known about them for a long time.

Second, although this clumpy structure might seem obvious in retrospect, this is usually fairly tacit knowledge. The clumps are not usually taught in the sense that a teacher says "we're going to work on the HIJK clump today." Third, the interesting grain size in this example is down a level from the structure that has a name. We call the whole thing "the alphabet," but we don't have names for the clumps. Finally, it is worth noting that the knowledge structure here retains some of the stamp of its origins. It's possible that, to some extent, this clumpy structure is a result of the alphabet song.

An Example From Physics

Now I want to give an example to illustrate how this focus on the "nearly obvious" can look in the case of physics learning. To begin, imagine that we are looking at a sheet of paper that is filled with physics equations. As physicists, we can very quickly recognize a great deal of structure in this sheet. For example, we will recognize individual symbols and we might recognize some entire equations

But the question is whether there is some sort of nearly obvious structure in this page of expressions, things we see for which we don't have any immediate names. I believe that there is a great deal of such tacit structure in these symbols and I'm going to talk about

a particular kind of structure that I call *symbolic forms* [4, 5]. A symbolic form is an association between two components. First, each symbolic form includes a *conceptual schema*. The conceptual schema is a way a way of understanding a physical situation that involves a few entities and some simple relations among those entities. The second component is a symbol template – it's a template that specifies how exactly to express the conceptual schema in a symbolic expression. Or, stated in crude terms, a symbolic form is an "idea" plus a specification of how exactly to express that idea in an equation.

balancing	$a = b$
parts-of-a-whole	$\square + \square + \square$
base±change	$\square \pm \Delta$

FIGURE 4. A page of physics expression

In previous work, I have discussed 21 distinct symbolic forms [4, 5]. For illustration, Figure 4 lists some symbolic forms along with the associated symbol templates. First, in the *balancing* form, a physical situation is schematized as involving two influences that are equal and opposite. The symbol template for this form specifies that, to express the idea of "balancing," we write two expressions, one for each influence, that are separated by an equal sign. In the *parts-of-a-whole* form, a situation is understood as involving a whole that is composed of two or more parts. The symbol template specifies that the way you express this particular idea is by writing two or more terms of some sort, separated by plus signs. Finally, in *base+change*, a base quantity is increased by the amount of a change. A thing to notice across the last two examples is that the same symbol structure can have multiple meanings.

Symbolic forms and the grand theory

Now what I want to argue is that my analysis of our understanding of symbol structure in terms of symbolic forms targets the right knowledge, and at the right grain size, so that we can begin to see how it might fit within the grand theory. To provide a sense for how that would work, I'm going to talk briefly about two kinds of precursor knowledge to symbolic forms, what diSessa calls "p-prims," and what Greeno called "patterns in arithmetic word problems."

Phenomenological Primitives

Andrea diSessa describes a portion of commonsense physics knowledge that he calls the sense-of-mechanism [6]. The sense-of-mechanism consists of knowledge elements called "phenomenological primitives" or just "p-prims" for short. They are called "primitives" because elements of the sense-of-mechanism form the base level of our intuitive explanations of physical phenomena. The idea of diSessa's program is to identify these primitive pieces of knowledge – those are the p-prims. For example, in a p-prim that diSessa calls *Ohm's p-prim*, a situation is schematized as involving an effort that works against a resistance to produce some result.

In his published work, diSessa identifies dozens of p-prims, and suggests the existence of many more. For example, there are p-prims, such as Ohm's p-prim, that have to do with force and agency. There also are p-prims – such as one that diSessa calls *dynamic balance* – that have to do with balance and equilibrium. And there are p-prims that apply to constraint phenomena, such as *blocking, supporting,* and *guiding*.

Patterns in Arithmetic Word Problems

In the early 1980s, a number of mathematics education researchers hypothesized that young students are sensitive to a certain class of patterns in simple arithmetic word problems. For example, Riley, Greeno, and Heller [7] described four patterns, which they called *change, combine, equalization,* and *compare*. Change problems, for example, are problems in which some amount is added onto a base quantity to increase the size of that base quantity. In contrast, in *combine* problems, two given numbers A and B are put together, to form a new third quantity. So the *change* pattern should sound similar to the *base+change* symbolic form, and *combine* should sound similar to *parts-of-a-whole*. And these patterns share the property, with symbolic forms, that the same mathematical expressions can have multiple meanings.

Putting the Story Together

The point is that, looking across this research, we can see some small threads of the grand theory begin to emerge (refer to Figure 5). For symbolic forms such as *balancing*, there might be a thread back to p-prims and the sense-of-mechanism. And for forms such as *parts-of-a-whole* and *base+change*, there might be threads back to patterns in arithmetic word problems. It is even possible that we might be able to trace Needham and Baillargeon's notion of *support* into the sense-of-mechanism.

This is only a small slice through the grand theory. Here I am tracing knowledge of a very particular sort; it's all moderately abstract schemas that have just a few entities in them. Some threads in the grand theory are going to look very different. For example, we certainly have knowledge of specific phenomena. We know what happens when you mix vinegar and backing soda. And we know things about specific classes of objects, like strings. We probably acquire some of that knowledge early and it gets refined during the progression to physics expertise. So not all threads in the grand theory will necessarily bear a close resemblance to this one. But this is the sort of game I think we should play.

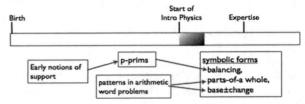

FIGURE 5. Some threads in the grand theory.

CONCLUSION

In summary, I stated that, even if it might be an impossible ideal, we should think of ourselves as working on a grand theory that describes changing knowledge structures throughout the life cycle of a physicist. I said that this required us to broaden our focus somewhat to look at more of the lifespan, and across a broader range of tasks. I said we could look to other fields to do some of the work for us, especially at the left side of the life cycle. And finally I tried to show how tweaking our stance to focus on the nearly obvious could allow us to do our work so that it fits more neatly within the grand program.

REFERENCES

1. J. Clement, "Use Of Physical Intuition And Imagistic Simulation In Expert Problem Solving," in *Implicit And Explicit Knowledge*, edited by D. Tirosh, Ablex, Hillsdale, NJ, 1994, pp. 204-244.
2. C. Singh, "When Physical Intuition Fails," *American Journal of Physics* **70** (11), pp. 1103-1109 (2002).
3. A. Needham and R. Baillargeon, "Intuitions About Support In 4.5-Month-Old Infants," *Cognition* **47** (2), pp. 121-148 (1993).
4. B. Sherin, "How Students Understand Physics Equations," *Cognition and Instruction* **19** (4), pp. 479-541 (2001).
5. B. Sherin, "Common Sense Clarified: Intuitive Knowledge And Its Role In Physics Expertise," *Journal of Research in Science Teaching* **33** (6), pp. 535-555 (2006).
6. A. A. diSessa, "Toward An Epistemology Of Physics," *Cognition and Instruction* **10** (2 & 3), pp. 165-255 (1993).
7. M. S. Riley, J. G. Greeno, and J. I. Heller, "Development Of Children's Problem-Solving Ability In Arithmetic," in *The Development Of Mathematical Thinking*, edited by H. P. Ginsberg, Academic Press, New York, 1983, pp. 153-196.

Conceptual Dynamics in Clinical Interviews

Bruce Sherin, Victor R. Lee, and Moshe Krakowski

School of Education and Social Policy, Northwestern University, Evanston, IL 60208 USA

Abstract. One of the main tools that we have for the study of student science conceptions is the clinical interview. Research on student understanding of natural phenomena has tended to understand interviews as tools for reading out a student's knowledge. In this paper, we argue for a shift in how we think about and analyze interview data. In particular, we argue that we must be aware that the interview itself is a dynamic process during which a sort of conceptual change occurs. We refer to these short time-scale changes that occur over a few minutes in an interview as *conceptual dynamics*. Our goal is to devise new frameworks and techniques for capturing the conceptual dynamics. To this end, we have devised a simple and neutral cognitive framework. In this paper, we describe this framework, and we show how it can be applied to understand interview data. We hope to show that the conceptual dynamics of interviews are complex, but that it nonetheless feasible to make them a focus of study.

Keywords: cognitive science, learning theory, conceptual change
PACS: 01.40.-d, 01.40.Ha

REVISITING THE "COHERENCE" QUESTION

An enormous amount of research has focused on the intuitive conceptions that students possess about science topics prior to formal instruction in those topics. Despite this significant body of research some of the most important questions are still in dispute. Perhaps the most fundamental of these questions is the coherence question. On one side of the debate are researchers that believe that students possess coherent models or even theories prior to instruction [1, 2] . On the other side of the debate are researchers that believe students possess a large number of pieces of knowledge that are employed in a sometimes inconsistent manner [3].

Although this question is still in hot dispute, we believe that, on one level, the answer to the coherence question is obvious. That answer is: It depends. It depends on…

1. **… the person and domain**. It simply must be the case that some individuals have coherent understandings in a domain while others do not, due to differences in prior experiences and other cognitive variables, such as age.
2. **… how knowledge is elicited**. The conditions and techniques for eliciting knowledge will affect what coherences we can observe in reasoning. Slight variations in interviews can produce substantial changes in apparent coherence of student understanding [4-6].
3. **…our definition of 'coherence.'** Different meanings of coherence have been applied, and they are, at times, in conflict. For example, diSessa and colleagues [7] have contrasted what they call *contextual* and *relational* coherence.

Given these dependencies, we believe that a change in tactic is necessary. We should not ask, generally speaking, whether students have coherent models. Instead, our goal must be to map out where, when, and what kinds of coherences are generated.

A FOCUS ON THE DYNAMICS OF CLINICAL INTERVIEWS

The above change in stance has many implications for research, some of which are methodological in nature. Research on student understanding of natural phenomena has tended to understand interviews as tools for reading out a student's knowledge. We believe it is more appropriate to think of coherences as arising throughout the course of an interviewing interaction. Thus, we must be aware that the interview itself as a dynamic process during which a sort of conceptual change will occur. We refer to these short time-scale changes that occur over a few minutes in an interview as *conceptual dynamics*. The goal of our

CP951, *2007 Physics Education Research Conference*, edited by L. Hsu, C. Henderson, and L. McCullough
© 2007 American Institute of Physics 978-0-7354-0465-6/07/$23.00

project is to devise new frameworks and techniques for capturing conceptual dynamics in interviews.

To understand the challenge that we face as researchers, consider the cartoon shown in Figure 1, which depicts a short exchange in a clinical interview. The person on the left in each pane is a student, and the person on the right is the interviewer. The shapes inside the student's head represent the knowledge resources that the student has available.

In the first pane, the interviewer asks a question. The student then responds in the second pane, drawing on a subset of his knowledge resources in order to construct an answer. In the third pane, the interviewer asks a follow-up question. Then, in the fourth pane, the student draws on a different subset of his knowledge resources, and constructs a different response. As cognitive researchers, we are interested in the knowledge possessed by the student – the shapes inside his head. But, as data, all that we have available are the behaviors of the interviewer and the student, which include their utterances, gestures, and any drawings that they make.

FIGURE 1. An exchange in a clinical interview.

THEORETICAL FRAMEWORK

Given the situation depicted in Figure 1, the challenge we face is to "see through" the interview to the knowledge possessed by the student. As researchers, we must seek to produce an analysis of the knowledge resources that, in part, generate the interview dynamics we observe. One challenge in conducting such an analysis is that there are many elements and types of knowledge employed during a clinical interview. However, we believe it is possible to make progress using a relatively simple and neutral framework. The major constructs of this framework are:

Nodes: We think of knowledge as consisting of a large number of elements we call *nodes*. We intend our notion of node to encompass mental elements of many different types and at multiple levels of abstraction.

Mode: A *mode* is a recurrent pattern of nodes that is activated for a particular class of cognitive tasks. The same node may participate in more than one mode. (Refer to Figure 2.)

Dynamic Mental Construct (DMC): A *DMC* is a constructed explanation that is produced by the reasoning that occurs within a mode. It can be highly context sensitive and it may have only momentary stability.

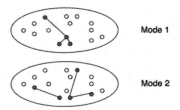

FIGURE 2. The same node may participate in more than one mode.

In terms of this framework, student reasoning during an interview is understood as follows: The interviewer asks a question and this leads to the activation of a mode. The student then reasons within this mode, assembling the nodes into a DMC (a model or answer to the question that was asked). The interviewer then asks another question. This can lead to more reasoning within the same mode, or to the activation of another mode.

DATA SOURCES

We have conducted ~160 clinical interviews with middle school students on a variety of science topics including: human biology, chemical reactions, optics, climate, energy transfer, and seasonal temperature variation.

AN EXAMPLE: LESLIE ON THE SEASONS

We now present a short excerpt to illustrate some of the interview phenomenology that we seek to understand and analyze. We present this to demonstrate conceptual dynamics and to show our framework in use. In this excerpt a student, Leslie, is asked to explain the seasons – why it is warmer in the summer and colder in the winter. The result is an extended response, over which her reasoning gradually unfolds.

TABLE 1. Analysis of the excerpt.

	Nodes	Dynamics	Current DMC
(1)	• Daylight Savings Time • Days are longer in the summer • More source ⇒ more effect	Skimming the mode throwing out first relevant elements. First assembly of a fledgling DMC.	Days are longer in the summer. Longer days mean more daylight which makes it warmer.
(2)	• Axis (word) • Earth moves	Skimming the mode throwing out relevant knowledge	Earth's movement and axis somehow are related to the seasons.
(3)	• Different seasons in different locations. • India is hot • India is near the equator • The equator is hot • Sun is a source	More skimming of the mode. She strikes out in a new direction now construct a DMC related to climate phenomena.	It's colder up north because there's less sun, warmer near the equator because there is more sun.
(4)	• Image of sunny winter wonderland • Snow/ice up north • Snow/ice are cold • More source ⇒ more effect	Leslie recalls a vivid image of a scene that shows a lot of snow but also a lot of sun. The cuing of this new node leads to a shift. Note how she corrects herself in mid sentence.	It's colder up north because of the snow and ice.

Interv. *First thing I want to know is why is it warmer in the summer and colder in the winter?*

Leslie **(1)** Well, umm, you know times savings? ... daylight savings time in the summer we have more time, like, with, like, daylight and that's why it gets warmer. **(2)** And with the circulation of the earth and the axis that it's on just has to do with like summer and winter. **(3)** And it depends on where we are on the earth, like if you look at, umm, India, it's toward the equator, you know? So it's like always hot. So if you go up north then it gets colder **[4]** because there's just, like, I can't really say less sun, but it kind of has to do with that and there's just a lot of snow and, like ice cause it's colder up there.

Leslie's utterance is broken into four chunks, and our analysis of each of those chunks is given in Table 1. In the first chunk, Leslie begins by mentioning "daylight savings time," and she then constructs her first fledging DMC: days are longer in the summer, and longer days mean more daylight, which makes it warmer. In the second chunk, she just mentions two nodes – *Axis* and *Earth moves* – without immediately incorporating either of them into a DMC. We believe that her mention of "axis" here signals no more than that she knows that this word is somehow relevant to explaining the seasons.

In chunk (3), Leslie seems to strike out in a new direction. She begins by saying "it depends on where we are on Earth." Then she goes on to construct a DMC that is more relevant to climate variation with latitude than to the seasons. She says, essentially, that it's colder up north because there is less sun and warmer near the equator because there is more sun. In

this chunk she also specifically mentions India as a location that is both hot and near the equator.

Finally, in chunk (4), Leslie seems to interrupt and correct herself, contradicting the DMC she constructed in chunk (3). She starts to say that it's colder up north because there is less sun, but then she says, instead, that it is colder because of snow and ice: "I can't really say less sun, but it kind of has to do with that and there's just a lot of snow and, like ice cause it's colder up there." As the interview proceeded, it became clear what Leslie was thinking here and why she corrected herself. It seems that she has a vivid memory of seeing photographs of a sunny "winter wonderland," showing a landscape covered in snow, but illuminated with bright sunlight. Based on this image, she concluded that cold regions could still have a great deal of sunlight.

We want to highlight a few features of the brief analysis summarized above. First, one might have thought that it would be extremely difficult to see conceptual dynamics as they unfold during an interview. But we hope that this excerpt makes it clear that, at least in some cases, it is possible to see student reasoning as it unfolds. Within this brief episode, we can see Leslie's thinking as it. Furthermore, although the dynamics are complex, they are not completely unfathomable; it is possible to make some reasonable guesses about what Leslie is thinking and why.

Finally, we want to emphasize the importance of attending to as much of Leslie's knowledge as possible, and even to relatively idiosyncratic elements of her knowledge. Many analyses of student reasoning during interviews reduce their knowledge to one of a small number of mental models. But, in Leslie's case, it is clear that her reasoning is being driven by more narrow and idiosyncratic elements of knowledge, such as her knowledge that India is hot, and her image of a sunny winter wonderland. If we were to ignore these more idiosyncratic elements of knowledge then we would not be in a position to understand why Leslie's reasoning unfolds as it does.

PATTERNS OF CONCEPTUAL DYNAMICS IN INTERVIEWS

Clinical interviews vary greatly in character, and can unfold in a complexly variable manner. However, there are some rough patterns and regularities that can be observed.

Mode skimming. When a question is first asked and a mode activated, sometimes a student will skim the surface of the mode. They will just list out ideas that might be relevant, without any attempt to fit them together into a DMC.

DMC construction and model building. Once some relevant nodes are selected, a student might engage in an extended process of trying to fit together those nodes into a DMC that has the desired properties.

DMC shifts. Many kinds of interviewing events can lead to a shift in the current DMC. These can be student-driven or interviewer-driven. Sometimes, as in the interview with Leslie, these shifts can occur just as the student's thinking unfolds. But they may also occur as a result of more dramatic events, such as a challenge from an interviewer. They may also occur when a student remembers a new and potentially relevant piece of knowledge, such as Leslie's image of a snowy winter wonderland.

Mode shifts. There can be a shift in the mode that is active, leading to a more dramatic change in DMC.

Attractor DMCs. Some DMCs can act as attractors for students; once discovered, these DMCs will be particularly appealing. For example, in the case of the seasons, one prominent explanation is what we call the closer-farther explanation. This explanation goes as follows: The Earth orbits the sun in such a way that it is sometimes closer to the sun and sometimes farther from the sun (perhaps an ellipse). Summer corresponds to the time when the Earth is closer, and winter corresponds to the time when it is farther [e.g., 8, 9]. This explanation, though incorrect, has many pleasing properties. Many people have heard that the Earth's orbit is elliptical. The fact that this explanation incorporates this technical bit of information means that it might acquire, by association, some of the weight of authority. And there is much that this explanation can predict that *is* correct. In particular, it predicts that the cycle of the seasons will be repeated with a period of one year, and it links this cycle to the Earth's orbit.

CONCLUSION

One of the main tools that researchers have for the study of students' science conceptions is the clinical interview. In this paper, we argued for a shift in how we think about and analyze interview data. In particular, we argued that it is important that we understand interviews as dynamic interactions, and that these interactions must themselves be a focus of our study. To this end, we described a simple and neutral cognitive framework, and we showed how that framework can be applied to understand interview data. We hope that, as a result, we have shown that the conceptual dynamics of interviews are complex, but it is nonetheless feasible to make them a focus of our analysis.

REFERENCES

1. Vosniadou, S. and W.F. Brewer, *Mental models of the earth: A study of conceptual change in childhood.* Cognitive Psychology, 1992. **24**(4): p. 535-585.
2. McCloskey, M., *Naive theories of motion*, in *Mental models*, D. Gentner and A. Stevens, Editors. 1983, Erlbaum: Hillsdale, NJ. p. 289-324.
3. diSessa, A.A., *Toward an epistemology of physics.* Cognition and Instruction, 1993. **10**(2 & 3): p. 165-255.
4. Nobes, G., et al., *Children's understanding of the earth in a multicultural community: mental models or fragments of knowledge?* Developmental Science, 2003. **6**(1): p. 72-85.
5. Vosniadou, S., I. Skopeliti, and K. Ikospentaki, *Modes of knowing and ways of reasoning in elementary astronomy.* Cognitive Development, 2004. **19**(2): p. 203-222.
6. Siegal, M., G. Butterworth, and P.A. Newcombe, *Culture and children's cosmology.* Developmental Science, 2004. **7**(3): p. 308-324.
7. diSessa, A.A., N.M. Gillespie, and J.B. Esterly, *Coherence versus fragmentation in the development of the concept of force.* Cognitive Science, 2004. **28**(6): p. 843-900.
8. Newman, D. and D. Morrison, *The conflict between teaching and scientific sense-making: The case of a curriculum on seasonal change.* Interactive Learning Environments, 1993. **3**: p. 1-16.
9. Sadler, P.M. *Alternative conceptions in astronomy.* in *Second international seminar on Misconception and Educational Strategies in Science and Mathematics.* 1987. Ithaca, NY: Cornell University Press.

Physics Learning In The Context Of Scaffolded Diagnostic Tasks (I): The Experimental Setup

Edit Yerushalmi[*], Chandralekha Singh[†], Bat Sheva Eylon[*]

[*]Department of Science Teaching, Weizmann Institute of Science; Rehovot, Israel
[†]Department of Physics and Astronomy, University of Pittsburgh, Pittsburgh, PA 15213, USA

Abstract. For problem solving to serve as an effective learning opportunity, it should involve deliberate reflection, e.g., planning and evaluating the solver's progress toward a solution, as well as self-diagnosing former steps while elaborating on conceptual understanding. While expert problem solvers employ deliberate reflection, the novices (many introductory physics students) fail to take full advantage of problem solving as a learning opportunity. In this paper we will focus on self-diagnosis as an instructional strategy to engage students in reflective problem solving. In self-diagnosis tasks students are explicitly required to carry out self diagnosis activities after being given some feedback on the solution. In this and a companion paper, we will present research exploring the following questions: How well do students self-diagnose, if at all, their solutions? What are the learning outcomes of these activities? Can one improve the act of self-diagnosis and the resulting learning outcomes by scaffolding the activity?

Keywords: problem solving, reflection, alternative assessment, self-diagnosis.
PACS: 01.40.gb, 01.40.HA

INTRODUCTION

One feature that characterizes experts problem solvers is continuous evaluation of their progress [1]. They monitor their progress towards a solution asking themselves implicit reflective questions such as: "What am I doing?", "Why am I doing it?", and "How does this help me?" [2]. Experts are recognized also by a strategic approach: first carrying out a qualitative analysis of the situation and then developing a plan to solve the problem on the basis of this analysis [3]. While self monitoring is directed mainly towards arriving at a solution, it might also involve self-diagnosis directed towards elaboration of the solver's conceptual understanding, knowledge organization and strategic approach.

In the context of problem solving it is beneficial to foster diagnostic self explanations. It has been shown that the activity of self-explanation leads to significant learning gains [4] and can be enhanced through interventions that require students to present their explanations [5], and encourage students to give justifications via peer interaction or human computer interaction.

In physics education, one common instructional strategy to enhance an expert-like problem solving approach that allows learning from problem solving is that of cognitive apprenticeship [6]. This approach incorporates three elements:

(1) modeling, (2) coaching, and (3) scaffolding and fading. In this approach, "modeling" means providing examples to demonstrate and exemplify the skills that students should learn. "Coaching" means providing students opportunity, and practice, with guidance and feedback so that they learn the necessary skills. "Scaffolding and fading" refers to decreasing the support and feedback consistent with students' needs to help them develop gradual self-reliance.

In the context of problem solving, this approach is often used with motivating realistic problems that need an expert-like approach mimicking the culture of expert practice. Students often work collaboratively [7] or with a computer [8] where they must externalize and explain their thinking while they solve a problem. In many of these interventions, students receive modeling of a problem-solving strategy [9,10] that externalizes the implicit problem solving strategies used by experts, and they are required to use it. Common steps amongst different strategies include asking students to: (1) Describe the problem, (2) Plan and construct a solution, (3) Check and evaluate their solution. These strategies have been shown to improve students' problem solving skills (planning and evaluating rather than searching for the appropriate

CP951, *2007 Physics Education Research Conference*, edited by L. Hsu, C. Henderson, and L. McCullough
© 2007 American Institute of Physics 978-0-7354-0465-6/07/$23.00

equation without reflection) as well as their understanding of physics concepts [11].

One challenge these approaches often face is that the assessment is traditional, focused on product rather than the process, thus undermining the intended outcomes. The negative impacts of traditional assessment include over-emphasis on grades, and under-emphasis on feedback to promote learning. Thus, traditional assessment approaches are lacking in formative assessment. Black and Wiliam [12] suggest that for formative assessment to be productive, students should be trained in self assessment so that they can understand the main purposes of their learning and thereby grasp what they need to do to succeed. Thus, tests and homework can be invaluable guides to learning, but they must be clear and relevant to learning goals. The feedback should give guidance on how to improve, and students must be given opportunity and help to work on the improvement.

Following these guidelines we constructed a self-assessment task in which students are required to present a diagnosis (namely, identifying where they went wrong, and explaining the nature of the mistakes) as part of the activity of reviewing their quiz solutions. We shall call these tasks "self-diagnosis tasks". Indeed, an integral activity in physics problem solving should be reviewing the solution that the learner has composed in order to improve it or learn from it. For instance, this activity can occur after comparing a final result of a problem solution to the back of the book answer, as well as when students get their graded work back. The notion that comparing solutions composed by the learners to a worked out example is conducive to learning is actually common amongst physics instructors.

In traditional situations in which students review their solutions, self-diagnosis is not guaranteed to occur. Students may or may not self-diagnose the solution, or it might occur only implicitly. In contrast, in self-diagnosis tasks, students are required explicitly to diagnose their solutions.

SELF-DIAGNOSIS TASKS

Self-diagnosis tasks can be distinguished by:
1. Instructions on how to carry out the diagnosis, e.g. level of detail laying out for students the possible deficiencies in their approach towards the solution and in its implementation.
2. Resources available to the students while diagnosing their solutions (e.g. information provided about the correct problem solution, notebooks and textbook).
3. Feedback: Provision of customized feedback (diagnostic information about the solution).

We report an in-class study focused on three types of interventions that require students explicitly to self-diagnose their solutions, yet differ in the instructions and resources students receive (see Table 1).

In all interventions students first solved realistic, motivating quiz problems (an example is shown in Figure 1). These kinds of problems are sometimes referred to in the literature as "context rich problems"[9] that have a context and motivation connected to reality, have no explicit cues (e.g. "apparent weight"), require more than one step to solve and may contain more than the information needed (e.g. the car's mass). Context Rich Problems require students to analyze the problem statement, determine which principles of physics are useful and what approximations are needed (e.g. smooth track), and plan and reflect upon the sub-problems constructed to solve the problem.

A friend told a girl that he had heard that if you sit on a scale while riding a roller coaster, the dial on the scale changes all the time. The girl decides to check the story and takes a bathroom scale to the amusement park. There she receives an illustration (see below), depicting the riding track of a roller coaster car along with information on the track (the illustration scale is not accurate). The operator of the ride informs her that the rail track is smooth, the mass of the car is 120 kg, and that the car sets in motion from a rest position at the height of 15 m. He adds that point B is at 5m height and that close to point B the track is part of a circle with a radius of 30 m. Before leaving the house, the girl stepped on the scale which indicated 55kg. In the rollercoaster car the girl sits on the scale. Do you think that the story she had heard about the reading of the scale changing on the roller coaster is true? According to your calculation, what will the scale show at point B?

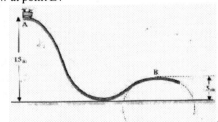

FIGURE 1. Context rich problem.

In some intervention groups, the problem statement included guidelines for how students should present their problem solutions, while in others instructors merely discussed with students these guidelines early in the semester (cf. Fig. 2).

In all intervention groups, in the recitation following the quiz, the instructor gave his students a photocopy of their solution, and asked them to diagnose mistakes in their last week's quiz solution. Students were credited 50% of quiz grade for the diagnosis. The instructor also motivated them by saying that self diagnosis will help them learn.

> **Problem description:** Represent the problem in Physics terms: Draw a Sketch, List known and unknown quantities, target variable
>
> **Solution construction:** Present the solution as a set of sub problems, In each sub problem write down:
>   The unknown quantity you are looking for
>   The physics principles you'll use to find it
>   The process to extract the unknown
>
> **Check answer:** Write down how you checked whether your final answer is reasonable

FIGURE 2. Guidelines for presenting problem solution.

In the 1st intervention group students received minimal guidance, namely, they were asked to circle their photocopied solutions and say what they did wrong in that part, aided by their notes and books only, without being provided the solution.

In the 2nd intervention group the students were also provided with a correct solution that the instructor handed out during the self-diagnosis activity (cf. Fig. 3). This solution followed the guidelines for presenting a problem solution (cf. Fig. 2).

In the 3rd intervention group the instructor discussed the correct solution with the students and they were required to fill in a self diagnosis rubric (cf. Fig. 4). The rubric was designed to direct students' attention at two possible types of deficiencies: deficiencies in approaching the problem in a systematic manner ("general evaluation" part) and deficiencies in the physics applied.

Besides the intervention groups, we also had a control group in which the instructor discussed with the students the solution for the problem, but they were not required to engage in a self–diagnosis task.

Solution

1. Description of the problem

Knowns:
The height of release: $h_A=15m$
Speed of the car at point A: $v_A=0$
The height at point B: $h_B=5m$
The radius at point B: $R_B=30m$
The mass of the car: M=120 kg
The mass of the girl: m= 55kg
Target quantity: N_B = Normal force at point B.
Assumptions: the friction with the track is negligible

2. Constructing the solution

Plan:

During the motion of the girl along the curved track, the magnitude of her velocity as well as the radial acceleration changes from point to point. If the radial acceleration changes, we infer that the net force acting in the radial direction on the girl changes as well. The net force on the girl is the sum of 2 forces acting on her: the force of gravity and the normal force. To calculate the normal force at point B we can use Newton's 2nd Law $\Sigma \vec{F} = m \cdot \vec{a}$; however we will need to know the acceleration at this point. To calculate the centripetal acceleration at B, $a_R = v_B^2/R$ we need to know the speed at point B.
We will calculate the velocity of the girl at point B using the law of conservation of mechanical energy between point of departure A and point B (the mechanical energy is conserved since the only force that does work is the force of gravity which is a conservative force. The normal force does no work because it is perpendicular to the velocity at every point on the curve).

Sub-problem 1: calculating the velocity at point B

We'll set ground as the reference for potential energy and compare the total mechanical energies of the girl and car at points A and B: $E_A = E_B$

Since the speed is zero at point A the kinetic energy is zero at that point. Therefore we get: $PE_A = KE_B + PE_B$

$$\Rightarrow (m+M)gh_A = (m+M)gh_B + \frac{1}{2}(m+M)v_B^2$$

$$\Rightarrow gh_A = gh_B + \frac{v_B^2}{2}$$

We can calculate the speed at B in the following way:
$$v_B^2 = 2gh_A - 2gh_B = 2(10m/s^2)(15m - 5m) = 200\,m^2/s^2$$

Sub problem 2: calculating the normal force at B

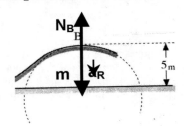

Using Newton 2nd law
$\Sigma \vec{F} = m\vec{a}_R$
$N_B - mg = -ma_R$
$N_B - mg = -mv_B^2/R_B$
$N_B = mg - mv_B^2/R_B$
$\Rightarrow N_B = (55kg)(10m/s^2) - (55kg)(200(m/s)^2/(30m)$
$\Rightarrow N_B = 183.33N$

Final result: when the car crosses Point B at the track, the scale indicates 183.3 N

3. Reasonability check of the final result:
- limiting cases of the parametric solution
$N_B = mg - mv_B^2/R_B$:
At rest (v=0): N = mg,
On horizontal surface (R → infinity): N = mg

FIGURE 3. Sample solution aligned with guidelines.

General evaluation	Performance level	Explain what is missing?
Problem construction description	Full Partial Missing	• In sketch • Known /unknowns
Solution construction description	Full Partial Missing	• Sub problem's unknown • Principles used
Check answer	Full Partial Missing	• Possible checks for reasonability

Circle and number mistakes you find in the solution. Fill in the following rubric

Mistake #	Mark x if mistake is in:			Explain mistake
	Physics	Math	Other	
1				
...				

FIGURE 4. self diagnosis rubric

TABLE 1. Summary of Different Interventions

Control	Rubric + TA outline	Sample solution	Notes + text books
	Student solve a quiz problem		
Instructor discusses the correct solution with the students	Instructor outline the correct solution with the students	Instructor provides written sample solution	
	50% of quiz credit for this diagnosis Motivation: diagnosis will help you learn		
NO self - diagnosis	Students are asked to circle mistakes in their photocopied solutions and fill in a self diagnosis rubric.	Students are asked to circle mistakes in their photocopied solutions and explain what they did wrong	Students are asked to circle mistakes in their photocopied solutions and explain what they did wrong
Grading solution	Grading the solution and the diagnosis		

The following table summarizes the various sequences in our study.

TABLE 2. Sequences in the Study:

Pretest	Pre problems, similar in complexity but not isomorphic to Intervention problems.
Intervention	Initial Training
	Problem-Solving, Diagnosis, Repeated twice
Posttest	Isomorphic (to Pre + Intervention) and far transfer problems both in midterm and final exams, FCI

The study was carried out at two sites:
1. Israel, 11th grade high school classes (algebra based, studying full year advanced high-school mechanics). 120 students, 3 instructors.
2. US, introductory college level (1 semester, algebra based, pre-meds), 240 students, 1 Instructor, 2 TAs

The above setup allowed us to explore questions such as: What kind of mistakes are students able to diagnose? How do students' solutions compare to their diagnosis? Do students' final exam grades correlate with their solutions or their diagnosis performed in the middle of the semester? How do students in different treatment groups compare in their ability to diagnose their mistakes?

Answering these questions would allow us to determine whether self-diagnosis is indeed helpful for students' learning. If it does help, how does it advance students' learning, and how could one modify the scaffolding to improve the outcomes? These questions will be the focus of the companion paper.

ACKNOWLEDGMENTS

This study was supported by ISF grant 1283/05 (EY, BE) and NSF grant DUE-0442087 (CS).

REFERENCES

1. J. Larkin, J. McDermott, D.P. Simon, and H.A. Simon, *Science* **208**, 1335 (1980).
2. A. H. Schoenfeld, in *Handbook of Research on Mathematics Teaching and Learning*, edited by D. A. Grouws, Macmillan, New York, 1992. p. 334.
3. J.I., Heller, and F. Reif, *Cognition and Instruction* **1**, 177 (1984).
4. M.T.H. Chi, , N. De Leeuw, M.H., Chiu, C. LaVancher,. *Cognitive Science*, **18**, 439 (1994)
5. V.,Aleven, K. R., Koedinger, & K. Cross, in *Artificial Intelligence in Education, Open Learning Environments proceedings of AIED-99*, edited by S. P. Lajoie and M. Vivet, IOS Press, Amsterdam, 1999, p.199.
6. Collins, J. S. Brown, and S. E. Newman, in *Knowing, learning, and instruction: Essays in honor of Robert Glaser*, edited by L. B. Resnick, Erlbaum, Hillsdale, NJ, 1989, p. 453
7. P. Heller and M. Hollabaugh, *American Journal of Physics* **60**, 637 (1992).
8. F. Reif, and L. Scott, *American Journal of Physics* **67**, 819 (1999).
9. A. Van Heuvelen, *American Journal of Physics* **59**, 898 (1991).
10. W. J. Leonard, R. J. Dufrense, and J. P. Mestre, *American Journal of Physics* **64**, 1495 (1996).
11. K. Cummings, J. Marx, R. Thornton, and D. Kuhl, *Physics Education Research, Supplement to American Journal of Physics* **67**, S38 (1999).
12. P. Black, & D. Wiliam, *Phi Delta Kappan* **80**, 139-148 (1998).

Physics Learning in the Context of Scaffolded Diagnostic Tasks (II): Preliminary Results

Chandralekha Singh[†], Edit Yerushalmi[*], Bat Sheva Eylon[*]

[†]Department of Physics and Astronomy, University of Pittsburgh, Pittsburgh, PA 15213, USA
[*]Department of Science Teaching, Weizmann Institute of Science; Rehovot, Israel

Abstract. In a companion paper we presented self-diagnosis tasks in which students are explicitly required to self diagnose their problem solutions after being given some feedback. In this paper we suggest a rubric to evaluate diagnosis and exemplify its use in two case studies. We present preliminary results regarding how students' performance on the self-diagnosis tasks relates to their performance in solving problems and to their progress during the course. In preliminary analysis, we find that the correlation between students' self-diagnosis grades and their performance in the mid-semester quiz was very low (0.16), the correlation between the grades in the mid-semester quiz and the final exam grades was also low (0.21) while the correlation between the self-diagnosis grades in the mid-semester quiz and the final exam grades was reasonably high (0.68). We suggest that these results can be explained by the hypothesis that the self-diagnosis grades measure the slope of students' learning curve.

Keywords: problem solving, reflection, alternative assessment, self-diagnosis.
PACS: 01.40.gb, 01.40.HA

INTRODUCTION

In the companion paper, we laid out an experimental setup designed to explore physics learning in the context of scaffolded diagnostic tasks. We want to investigate whether scaffolded self-diagnosis is helpful in students' learning, what happens in the self-diagnosis that may be helpful in students' learning, and how can we improve the scaffolding. We described three interventions differing in the nature of scaffolding provided.

In this paper we develop a rubric to evaluate diagnosis and exemplify its use in two case studies. We present preliminary results regarding how students' performance on self-diagnosis tasks relates to their performance in solving problems. At this stage we consider the three interventions together to identify general trends that go beyond the particular scaffolding provided.

We focus on the study we conducted in the US in the introductory algebra based course for pre-meds, with 240 students, one instructor and two teaching assistants. We report on the preliminary analysis of seventeen students' self diagnosis of a mid-semester quiz that was given in the recitation in week 6 of a 15 week semester. There were roughly equal number of students from each intervention group.

WHAT IS A GOOD DIAGNOSIS?

The characterization of a good diagnosis is based on defining what is a good solution. Aligned with the cognitive apprenticeship approach in which a principal component is the externalization of an expert strategy to problem solving (PS), we focus the evaluation of students' solution and diagnosis both on the application of physics principles/concepts, as well as on the application of systematic PS approach.

We developed a rubric (shown in table 1) that spans possible deficiencies in these two categories. Similar rubrics adapted to lab assessment [1] and PS assessment [2] can be found in the literature. For example, for the problem shown in the companion paper, appropriate principles that need to be invoked include conservation of mechanical energy and Newton's second law. Yet many students did not invoke them. Moreover, many invoked inappropriate principles such as kinematics equations in uniform acceleration motion, or equilibrium application of Newton's second law. Useful description for this problem would include, e.g., drawing a free body diagram, an acceleration vector, and coordinate axes, defining the reference level for the potential energy, etc. An explicit strategy would include defining an appropriate target quantity (i.e., normal force that the scale exerts upon the girl), defining appropriate intermediate variables and explicitly stating in words and/or in a generic mathematical form the principles used to find these variables. Common deficiencies in this category would be laying out surplus equations, writing derived equations related to special cases, with no reference to the principles they are derived from, and ignoring intermediate or final results that make no sense.

TABLE 1: Summary of Diagnostic Components:

Diagnostic components
Application of Physics principles/concepts
1. appropriate principles that were not invoked
2. inappropriate principles that were invoked
3. principles invoked yet applied incorrectly
Application of systematic PS approach
4. Useful description (re-stating the problem in terms of knowns and unknowns, useful variables, drawing, etc.)
5. Explicit strategy to solve the problem (explicitly defining the physics concepts or principles that will be used to find intermediate variables)
6. Explicit checking of the final answer

A good self-diagnosis should reflect the student's realization of these deficiencies. In particular it should allow the students to learn the lesson from these deficiencies thus preventing them from doing similar mistakes in the future. To achieve this goal students need: 1) to identify where deficiencies have occurred; 2) to explain or/and correct the deficiencies; and 3) to acknowledge the nature of the deficiencies (physics, math,...). With these considerations in mind we evaluated students' self-diagnosis.

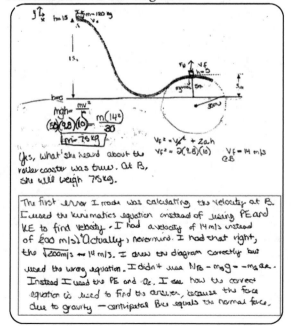

FIGURE 1. Cathy: solution & self-diagnosis.

In Figure 1 and Figure 2 we present two case studies of solutions of diagnoses differing in quality.

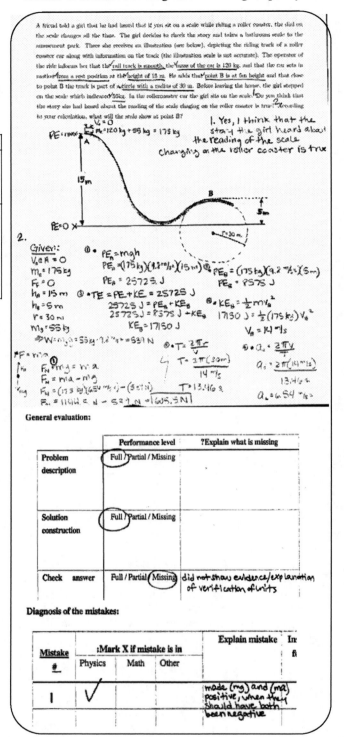

FIGURE 2. Beth: solution & self-diagnosis.

Cathy (all names are pseudonyms), for example, carried out a relatively good self-diagnosis on a relatively poor solution (see figure 1). Cathy belonged to the 2nd intervention group in which students were provided with a correct solution that the instructor handed out during the self-diagnosis activity. These students were asked to circle mistakes in their photocopied solutions and write what they did wrong in each circled part.

Cathy drew a free body diagram (FBD) but it appears that she was thinking of equilibrium application of Newton's 2nd law, taking centripetal force as a physical force. Moreover, the FBD was a surplus information in her solution as she did not use the FBD for anything else. She invoked a derived equation to find the final speed at point B. She used equation $mgh = mv^2/r$ that has energy on one side of the equation and force on the other side, and as she did not carry out dimensional analysis she did not realize this equation is dimensionally incorrect.

Surprisingly, in her self diagnosis she focused on the principles to be used, realized all the principles she misused, and which principles should be used instead. For example, she realized she was using the equation of kinematics to find v_B, rather than the conservation of mechanical energy. She also realized the strength in her approach to the problem solution. For example she realized she had a relatively good FBD that she did not use.

Beth's solution on the other hand, is an example of a little better solution, yet poor diagnosis. Beth belonged to the 3rd intervention group in which the instructor discussed the correct solution with the students and they were required to fill in a self diagnosis rubric

Her solution has FBD, although she did not draw the axes. She invoked both the conservation of mechanical energy as well as Newton's second law, although she confused the signs and used a wrong mass to calculate the normal force.

She also invoked the expression for the period of revolution T in uniform circular motion, a legitimate principle that is not appropriate in the specific problem, as well as a non-legitimate expression for the acceleration "a" in uniform circular motion as $a_c = 2\pi v/T$, resulting together in a correct equation. Her presentation is organized, but she did not check whether her results make sense.

In contrast, her diagnosis is poor. Beth was in the intervention group that was instructed to diagnose using a rubric. In the first part of the rubric, students were required to mark their performance level for "problem description" and "solution construction". "Problem description" stands for representing the problem in Physics terms (drawing a sketch, determining target variables, etc.), and "solution construction" stands for presenting a set of sub

problems differing in intermediate variables as well as physics principles used to find them. Beth marked both of those categories as "full", not realizing she treated the circular motion as uniform one without justifying why it may work at point B. While she noticed her sign confusion, and even classified it as a physics mistake, she did not explain why this is wrong and which principles/concepts she had applied inappropriately.

TABLE 2. Partial rubric for evaluating a solution:

	Positive	Negative	+	-
Invoking principles	Appropriate principles/concepts 1. Conservation of mechanical energy; 2…	Non appropriate principles/concepts a. Kinematics equations in uniform acceleration motion; b…. Non-legitimate principles/concepts		
Application	1. Setting KE to zero when the girl+cart are released; 2….	a. The centripetal force is one of the physical forces …		
Plan	Appropriate target (i.e., F_N) and intermediate (v_B, a_c) quantities chosen; 2…	a. surplus equations or intermediate variables are written down; b…		
Presentation	1. Invokes a FBD; 2…	b. not stating in words or a generic form the principles used to solve sub-problems		

TABLE 3. Partial rubric for evaluating self-diagnosis

	Things the students realized regarding his/her solution	
Invoking principles	Realizes that s/he was missing one of the following principles in solving the problem, e.g.,: 1. Conservation of mechanical energy; 2… Realizes that s/he was using one of the following inappropriate principles in solving the problem, and explaining why this was inappropriate, e.g.,: a. Kinematics equations in uniform acceleration motion; b.	
Application	Realizes principle/concept mis applied, e.g. a) The centripetal force is not one of the physical forces; b) i. Realizing principles/concepts that were not applied correctly, ii. why the way they were applied was wrong or iii. how they could have been applied correctly e. g. iv. categorizes correctly to Physics/math/other error	
Plan	Realizes s/he was missing or providing inappropriate in the following way.: a. surplus equations or intermediate variables are written down b. Inappropriate intermediate variables chosen	
Presentation	Realizes that s/he was missing one of the following 1. FBD; 2. explicitly stating in words or a generic form the principles used to solve this sub-problems	

To transform the above qualitative analysis to a quantitative one and determine a grade for the quiz solution and for the self diagnosis, we constructed a problem specific rubric (Table 2,3). Table 2 lists some specific examples for each of the items in table 1. We marked each item that appeared in the solution of a student in a positive or negative column depending upon their suitability. The analysis was done by two researchers (EY & CS) and any disagreements were discussed and resolved. Based on this analysis, the solutions of all the 17 students were compared to each other and assigned a relative grade.

The above grading procedure using the rubric revealed that the above two examples (Cathy and Beth) reflect a more general pattern, namely, the level of performance on the mid-semester quiz was not a good predictor for the level of self-diagnosis. We note that it is true that if a student had solved the problem correctly, self-diagnosis of mistakes would not be required. However, none of the students we have graded so far were able to solve the context rich problem discussed here correctly. The multi-part problem was sufficiently challenging that all of these students made some mistakes.

Figure 3 shows a plot of the mid-semester quiz grades (1-10 scale) vs. self diagnosis grades (1-10 scale) for 17 students. The correlation coefficient between quiz and self diagnosis grades was very low (0.16).

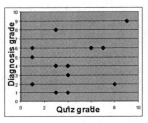

$$\rho = \frac{COV(X,Y)}{\sigma_y \cdot \sigma_x} = 0.16$$

FIGURE 3. Mid-semester quiz vs. self diagnosis grades.

One can now ask which of these two indicators, the solution or the self diagnosis, correlates better with student's final exam grades?

Our preliminary results suggest that the self-diagnosis grade for the mid-semester quiz was a better predictor of the final exam grades than the grade of the quiz solution itself (see figure 4). The correlation coefficient between the mid-semester quiz grades and the final exam grades (1-40 scale) was very low $\rho = 0.21$ and between self diagnosis and final grades was very high: $\rho = 0.68$.

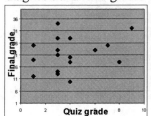 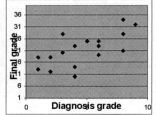

FIGURE 4. Left: Quiz vs. final exam grades. Right: Self diagnosis vs. final exam grades (N=17).

Moreover, comparison of the students we have analyzed in different intervention groups, shows that those students who received an example solution (i.e., received a paper copy of the solution or watched the TA discuss and write the solution outline on the blackboard) were better able to self-diagnose than those who only used the textbook and their notes.

POSSIBLE IMPLICATIONS

We hypothesize that the performance on the self-diagnosis tasks is an indicator of the slope of the learning curve of the students, namely, how well they will progress during the course. Hence, students who performed well on the self-diagnosis tasks are more likely to show superior performance towards the end of the course regardless of their performance at the beginning as measured by their grades in the mid-semester quiz. On the other hand, students with low performance on the self-diagnosis tasks may not progress significantly during the course and may not perform very well at the end of the course even if their performance on the mid-semester quiz was better than many others. An alternative interpretation of our preliminary finding is that the mid-semester quiz was too difficult and thus did not differentiate well between students' of different ability level.

Note that even the presumably "good diagnosis" of Cathy was not that "good". Although she mentioned which principles she did not invoke, she did not explain how she misapplied the ones she did use. However, the preliminary finding that those students who had a superior self-diagnosis performance according to our rubric had significantly better grades at the end of the course than those with poor self-diagnosis performance is encouraging. Scaffolded self-diagnosis activities should be used more often for teaching and learning.

ACKNOWLEDGMENTS

We wish to express our gratitude to the teachers who participated in the study: Korina Polinger, Philip Rojnikovski, Shoshana Ozeri, Jeremy Levy, Patrick Irvin, Shanti Wendler. This study was supported by ISF grant 1283/05 (EY, BE) and NSF grant DUE-0442087 (CS).

REFERENCES

1. E. Etkina, A. Van Heuvlen, S. White-Brahmia, D. T. Brookes, M.1 Gentile, S. Murthy, D. Rosengrant and A. Warren, Phys. Rev. ST Phys. Educ. Res. 2, 020103 (2006)
2. J. Doktor, K. Heller, P. Heller, T. Thaden-Koch and J. Li, "Robust Assessment Instrument for Student Problem Solving", Abstract, AAPT Summer Meeting, July 2007

SECTION II

Peer-Reviewed Papers

Investigating Peer Scaffolding in Learning and Transfer of Learning Using Teaching Interviews*

Bijaya Aryal and Dean A. Zollman

Department of Physics, Kansas State University, Manhattan, KS, 66506-2601

Abstract. We conducted teaching interviews with nine groups of students enrolled in an introductory level algebra-based physics course and consisted of two sessions—a learning session and a transfer session. The students were engaged in hands-on activities to learn various physics ideas in the learning session. We expected the students apply the physics learning to understand positron emission tomography (PET) in a transfer session. After providing worksheets, we asked the students to write their responses before and after the group discussion. To present the dynamics of group learning and the influence of peer scaffolding we compared the results of this study with our prior study [3] where students were individually engaged with a similar set of activities. Results suggest that peers were effective in activating and challenging each other's conceptual resources as well as facilitating transfer of learning. The results of this study also showed that students' performance was better when they were provided the direct hint instead of graduated hints. However, we found that the students gave the right answer with the wrong reasoning when a direct hint was provided, and they gave wrong answer with relatively better reasoning when the hints were graduated.

Keywords: Positron emission tomography, scaffolding, conceptual resources.
PACS: 01.40.Fk

INTRODUCTION

The research presented in this paper concentrates on revealing the nature of social influences to dynamics of learning. It discusses students' achievements in both learning and transfer of physics learning through group interaction with the aid of the hands-on activities. The instructional goal of the activities is to help students learn a range of physics ideas and apply those ideas in understanding the technology of positron emission tomography. A group teaching interview method [1], where students discuss with peers and respond to the questions, is used to collect data.

We describe the details of the students' activation of resources and the facilitation of transfer of learning through peer interaction. A resource-based transfer framework [2] is used to analyze qualitative data. To discuss the students' achievements in group learning we compare some of the results with that of our previous study [3] where the students of similar physics backgrounds participated individually using the same set of teaching activities.

The goal of research presented in this paper was to answer the following questions.

1. What is the effect of scaffolding provided to facilitate learning?

2. What is the effect of group interactions on activation of students' cognitive resources and facilitation of learning transfer using the physical models?

LITERATURE REVIEW

Literatures in psychology and education document the difference between the individual and group learning [4]. It is accepted that students' learning accomplishment is greater through group interaction than the individual efforts [5]. Various cognitive developmental and behavioral theories consider learning a social process and regard the role of social interaction in construction of knowledge. Vygotsky's [6] cognitive theory emphasizes the role of scaffolding in the cognitive development by introducing a notion of Zone of Proximal Development (ZPD). In a peer interaction context, the scaffolding comes from the cumulative efforts among the students of more or less equal capability. The group teaching interviews used in the research are based on Vygotsky's idea of learning within the ZPD.

One way to describe students' learning performance and ZPD is by counting the number of problem steps, number of hints provided and strength of hints [7]. In this study we count the number of hints, problem steps as well as the time required to complete the task to describe students' achievements in learning and transfer of learning.

METHODOLOGY

The participants of this study were students enrolled in an algebra-based physics course and were exposed to kinematics in their physics course. Nine groups were formed from 21 students (10 females and 11 males). They participated in two sessions of the teaching interview each about one hour long. A peer instruction format [8] was adapted in the interviews with minimal interviewer intervention.

The first session of teaching interview used series of hands-on activities where students learn physics through active engagement. One activity of the session involved the collision carts on a track. A barrier was placed in front of the track so that students were able to see the end of the track but not the location where the carts were released (Figure 1 (a)). They were asked to discuss the location where they started, both qualitatively and quantitatively. Another activity of the first session simulated a series of simulated explosions using the light activity (Figure 1(b)). The result of each "explosion" produced two light spots on the wall of the cylindrical enclosure. Students were asked to deduce the explosion location that produced the light pulses. Students observed several pairs of lights, recorded the positions of light on a circular graph paper, and noted any pattern or trend in a graph.

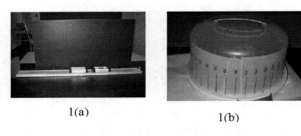

1(a)   1(b)

Figure 1: Activities used in the teaching interview

The interviewer guided a discussion of general ideas about positron emission tomography (PET) at the beginning of the second session. Students engaged in a series of problems related to the physics PET technology such as building a model of locating the exact position of electron-positron annihilation in a brain. Finally, they were asked to complete activities that enabled them to find the region of a tumor in which the annihilations were occurring.

The interviews were video and audio taped. The data analysis followed most of the steps with Colaizzi's [9] phenomenological analysis technique. Six physics education researchers independently analyzed several samples of interview data for the inter-rater reliability test. An agreement of 67% or above was established among the researchers in different categories.

RESULTS AND DISCUSSION

Peer Scaffolding in Learning

Expansion of ZPD Through Peer Scaffolding:

In a task of locating hidden events using the cart activity students' performance was categorized into three levels. The students at the 'quantitative' level could use the appropriate variables and put them in an equation to come up with the numerical results. The students at the 'qualitative' level used the appropriate reasoning identifying the correct variables without being able to use the variables in an equation and the students in the 'unsuccessful' group could not come up with a correct reasoning.

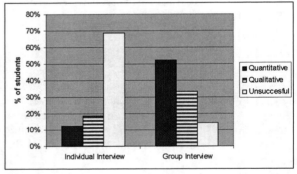

Figure 2: Students Performance in Locating Cart Release

Figure 2 indicates that student performance in the group teaching interviews was significantly better than that in individual teaching interviews. During group teaching interviews 11 out of 21 (52%) students were successful in quantitative task, 33% were successful in qualitative and 15% students were not able to use an appropriate reasoning. In our prior study [3] we had conducted the individual teaching interviews with 16 students where 11 students (68.75%) were not successful in making any kind of approach to solve the problem. Only two students (12.5%) were able to complete the task quantitatively and the remaining three students (18.75%) could complete the task with qualitative reasoning successfully.

Students' performance difference in different contexts can be explained on the theoretical basis of Vygotsky's zone of proximal development. The interviewer in the individual interviews scaffolded almost 69 % of the students to enable them to use appropriate reasoning whereas during the group interviews 72% of the students make appropriate reasoning through peer interaction. More students were successful through peer scaffolding than through interviewer scaffolding to do the quantitative task (53% versus 13%). The students constructed their knowledge socially with the help of others in both the cases the peers were more effective than the interviewer.

Activation of Appropriate Reasoning by Peers

Peers were capable of challenging each other's symmetry argument. For example, the students were asked to deduce the location of an explosion that produced lights on the wall of the cylinder. They used symmetry reasoning to predict locations of the hidden events where this type of reasoning was not relevant. This "central tendency" was more prominent during individual teaching interviews. The students who participated in the group interviews also held a similar idea but they changed it easily through peer scaffolding.

Most of the students (87%) who participated individually held the central tendency. Similarly, most of the student groups (7 out of 9 groups) who participated in the group interviews started out with the central tendency but with the completion of their discussion six groups (15 students) did not indicate central tendency. Only three groups (6 students) still had the central tendency.

In the context of the light activity we expected students to make the association of 'time' with 'location' to find the origin of explosion in the light activity. We noted that students transferred the idea of the cart activity to the light activity to associate 'time' with the 'location'. Six out of nine groups (14 out of 21 students) relied on time to deduce event location in the light activity. Two groups (5 students) relied on intensity of light and the last group (2 students) inferred the location based on the size of the light spot of the wall of the cylinder.

The peer scaffolding was found effective in activation of 'time' resource to locate events. Peer helped to suppress the association of 'location' with 'intensity' through several examples. They helped each other to transfer the idea of the cart activity to facilitate the association of the 'time' with 'location' in the light activity.

Effect of Sequencing Hints

Students described the cart motion by either associating the cart behavior with kinematics or magnetic terms. Before asking them to describe the motion of the carts they were asked if they had either seen the carts before and if so in what context. Or they were provided information that the carts were magnetic. The association of the carts motion with kinematics idea was dominant (12 out of 16 students) among the students who were asked if they saw the carts before. The association of the cart's motion with magnetic interaction was more popular (12 out of 21 students) among those students who were provided with the information that the carts were magnetic.

It is reasonable to argue that the students activated resources of different domains based on the types of scaffoldings. Activation of students' resources from a prior physics class was facilitated when they were asked if they saw the carts before. On the other hand the information of magnetic carts helped to trigger the students' prior experiences of magnetism, and they associated magnetic interaction with the cart's behavior.

We investigated the effect of direct hint versus indirect graduated hint. Students were asked to describe the least number and direction of gamma rays produced in the electron-positron annihilation process. None of the students applied the idea of momentum conservation themselves when the initial hint was phrased as 'momentum of system is zero' and several hints followed afterwards. To the other groups of students the question in the worksheet was phrased like the following: "When an electron and positron annihilate how many gamma rays in the least should be produced in order to conserve the momentum (hint: momentum of the electron-positron system was zero just before annihilation)". The result was that 15 out of 21 students (72%) of different groups individually came up with the idea immediately and wrote that there must be at least two gamma rays produced in the process to conserve momentum.

Peer Scaffolding in Transfer of Learning

We classified types of transfer into four categories. A student group is said to make spontaneous transfer if they correctly answer the PET problems and refer to the physical models immediately during their explanation. If they refer back the physical models upon being asked the basis of their answer, this group of students is considered to exhibit semi-spontaneous transfer. If they are successful in solving the PET problem but make the association of the physical models with the PET problem only after being asked if they had previously seen a similar situation, the students are said to be in non-spontaneous transfer class. They are said to be in no transfer class if they are not successful in their task of the PET problem.

A group was considered to transfer spontaneously if every student contributed in the problem solving and made associations with the physical models. If one student started the discussion and other students helped build upon each others' ideas, then we considered that the group as a whole transferred. On the other hand if a student in a group could solve the problem successfully and the other students of the group sought clarification to make sense of it, then the latter ones were considered in either the low level of transfer or no transfer class.

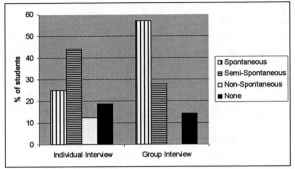

Figure 3: Transfer in individual and group interviews

Figure 3 indicates that the largest student population falls in the spontaneous class and semi-spontaneous comes second when students worked in groups. This is the reverse of our results with the individual interviews.

Students in the groups helped each other in triggering their association of the physical models with the PET problem. This association, which is considered as transfer of learning in this research, is facilitated through peer interaction. We regard that spontaneous and semi-spontaneous transfer are similar in terms of student achievements. The total percentage of these two in the individual interviews was almost 70% whereas that in the group interviews was 85%. This indicates that group interactions have enhanced the transfer of physics learning from the physical models to the PET problems.

CONCLUSION

Group teaching interviews were conducted to investigate the role of peer scaffolding to learning and transfer of learning to the understanding of PET. The results showed that the students' idea progressed significantly when they worked in groups. The students helped each other not only by challenging peer's ideas but also by providing them resources to conduct alternative reasoning. A large majority of the students (69%) could not complete a certain tasks by themselves. Group efforts resulted in the significant improvement in the students' achievements.

The change in students' responses with the change of sequence of hints or information was investigated. Some students were told that the carts were magnetic, and the others were asked if they had prior experience about the 'carts on the track' followed by a common question asking them to describe the motion of 'carts on the track'. The use of the idea of magnetic interaction was dominant in the former group and the use of the kinematics idea was dominant in the latter group while describing the motion of the carts. We also investigated the effect on sequencing of weak hints and strong hints in context of electron-positron annihilation. Students who were given the strong hint before the weak hint came up with the target idea more easily. Many of the students who got the hints in the opposite order could not get the target idea easily.

We discussed peer scaffolding in transferring physics learning to understand PET. The results of this study indicated that the students helped each other facilitate transfer of physics learning from the physical models to the PET problems. The students could trigger each other's ideas to associate the physical models with the related PET problems.

REFERENCES

* Supported by the National Science Foundation under grant DUE 04-2675.

1. M.Komorek and R.Duit, *International Journal of Science Education 26* (5), 619 (2004)
2. N. S. Rebello, D. A. Zollman, A. R. Allbaugh et al., in *Transfer of learning from a modern multidisciplinary perspective*, edited by J. P. Mestre (Information Age Publishing Inc., Greenwich, CT, 2005).
3. B. Aryal et al., 2006 *PERC Proceedings*, AIP Conf. Proc. No.883 (AIP, Melville, NY, 2007)
4. D. Johnson, R. Johnson, and K. Smith, *Active learning: Cooperation in the College Classroom*. (Interaction book, Edina, MN, 1998)
5. P. D. Cottle and G. E. Hart, Research in Science Education, 219 (1996).
6. L. S.Vygotsky, *Mind in society: The development of higher psychological processes*. (Harvard University Press, Cambridge, 1978).
7. T.Murray and I.Arroyo, in *International Conference on Intelligent Tutoring Systems* (London, UK, 2002).
8. E. Mazur, *Peer instruction: A user's manual*. (Prentice-Hall, Upper Saddle River, NJ, 1997).
9. L.Cohen and L.Manion, *Research methods in education*. (Routledge, New York, NY, 1994).

Student Perceptions of Three Different Physics by Inquiry Classes using the Laboratory Program Variables Inventory

Gordon J. Aubrecht, II

The Ohio State University Marion Campus

Abstract. Students from different versions of Physics by Inquiry courses (properties of matter, electric circuits, and astronomy by sight and optics) determined most and least characteristic aspects of their classes using the Laboratory Program Variables Inventory (LPVI), a Q-sort instrument. Students generally described these separate courses similarly, but with certain differences. We also compare student rankings with instructor rankings.

Keywords: Physics by Inquiry, Q-sort instrument, laboratory survey, physics education research
PACS: 01.40.Di, 01.40.Fk, 01.50.Qb

INTRODUCTION

The setting of Physics by Inquiry (PbI) classes at the Ohio State University (OSU) is the laboratory. Students do experiments as suggested by the text as well as creating their own experiments to test predictions they have made about nature's behavior.

Evaluation of laboratory classes often has been performed in a disorganized fashion and not been useful. Formative evaluation can be accomplished utilizing a form of Q-sort assessment that eschews affective information.

The Q-sort mechanism, devised by William Stephenson in 1935,[1] consists in organizing the statements or pictures into a ranking scheme. The assessment forces students to categorize the extent to which they think twenty-five descriptive statements characterize their laboratory class experience. They sort the statements from most to least characteristic of the course into boxes in bins of successive size 2, 6, 9, 6, 2 (forcing a "normal" distribution).

We have described our version of the instrument, the Laboratory Program Variables Inventory (LPVI), elsewhere. [2,3] This work presents a new example of the utility of instructor use of the LPVI, for comparing different groups' responses. Q-sort instruments can get first-hand data from large numbers of students in a short time; statements describe common lab activities. The LPVI measures what individual students think are the most and least characteristic features of classroom activities, not what they most like or most dislike about the course or the instructor. PbI uses guided inquiry, and students assessment of PbI classes reflect some important aspects of inquiry as conceived by instructors, while not supporting others

METHOD

Students in the Ohio State Physics by Inquiry (PbI) course were instructed to sort the 25 statements of the modified LPVI into five groups. Group I is considered most descriptive of the course; Group V is least descriptive. Students are forced to rank the statements into groups of size 2, 6, 9, 6, 2 by writing statement numbers into pre-designated boxes, forming a quasi-normal distribution. While the instructions imply that the statements are organized in groups, many students carefully sort and weigh each statement against the others before entering into boxes. We examine here the most descriptive and least descriptive statements as ranked by students in three different versions of the inquiry class—properties of matter (V.1 [4]), electric circuits (V. 2 [5]), and optics and astronomy by sight (V.1 and V.2 [4,5]). Data have been gathered on student rankings in these OSU PbI classes since Autumn 2004. The sample consists of responses of N = 174 students in properties of matter, N = 239 in electric circuits, and N = 53 in optics and astronomy (this class has fewer students overall, and I misplaced a large number of responses). In addition, 18 PbI instructors filled out the LPVI during this time period.

With these large numbers of students, nearly all statements are significant at the 99.99% level using chi-square. Only statements 2 and 3 in optics and astronomy in our sample are not significant at the 95% level (i.e., by convention indistinguishable from

CP951, *2007 Physics Education Research Conference*, edited by L. Hsu, C. Henderson, and L. McCullough
© 2007 American Institute of Physics 978-0-7354-0465-6/07/$23.00

chance). For the instructors, statements 3 and 4 are not significantly different from chance, but all others are significant above the 99.99% confidence level.

Filling the 25 statements by the five groups a matrix is created for each class; it is normalized by dividing by the appropriate N; the null result is extracted to give a matrix of variations (if student choices were random, all elements would be zero).

I also calculated an average "position" for each statement. For example, if each student chose the same statement as most descriptive, its position would be 1; if each chose one statement as least descriptive, its position would be 25. While "position" is not an exact measure, because not all students order the statements with painstaking precision, it is a proxy for the most and least descriptive statements (there is clearly no discriminatory power for the middle statements).

RANKING OF STATEMENTS

Do all three PbI classes' students see the courses (all taught the same way) as acting identically?

We adopted three different methods to determine the most descriptive and least descriptive statements. One, described above, is the average position; a second, denoted average score, was discussed in Ref. [3] and is formed by weighting groups I and V by 2 and II and IV by 1 ($2M_I + M_{II} - M_{IV} - 2M_V$); and the third, with equal weighting is denoted the matrix score ($M_I + M_{II} - M_{IV} - M_V$). The matrix score can be between 1 and -1, the average score between 2 and -2, and the position between 1 and 25. Tables 1 through 4 show the results for each group.

Each method results in very similar most and least descriptive statements in each table. It should be noted that the smallest position difference between statements listed and the closest lower statement is 1.16, 0.81, 0.45, and 1.06 for Tables 1 to 4, respectively, and

0.80, 0.89, 0.53, and 1.22, for Tables 1 to 4, respectively, for the least descriptive statements.

Taking averages for each category in Tables 1 to 4, we compare the four groups' responses in Table 5.

Table 5 shows a great deal of agreement among the groups. Statement 17 ("Students discuss their data and conclusions with each other."), statement 23 ("In discussion with the instructor, assumptions are challenged and conclusions must be justified."), and statement 13 ("The instructor or laboratory manual requires that students explain why certain things happen.") are perceived as strongly characteristic and statement 6 ("The instructor lectures to the whole class."), statement 24 ("Students usually know the general outcome of an experiment before doing the experiment."), and statement 7 ("Students are asked to design their own experiments.") are perceived as strongly uncharacteristic.

There is somewhat less importance attributed to statement 1 ("Students follow the step-by-step instructions in the laboratory manual.") and statement 5 ("Laboratory activities are used to develop concepts.") on the characteristic side and statement 21 ("Students identify problems to be investigated."), statement 14 ("Laboratory is used to investigate a problem that comes up in class."), and statement 4 ("Students are allowed to go beyond regular laboratory exercises and do experiments on their own.") on the uncharacteristic side.

Statements 16 ("Questions in the laboratory manual require that students use evidence to back up their conclusions.") and 22 ("During laboratory students check the correctness of their work with the instructor.") are interesting because of the difference between the groups. Instructors, electric circuit students, and optics and astronomy students believe that evidence is important, but properties of matter students rate it as less important than other aspects of the course.

TABLE 1. Properties of matter (N = 174)

	Statement	Matrix score	Statement	Average Score	Statement	Position
Most Charact- teristic	13	0.52	13	0.7	13	8.49
	23	0.46	23	0.63	23	8.99
	17	0.43	17	0.61	1	9.07
	5	0.41	1, 5	0.55	17	9.17
	1	0.38	22	0.47	5	9.22
	...	...	...	...	...	...
	21	-0.35	14	-0.53	14	16.8
Least Charact- teristic	14, 24	-0.46	24	-0.57	24	17.1
	4	-0.52	4	-0.72	4	17.6
	7	-0.6	7	-0.83	7	18.4
	6	-0.79	6	-1.41	6	21.3

TABLE 2. Electric circuits (N = 237)

	Statement	Matrix score	Statement	Average Score	Statement	Position
Most Charact-teristic	17	0.46	17	0.66	17	8.68
	13	0.44	23	0.63	13	9.05
	23	0.41	13	0.61	23	9.1
	16, 22	0.36	1, 16	0.48	1	9.76
	5	0.35	22	0.47	16, 22, 5	10.1
	. . .	. . .	. . .	. . .	. . .	. . .
	14	-0.33	21	-0.42	14	15.9
Least Charact-teristic	21	-0.35	14	-0.46	21	16
	7	-0.56	7	-0.76	7	18
	24	-0.61	24	-0.78	24	18.4
	6	-0.89	6	-1.69	6	22.8

TABLE 3. Optics and astronomy (N = 53)

	Statement	Matrix score	Statement	Average Score	Statement	Position
Most Charact-teristic	17	0.7	17	1.08	17	6.26
	23	0.68	23	0.92	13	7.38
	13	0.66	13	0.87	23	7.51
	16	0.55	16	0.68	16	8.4
	22	0.45	22	0.64	1, 22	9.55, 9.58
	. . .	. . .	. . .	. . .	. . .	. . .
	4	-0.4	4	-0.42	21	16.3
Least Charact-teristic	14, 21	-0.43	14, 21	-0.47	14	16.5
	7	-0.53	24	-0.64	24	17.9
	24	-0.55	7	-0.81	7	17.9
	6	-0.87	6	-1.7	6	22.9

TABLE 4. PbI instructors (N = 18)

	Statement	Matrix score	Statement	Average Score	Statement	Position
Most Charact-teristic	17	0.72	17	1.06	5, 17	6.67
	5	0.61	5	0.89	8, 23	8.06
	16	0.56	13	0.72	1	8.11
	8, 23	0.5	1, 23	0.67	16	9.17
	1	0.44	23	0.67	13	9.44
	. . .	. . .	. . .	. . .	. . .	. . .
	4	-0.33	4	-0.5	18, 7	17.2, 17.6
Least Charact-teristic	18	-0.44	14	-0.72	14	18.4
	7, 14, 21	-0.61	7, 21	-0.78	24	18.7
	24	-0.67	24	-0.89	21	19.2
	6	-1	6	-1.78	6	23.2

Electric circuit students and optics and astronomy students choose checking with the instructor as important aspects of the course, while properties of matter students do not rank this in the top five.

Instructors do not rate alternate explanations as encouraged, as their choice of statement 18 ("The instructor or laboratory manual asks students to state alternative explanations of observed phenomenon.") as uncharacteristic shows, but students do not recognize this aspect as strongly (it is also seen as uncharacteristic by students). As scientific consensus leads to just one explanation, it is not terribly surprising that the experiments lead to a single conclusion. Statement 8 ("During laboratory students record information requested by the instructor or the laboratory manual.") is also chosen among the top five only by instructors. Access to and reliance on these data is clearly more important to instructors than to students.

TABLE 5. Comparison of groups using their respective top five rankings (+: characteristic; -: uncharacteristic)..

Statement Number	PM N = 174	EC N = 237	OA N = 53	instructors N = 18
1	4+			3+
4	3-	5-		
5	5+			2+
6	1-	1-	1-	1-
7	2-	3-	3-	4-
8				4+
13	1+	2+	3+	5+
14	5-	5-	5-	4-
16		4+	4+	4+
17	3+	1+	1+	1+
18				5-
21		4-	5-	3-
22		5+	5+	
23	2+	3+	2+	2+
24	4-	2-	2-	2-

Table 5 presents only the top five among the ranked arithmetic averages of matrix score, average score, and position. To see how Table 5 is filled in, the averages for the top five PM characteristic statements (and corresponding rank) in the Table are first 13 (1.00), second 23 (2.00), third 17 (3.33), fourth 1 (4.00), and fifth 5 (4.33). The next highest-ranked average for PM students is statement 22 at an average of 5.67. The other entries in Table 5 are found in similar fashion. In EC, the next ranked statement after 22 (4.67) is 1 (5.00); in OA, the next ranked statement after 22 (5.00) is also 1 (6.33). The differences in PM and OA are quite large, while the difference in EC is quite small. The only other small difference is between OA uncharacteristic statements: the fifth least characteristic statement (14) averages 4.00, while unranked statement 4 averages 4.33, as small a difference as between statements 22 and 1 in EC.

There are many statements that are seen as strongly characteristic and strongly uncharacteristic that would likely please teachers like me who want students to see the utility of inquiry and to use inquiry in their own classes when they become teachers. Lectures don't happen. Conclusions are analyzed, justified, and explained by students. However, such teachers would probably want students to know they were able to do their own experiments extending what they are investigating. Ironically, in our OSU PbI courses, this sort of experimentation does happen, and often, but students apparently do not register it when it occurs. I have struggled with how to increase student recognition of their own experimentation but have not yet achieved it in my classes.

Because inquiry in PbI classes is guided, it is not surprising that students recognize that they are being led (by the teachers and the book) to the problems that are being investigated. The canonical understanding of nature is being sought, so students in PbI are led to that canonical explanation and alternate explanations, though essential to experimental science research, are not encouraged.

CONCLUSIONS

A benefit of Q-sort is that this can be used as a formative evaluation for the course (although because we give it at the end of the class, it is formative in the sense that future classes—not the current class—will benefit from any insights the instructor garners). The consonance between instructor goals and lab achievement in meeting these goals can be determined by comparing statements the instructor or instructors consider desirable and undesirable characteristics with student rankings. The choices of the student compared to those of the instructor could suggest ways to change the class so that students might converge to the instructor's choices.

This work supports the usefulness of Q-sort for assessing laboratory courses and shows that it can be used to provide instructors with formative assessment of their classes. The LVPI is a valuable tool for use because it gives instructors useful information about how students perceive what actually happened in a course without the need for lengthy classroom observations.

REFERENCES

1. Stephenson, W., "Technique of factor analysis," *Nature*, **136**, 297–299 (1935).
2. Lin, Y. , Demaree, D. , Zou, X., and Aubrecht, G. J., "Student Assessment Of Laboratory In Introductory Physics Courses: A Q-sort Approach," Proceedings of the 2005 Physics Education Research Conference, AIP, Melville, NY, 2006, 101-104.
3. Aubrecht, G. J., Lin, Y., Demaree, D., Brookes, D., and Zou, X., "Student Perceptions of Physics by Inquiry at Ohio State," Proceedings of the 2005 Physics Education Research Conference, AIP, Melville, NY, 2006, 113-116.
4. McDermott, L. C., *Physics by Inquiry*, vol. 1, Wiley, New York, 1995.
5. McDermott, L. C., *Physics by Inquiry*, vol. 2, Wiley, New York, 1995.

Humans, Intentionality, Experience And Tools For Learning: Some Contributions From Post-cognitive Theories To The Use Of Technology In Physics Education

Jonte Bernhard

Engineering education research group, ITN, Campus Norrköping, Linköping University, SE-601 74 Norrköping, Sweden

Abstract. Human cognition cannot be properly understood if we do not take the use of tools into account. The English word cognition stems from the Latin "cognoscere," meaning "to become acquainted with" or "to come to know." Following the original Latin meaning we should not only study "what happens in the head" if we want to study cognition. Experientially based perspectives, such as pragmatism, phenomenology, phenomenography, and activity theory, stress that we should study person-world relationships. Technologies actively shape the character of human-world relationships. An emergent understanding in modern cognitive research is the co-evolution of the human brain and human use of tools and the active character of perception. Thus, I argue that we must analyze the role of technologies in physics education in order to realize their full potential as tools for learning, and I will provide selected examples from physics learning environments to support this assertion.

Keywords: Labwork, experience, mediation, pragmatism, phenomenology, philosophy of technology, MBL.
PACS: 01.40.Fk, 01.40.Ha, 01.50.H-, 01.70.+w.

INTRODUCTION

In 1940 Müller [1] stated that:
"There is little evidence to show that the mind of modern man is superior to that of the ancients. His tools are incomparably better."
I argue here that to study cognition properly we must take an experiential perspective [2, 3] and study person-world relationships. To do this we must study the tools, i.e. technologies, used by humans since, as implied by Müller above and succinctly expressed by Mitcham [4], we *think through technology.*

COGNITION AND MIND

The English word "cognition" stems from the Latin "cognoscere," meaning "to become acquainted with," "know" or "to come to know." Thus, the meanings of the word's roots indicate that we should not merely consider "what happens in the head" when studying cognition. It is important to understand that "cognitive science" is not synonymous with "cognitivism" or "mentalism." What he calls the "myth of Descartes" and concepts of mind are more thoroughly discussed by Ryle [5]. Still and Costal [6] have summarized the dogma of cognitivism as

"the presumption that all psychological explanation must be framed in terms of internal mental representation, and processes (or rules) by which these representations are manipulated and transformed." (p. 2)

Instead I argue, following Dewey [7, 8], Peirce [9] and James [10] that cognition is related to capabilities for action [11, 12] and is itself a specific kind of action of the human body. This is stated by Rorty [13] as: "[we should not] view knowledge as a matter of getting reality right, but as a matter of acquiring habits of action for coping with reality." Similarly, Noë [14] argues that "perception is not something that happens to us, or in us, [but] something we do" and Wartofsky [15] argues that "[separating] *logos* from *praxis* is impossible" (p. 173).

Central to Dewey's theory of cognition is his principle of continuity, he states [3] that:

"there is no breach of continuity between operations of inquiry and biological operations and physical operations. 'Continuity' ... means that rational operations *grow out of* organic activities, without being identical with that from which they emerge" (p. 26).

Pragmatism provides us with a non-dualistic, non-representational model of an embodied mind. Lakoff and Johnson [16] (cf. [17]) purport the importance of Dewey and Merleau-Ponty [18] for the idea of the embodied mind. Related to this is the importance

CP951, *2007 Physics Education Research Conference,* edited by L. Hsu, C. Henderson, and L. McCullough
© 2007 American Institute of Physics 978-0-7354-0465-6/07/$23.00

Dewey saw in seeing acts as *dynamic* and *holistic* units. In his classic paper about the reflex arc concept [19] he criticized theories that turned the dynamic process of acting into a sequence of static and disjointed stimuli and responses.

In recent findings of cognitive science there is mounting evidence that the human brain co-evolved with the development of human culture and human use of tools [11, 20]. The neuroscientists Quartz and Sejnowski [21] summarize this as follows,

> "culture plays a central role in the development [together] with genes to build the brain that underlies who you are" (p. 58). They argue that "the central role of culture in our mental life reveals that intelligence isn't just inside the head" (p. 233). They further claim that "Our brains evolved to engage the world ... not to sit around passively. ...Brain functions ... are highly integrated and crosscut ... 'multiple' intelligences." (p. 249)

In line with the concept of "intentionality" stemming from Brentano [22] the theories of for example pragmatism, phenomenology [23] and phenomenography [2] emphasize that there is no detached thinking, seeing, learning etc. We always think of something, learning is always related to something etc. The term "Intentionality" implies that we must treat the human-world correlate as a single unity, like it is in the "experiential perspective."

Nardi and Kaptelinin [24] have attributed the term "postcognitivist" to theories of activity theory, distributed cognition, actor-network theory, and phenomenology, that are critical of the assumptions of "cognitivism." According to them a "major point of agreement among postcognitivist theories is the vital role of technology in human life [and that these] theories are highly critical of mind-body dualism." I here claim that pragmatism in the tradition of Dewey is missing in the list of "postcognitive" theories. Also, the term can give the mistaken impression that these theories post-dated "cognitivism." Instead, the term post should be seen as an expression of the currently increased interest in some "classic" cognitive theories.

MEDIATED ACTION AND PHILOSOPHY OF TECHNOLOGY

Human experience of our world is, as briefly mentioned in the introduction, shaped by physical and symbolic tools (mediating tools). The concept of mediation and mediating tools could be represented diagrammatically as:

Human ⇔ Mediating tools ⇔ World

Questions about the role of technology (artifacts) in everyday human experience include: How do technological artifacts affect the existence of humans and their relationship with the world? How do artifacts produce and transform human knowledge? How is human knowledge incorporated into artifacts? What are the actions of artifacts?

Tools play important roles in Dewey's philosophies of both education and technology [25]. In the socio-cultural theory and in activity-theory, which is rooted in the thinking of Vygotsky, "tool" and "mediation" are key concepts [12, 20, 26]. Miettinen [27] has pointed out the similarities between the thinking of Dewey and Vygotsky regarding tools and mediation.

FIGURE 1. A model showing the concept of mediation adapted and modified from Vygotsky [26] and Cole [20]: the triadic relationship between *subject – mediating tools – object* illustrating that the relationship is transformed by mediation.

The philosopher of technology Don Ihde synthesized non-foundational phenomenology and pragmatism in an approach dubbed postphenomenology [28]. According to him perception is co-determined by technology. In science instruments do not merely "mirror reality," but mutually constitute the reality investigated. The technology used places some aspects of reality in the foreground, others in the background, and makes certain aspects visible that would otherwise be invisible [29]. Neglecting the role of technology in science leads to naïve realism or to naïve idealism [29, 30]. Ihde developed the following schematic distinctions regarding mediated intentional relationships between humans and their world:

Embodiment: (Human ⇔ Technology) ⇔ World
Hermeneutic: Human ⇔ (Technology ⇔ World)
Alterity: Human ⇔ Technology (⇔ World)

In embodiment relationships we are normally unaware of the technology. In hermeneutic relationships some kind of interpretation is involved, hence the term hermeneutic. In both embodiment and hermeneutic relationships experience is transformed by the mediating technology. In alterity relationships humans are not related to the world through a technology, or to a world-technology complex, but to a technology.

TECHNOLOGY IN LABS

'Microcomputer Based Laboratory' (MBL) activities are examples of the use of "interactive technology" as a tool for learning in physics education [31]. In MBL activities students do experiments using various sensors (e.g., force, motion, temperature, light or sound sensors) connected to a computer via an interface. The arrangement provides a powerful system for *simultaneous* collection, analysis and display of experimental data, sometimes referred to as *real-time* graphing. The PER and lab-based curricula "Tools for scientific thinking" and "Real-Time Physics" have proven effective in fostering a functional understanding of physics [32], and in the "experientially based physics" project MBL has proven to be effective in a Swedish context [33], achieving normalized gains in the FMCE-test of 61%.

However, I have shown that the same sensor-computer-technology ("probeware") used in MBL can also be implemented in ways that lead to low achievements in conceptual tests, thus refuting technological determinism. My findings indicate that the form of the educational implementation is crucial [33], i.e. we must look at how the intentional *Human-Technology-World* relationship is established.

Nevertheless, as noted by Ihde [29] and Kroes [34], for example, observation is not generally regarded as problematic in positivist approaches and from the anti-positivist perspective, the praxis-ladenness of observations tends to be overlooked. Kroes expresses this as follows:

> "[i]n [the traditional] view, the physicist is essentially a passive observer in experiments: once the stage is set he just observes (discovers) what is going to happen."

Figure 2 illustrates two common views of technology in education. In these views the *Human – World* relationship is not seen as being affected.

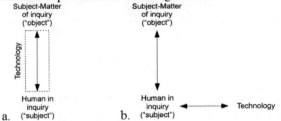

a.   b.

FIGURE 2. a) A 'Transmissive' view of technology where technology is seen merely as a vehicle for information **b)** An 'Auxiliary' view where technology is seen merely as a provider of information or support.

According to "Variation theory," developed by Marton and co-workers [35], we learn through the experience of difference, rather than the recognition of similarity. In this theory the experience of discernment, simultaneity (synchronic and asynchronic) and variation are necessary conditions for learning.

It is not possible in this short paper to present a full phenomenological analysis [23, 36] of the role of technology and the *Human-Technology-World* relationships that the learning environment affords [37]. However, I will briefly discuss an example from one of the earlier tasks in a typical MBL-lab. In this task students are asked to walk a trajectory that matches a given velocity-time graph. While moving the student, and his/her peers, can see the experimental graph produced in *real-time* (see figure 3). Prior to this, students have solved tasks involving position-time graphs.

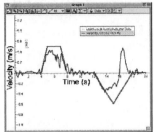

FIGURE 3. Example of a task that students attempt in a MBL-lab. Displayed is a v(t)-graph with a curve that the students are asked to recreate together with an experimental graph produced by a student.

As mentioned above, the way technologies are implemented shapes figure-background relationships, and the variations that can be discerned. Wartofsky [15] expressed this as follows (p. 204):

> "I take the artifacts (tools and languages) to be objectifications of human needs and intentions ... *already* invested with cognitive and affective content."

What the technology does in this task is to bring velocity to the fore, i.e. it enters in the focal awareness [38] of students. Other features of the situation, physical as well as non-physical, are not highlighted, i.e. some discernment has already occurred. It is also important that velocity is established as a relationship to objects and events in the world (cf. [39]). In order to complete the assignment, students have to understand this and also make important conceptual distinctions.

I have examined labs that use "probeware" and have lead to either low or high achievements. In high-achieving labs the technology is used to bring important concepts and relationships into students' focal awareness, i.e. it is used as a "cognitive tool." A preliminary analysis of the critical aspects of "probeware" use have been presented previously [33] and a paper containing an in depth analysis based on variation theory and the philosophy of technology is forthcoming (cf. [40]).

CONCLUDING REMARKS

In this short essay it has only been possible to give a brief account of *some* theories that could be applied in analyzing the role of technologies in physics education. Important theories, such as those for tool-use by Heidegger and actor-network theory by Latour, have not been discussed due to space limitations.

In conclusion I contend that to use technologies as learning tools we must understand their cognitive role(s), identifying which aspects of the world are brought into focus, thus making learning possible. By understanding this, and the active nature of perception, we can use technologies to their full potential.

ACKNOWLEDGEMENTS

Grants from the Swedish Research Council are gratefully acknowledged.

REFERENCES

1. R.H. Müller, "American Apparatus, Instruments, and Instrumentation," Ind. and Eng. Chemistry: Anal. Ed., **12**, 571-630 (1940).
2. F. Marton, "Phenomenography: Describing conceptions of the world around us," Instr. Sci., **10**, 177-200 (1981).
3. J. Dewey, "Experience and Nature," in *John Dewey: The Later Works (Vol. 1)*, J.A. Boydston, ed., Southern Illinois University Press, Carbondale, 1981 (1925).
4. C. Mitcham, *Thinking through Technology: The Path between Engineering and Philosophy*, The Univ. of Chicago Press, Chicago, 1994.
5. G. Ryle, *The Concept of Mind*, The Univ. of Chicago Press, Chicago, 1949.
6. A. Still and A. Costall, *Cognitive psychology in question*, Harvester Press, Brighton, 1987.
7. J. Dewey, "How We Think: A Restatement of the Relation of Reflective Thinking to the Educative Process," in *John Dewey: The Later Works (Vol. 8)*, J.A. Boydston, ed., Southern Illinois Univ. Press, Carbondale, 1986 (1933).
8. J. Dewey, "Logic: The Theory of Inquiry," in *John Dewey: The Later Works (Vol. 12)*, J.A. Boydston, ed., Southern Illinios Univ. Press, Carbondale, 1986 (1938).
9. C.S. Peirce, "How To Make Our Ideas Clear," Popular Science Monthly, **12**, 286-302 (1878).
10. W. James, *The Principles of Psychology*, Henry Holt, New York, 1890.
11. M. Cole and J. Derry, "We Have Met Technology and It Is Us," in *Intelligence and Technology: The Impact of Tools on the Nature and Development of Human Abilities*, R.J. Sternberg and D.D. Preiss, eds., Lawrence Erlbaum, Mahwah, 2005.
12. J.V. Wertsch, *Mind as Action*, Oxford Univ. Press, Oxford, 1998.
13. R. Rorty, *Objectivity, Relativism, and Truth: Philosophical Papers Volume 1*, Cambridge Univ. Press, Cambridge, 1991.
14. A. Noë, *Action in Perception*, The MIT Press, Cambridge, 2004.
15. M.W. Wartofsky, *Models: Representation and the Scientific Understanding*, Reidel, Dordrecht, 1979.
16. G. Lakoff and M. Johnson, *Philosophy in the Flesh: The Embodied Mind and Its Challenge to Western Thought*, Basic Books, New York, 1999.
17. M. Johnson and T. Rohrer, "We Are Live Creatures: Embodiment, American Pragmatism, and the Cognitive Organism," in *Body, Language, and Mind*, J. Zlatev, et al., eds., Mouton de Gruyter, Berlin, in press.
18. M. Merleau-Ponty, *Phenomenology of Perception*, Routledge, London, 1962 (1945).
19. J. Dewey, "The Reflex Arc Concept in Psychology," Psych. Rev., **3**, 357-370 (1896).
20. M. Cole, *Cultural psychology: A once and future discipline*, Harvard Univ. Press, Cambridge, 1996.
21. S. Quartz and T. Sejnowski, *Liars, lovers, and heroes: what the new brain science reveals about how we become who we are*, William Morrow, New York, 2002.
22. F. Brentano, *Psychology from an Empirical Standpoint*, Routledge, London, 1995 (1874).
23. D. Ihde, *Experimental Phenomenology: An Introduction*, State Univ. of New York Press, Albany, 1986 (1977).
24. B.A. Nardi and V. Kaptelinin, *Acting with Technology: Activity Theory and Interaction Design*, MIT Press, Cambridge, 2006.
25. L.A. Hickman, *John Dewey's Pragmatic Technology*, Indiana Univ. Press, Bloomington, 1990.
26. L.S. Vygotsky, *Mind in society: The development of higher psychological processes*, Harvard Univ. Press, Cambridge, 1978.
27. R. Miettinen, "Artifact Mediation in Dewey and in Cultural-Historical Activity Theory," Mind, Culture & Activity, **8**, 297-308 (2001).
28. E. Selinger, ed., *Postphenomenology: A Critical Companion to Ihde*, State Univ. of New York Press, Albany, 2006.
29. D. Ihde, *Instrumental Realism: The Interface between Philosophy of Science and Philosophy of Technology*, Indiana Univ. Press, Bloomington, 1991.
30. D. Ihde and E. Selinger, eds., *Chasing Technoscience: Matrix for Materiality*, Indiana Univ. Press, Bloomington, 2003.
31. R.F. Tinker, ed., *Microcomputer-Based Labs: Educational research and Standards*, Springer, Berlin, 1996.
32. D. Sokoloff, P. Laws, and R. Thornton, "RealTime Physics: active learning labs transforming the introductory laboratory," Eur. J. Phys, **28**, S83-S94 (2007).
33. J. Bernhard, "Physics learning and microcomputer based laboratory (MBL): Learning effects of using MBL as a technological and as a cognitive tool," in *Science Education Research in the Knowledge Based Society*, D. Psillos, et al., eds., Kluwer, Dordrecht, 2003, pp. 313-321.
34. P. Kroes, "Physics, Experiments, and the Concept of Nature," in *The Philosophy of Scientific Experimentation*, H. Radder, ed., Univ. of Pittsburgh Press, Pittsburgh, 2003, pp. 68-86.
35. F. Marton and A.B.M. Tsui, eds., *Classroom Discourse and the Space of Learning*, Lawrence Erlbaum, Mahwaw, 2004.
36. D. Ihde, *Technics and Praxis*, Reidel, Dordrecht, 1979.
37. J. Gibson, *The Ecological Approach to Visual Perception*, Houghton Mifflin, Boston, 1979.
38. F. Marton and S. Booth, *Learning and Awareness*, Erlbaum, Mahwah, 1997.
39. A. Tiberghien, "Labwork activity and learning physics - an approach based on modeling," in *Practical work in science education*, J. Leach and A. Paulsen, eds., Roskilde Univ. Press, Fredriksberg, 1998, pp. 176-194.
40. J. Bernhard, "Thinking and learning through technology," The Pantaneto Forum, **27** (2007).

Improving Students' Conceptual Understanding of Conductors and Insulators

Joshua Bilak and Chandralekha Singh

Department of Physics and Astronomy, University of Pittsburgh, Pittsburgh, PA, 15260

Abstract. We examine the difficulties that introductory physics students, undergraduate physics majors, and physics graduate students have with concepts related to conductors and insulators covered in introductory physics by giving written tests and interviewing a subset of students. We find that even graduate students have serious difficulties with these concepts. We develop tutorials related to these topics and evaluate their effectiveness by comparing the performance on written pre-/post-tests and interviews of students who received traditional instruction vs. those who learned using tutorials.

INTRODUCTION

Conductors and insulators are taught at increasing levels of mathematical sophistication starting from algebra- and calculus-based introductory physics (or high school physics) to graduate level physics courses [1, 2, 3, 4]. Considering the frequent appearance of these topics in electricity and magnetism (E&M) courses at various levels, one may assume that physics graduate students have a clear understanding of these concepts. Here, we investigate the conceptual difficulties that introductory students, undergraduate physics majors, and physics graduate students have with concepts related to conductors and insulators covered at the introductory level. We also develop tutorials and evaluate their effectiveness by comparing the performance of students who learned these topics using traditional and tutorial instructions.

First, conceptual multiple-choice questions were developed and administered to introductory students and physics graduate students and interviews were conducted with some students. We then developed tutorials and pre-/post-tests on these topics at the level of introductory physics. We performed a controlled study and gave the pre-/post-tests in equivalent introductory physics courses before and after traditional or tutorial instruction. These tests require written responses often asking students for drawings and explanations. Interviews with introductory students and physics majors were conducted using a "think-aloud" protocol to gain further insight into their reasoning and the origins of their difficulties. In this protocol, students are asked to talk aloud as they work through the material without interruption and they are asked for clarification of the points they had not otherwise made clear at the end. Some physics majors worked on the tutorials while thinking aloud and were given

the pre-/post-tests while other physics majors worked through the pre-/post-tests only without working on the tutorials. In some interviews, introductory students and physics majors were shown relevant experimental setups, asked to predict what should happen in certain situations and then asked to perform the experiment to check their prediction and reconcile the differences between their original prediction and observation if they did not match. We briefly discuss some findings.

RESULTS AND DISCUSSION

We first discuss the performance of introductory students and graduate students on three conceptual multiple-choice questions. Their performance shows that neither group has an adequate comprehension of these concepts. While 235 introductory students in three different classes were administered these questions after instruction as part of a recitation quiz, 19 first year graduate students worked on them as part of a placement quiz for entry to a Jackson-level E&M course. Graduate students were told ahead of time that their placement quiz will be on undergraduate level E&M. Both graduate and introductory students made very similar mistakes and the commonality suggests that these concepts covered in introductory physics are indeed challenging. The questions described below deal with the charge distribution on a conductor with a charge outside, the effect of grounding a conductor with a charge nearby, and the electric field inside a hollow insulator due to a charge outside:

(1) A small aluminum ball hanging from a thread is placed at the center of a tall metal cylinder and a positively charged plastic rod is brought near the aluminum ball in such a way that the metal wall is in between the rod and the ball. Which one of the following is the correct qualitative picture of the charge distribution in this

CP951, *2007 Physics Education Research Conference*, edited by L. Hsu, C. Henderson, and L. McCullough
© 2007 American Institute of Physics 978-0-7354-0465-6/07/$23.00

situation?

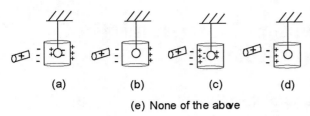

(a)   (b)   (c)   (d)

(e) None of the above

(2) A positively charged plastic rod is taken close to (but not touching) a neutral aluminum ball hanging from a thread. The aluminum ball is grounded by connecting it with a wire that touches the ground. Which one of the following statements is true about the force between the rod and the ball?

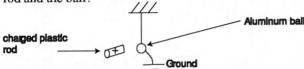

(a) The aluminum ball will not feel any force due to the charged rod.
(b) The aluminum ball will be attracted to the charged rod and the force of attraction is the same as that without the grounding wire.
(c) The aluminum ball will be attracted to the charged rod and the force of attraction is more than that without the grounding wire.
(d) The aluminum ball will be attracted to the charged rod and the force of attraction is less than that without the grounding wire.
(e) The aluminum ball will be attracted or repelled depending upon the magnitude of the charge on the rod.

(3) A small aluminum ball hanging from a thread is placed at the center of a tall cylinder made with a good insulator (Styrofoam) and a positively charged plastic rod is brought near the ball in such a way that the Styrofoam wall is in between the rod and the ball. Which one of the following statements is true about this situation?

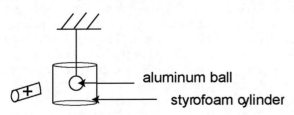

(a) The aluminum ball will not feel any force due to the charged rod.
(b) The aluminum ball will be attracted to the charged rod and the force of attraction is the same as that without the Styrofoam cylinder.
(c) The aluminum ball will be attracted to the charged

rod but the force of attraction is more than that without the Styrofoam cylinder.
(d) The aluminum ball will be attracted to the charged rod and the force of attraction is less than that without the Styrofoam cylinder.
(e) None of the above.

TABLE 1. The average percentage of 235 introductory physics students and 19 physics graduate students who chose options (a)-(e) on the three multiple-choice questions. The correct response for each question is italicized.

Q	Group	a	b	c	d	e
1	Introductory Students	44	*23*	26	6	1
	Graduate Students	32	*37*	16	5	11
2	Introductory Students	21	15	*44*	16	1
	Graduate Students	16	32	*53*	0	0
3	Introductory Students	27	1	12	*46*	1
	Graduate Students	32	26	0	*42*	0

Table 1 shows that these questions were difficult and the performances of the introductory students and graduate students on each of these questions is not much different with neither group performing better than 53% on any of them. While it is true that the introductory students had learned these concepts recently so they may be fresh in their minds, the questions are sufficiently conceptual that we expected the graduate students to perform significantly better than they actually did.

Question (1) probes student's knowledge of the charge distribution via induction due to a charge outside the cavity. The correct response is option (b). The electric field inside the conductor is zero with no charge separation on the aluminum ball. Approximately one-fourth of the introductory students and one-third of the graduate students provided the correct response. The difficulties were common for both groups and 70% of the introductory students and half of the graduate students believed that there will be a charge separation on the aluminum ball as well (options (a) and (c)). During individual interviews with introductory students, those who chose this option explained that the negative and positive charges on the wall of the hollow conductor will exert forces on the electrons in the aluminum ball and induce a charge separation in the ball. When asked explicitly about the electric field in the hollow region, these students often believed that there was a non-zero electric field inside. When asked explicitly about the charges that produce the electric field in the hollow region, some students claimed that the charge on the plastic rod outside will also contribute to the field inside in addition to the induced charges. Others claimed that it was only the charges induced on the walls of the hollow region that produce the field inside because the electric field due to the outside charge on the plastic rod was shielded by the conductor. Thus, although some of the students used the concept of shielding the inside of the conductor from the charges outside, they did not realize that shielding implies that the net electric field inside the conductor is zero due to the cumulative effects of the charges on the plastic rod and those induced on the outer walls of the conductor. Interviews suggest that some students chose options (a) or (c) because they believed incorrectly that the situation given is similar to the one with a charge inside a conductor cavity (perhaps because there was an aluminum ball in the cavity). Unlike the given situation, the charge inside a cavity can produce a field inside the cavity in addition to inducing charges on the cavity walls.

In Question (2), if the aluminum ball with a positively charged rod nearby is grounded, free electrons come from the ground and neutralize the positive charge on the aluminum ball so that the net attraction between the charged rod and the aluminum ball is more than when there was no grounding wire (option (c)). Approximately half of the students from both groups did not provide the correct response. The incorrect responses suggest that one-third of the graduate students believed that the grounding wire will not affect the force between the aluminum ball and the charged plastic rod whereas 16% of them believed that a grounded object cannot be charged. The introductory students had similar difficulties. In a pretest given to introductory students before college instruction about these concepts (in which they had to draw diagrams and explain their reasonings), students were asked a similar question involving a grounded steel ball in the vicinity of a positively charged comb (see figure below). Some students who had high school instruction in these concepts drew incorrect diagrams such as the following, in which protons travel from the ball to the ground and the electrons on the steel ball spread out uniformly on its surface while the comb is still nearby:

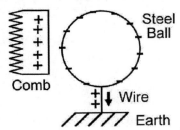

Written explanations and diagrams drawn on the pretest (before college instruction) and post-test (after college instruction) given to introductory students and physics majors in the free-response format and interviews with some of them suggest that students had difficulty with the role of grounding and reasoning about the changes in the charge distribution due to grounding. A major problem with this concept of grounding dealt with inaccurate characteristics attributed to the Earth. Since the Earth is a conductor, some students believed that it is negatively charged. This notion caused some students to draw positive charges on the steel ball shifted downward due to the attraction from the negatively charged Earth as in the following drawing by a student who answered the question while "thinking-aloud":

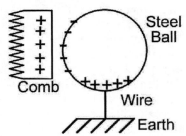

Question (3) asks how an aluminum ball, encased in an insulator (Styrofoam), will be affected by the presence of a positively charged rod outside. Since Styrofoam does not have free electrons, the atoms in the Styrofoam will become polarized and the electric field inside the Styrofoam will be smaller than that without the Styrofoam. Therefore, the force on the free electrons in the aluminum ball inside the Styrofoam (and hence the separation of charges on the aluminum ball) will be smaller than the case without the Styrofoam. The force of attraction between the plastic rod and the ball will be less than that without the Styrofoam (option (d)). There is no significant difference between the performance of the introductory students and graduate students on this question, with both groups obtaining less than 50%.

Surprisingly, approximately one-third of the graduate students invoked the notion of shielding and believed that no force would be felt by the aluminum ball from the plastic rod outside. Introductory physics students and physics majors participating in the think-aloud protocol often explained that an insulator will insulate the inside of the cylinder from the outside and prevent the electric field due to the charges on the plastic rod from penetrating the walls and reaching the inside. Written responses requiring explanations and interviews suggest that many students took the word "insulator" literally. These students incorrectly claimed that a conducting wall around a cavity is not nearly as effective as an insulating wall in preventing the electric field from the external sources from penetrating inside in analogy with an insulating wall required to prevent heat transfer. Asking students for the microscopic mechanism for zero field, i.e., asking them to account explicitly for the charges whose electric fields will add up vectorially to make the net electric field zero everywhere inside the insulator turned out to be difficult for them. Most students had not thought about why an electrical insulator will insulate the inside from the outside charges, but believed in it intuitively. In some interviews, after making the prediction that the inside of the Styrofoam is not influenced by the charges outside, students were asked to conduct the experiment, they observed that the aluminum ball inside the Styrofoam is attracted to the plastic rod and the electric field inside the Styrofoam cannot be zero. However, this observation alone was not sufficient to help them formulate correct reasoning about this situation. We plan to evaluate student performance on these questions with the word insulator replaced with non-conductor.

These universal difficulties that persist even at the graduate level inspired us to develop tutorials and pre-/post-tests about these concepts. The details of the tutorials that use a guided approach to helping students reason about these concepts will be provided elsewhere. We performed a controlled study involving different introductory physics courses to evaluate the effectiveness of the tutorials and gave the pre-/post-tests to students (before and after college instruction) in classes using traditional and tutorial instructions. The pre-tests and post-tests required students to explain their reasoning and draw diagrams. We also interviewed some students. Incidentally, there was no significant difference in the performance of students on the pre-test who had high school instruction in conductors and insulators vs. those who did not. The graph in Figure 1 below shows that there was no significant improvement in introductory students' performance on the post-test compared to the pre-test when students were taught traditionally using lecture alone (lec. group in the graph). The grading for these free-response questions was done using a scale where each incorrect response was coded as 0, partially correct response as 1 and a correct response as 2. The pre-test scores for both groups (lec. and tut.) were combined because there was no significant difference. The tutorial group performed significantly better on the post-test.

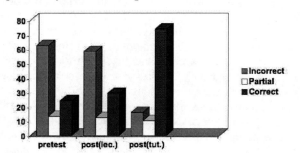

FIGURE 1. The average percentage of post-test responses that were answered incorrectly, partially correctly, or correctly.

For example, when asked to show the distribution of the induced charges on a conducting shell with a point charge outside the shell (similar to the multiple-choice question (1)), the students taught traditionally (via lecture only) continued to have trouble on the post-test. Compared to the traditionally taught students, 24%

more students who learned using the tutorials depicted the charge configuration correctly. Many students in the traditional group continued to draw induced charges on the inner walls of the conducting shell.

On the pre-/post-test free response questions that were similar to the multiple-choice question (2) with the grounding wire, many traditionally taught students displayed difficulties similar to those on the pre-test. They still drew positive charges leaving the conductor via the grounding wire or shifted the positive charge downward assuming that free electrons in the Earth will attract the positive charges on the conductor. Although some students realized that the order in which the grounding wire and the charged comb are removed is important in order to make the conductor develop a net charge via this process, they often did not provide an adequate explanation for how this is the case. On the other hand, the tutorial group performed significantly better on the post-test questions about grounding. Not only could many of these students draw the final charge configuration on the conductor after induction and grounding processes, they drew and justified the proper charge configuration.

CONCLUSIONS

We find that the concepts related to conductors and insulators covered in introductory physics are very challenging not only for the introductory physics students but also for the physics graduate students. Since homeworks or exams in upper-level courses seldom require students to make qualitative inferences from quantitative tools, students do not develop a good grasp of the underlying concepts or learn to apply the formalism to physical situations. The conceptual questions discussed here often require long chains of systematic reasoning. These difficulties motivated the development of tutorials about these concepts which were evaluated using a controlled study in which students were given a pre-test and post-test before and after the traditional and tutorial instructions. The tutorials have been used by introductory students and physics majors and are helpful in improving students' understanding of these concepts.

ACKNOWLEDGMENTS

We are grateful to the NSF for award DUE-0442087.

REFERENCES

1. D. Maloney, T. O'Kuma, C. Hieggelke, A. V. Heuvelen, Am. J. Phys. **69**, S12 (2001).
2. V. Otero, Proceedings of the International School of Physics, eds. Vicentini and Redish, (IOS Amsterdam), Varenna, Italy, 1-33, (2003).
3. C. Singh, Am. J. Phys., **74**(10), 923-936, (2006).
4. L. Ding, R. Chabay, B. Sherwood, R. Beichner, Phys. Rev. ST PER **1**, 10105, (2006).

Epistemic Games in Integration: Modeling Resource Choice

Katrina E. Black[*] and Michael C. Wittmann[*¶†]

[*]Department of Physics and Astronomy, University of Maine, Orono, ME 04469, USA
[¶]College of Education and Human Development, University of Maine, Orono, ME 04469, USA
[†]Center for Science and Mathematics Education Research, University of Maine, Orono, ME 04469, USA

Abstract. As part of an ongoing project to understand how mathematics is used in advanced physics to guide one's conceptual understanding of physics, we focus on students' interpretation and use of boundary and initial conditions when solving integrals. We discuss an interaction between two students working on a group quiz problem. After describing the interaction, we briefly discuss the procedural resources that we use to model the students' solutions. We then use the procedural resources introduced earlier to draw resources graphs describing the two epistemic game facets used by the students in our transcript.

Keywords: Physics education research, differential equations, air resistance.
PACS: 01.40Fk

INTRODUCTION

We find that students typically use one of two methods to apply boundary conditions to integrals that must be performed when solving first-order, separable differential equations (FOSDEs) in a physics setting. In one method, an undetermined integration constant ("+c") is added following integration and the boundary condition is used to find the value of the constant. In the other method ("limits"), the boundary or initial conditions are used to choose appropriate limits of integration. Although these methods lead to physically equivalent solutions, the choice of method, whether tacit or deliberate, is found to affect students' abilities to arrive at complete descriptions of the physics. We found that in physics-like scenarios, students who used the "+c" method did not use the boundary condition to find the value of the integration constant; rather, they left the constant unspecified. We present a detailed transcript of two students discussing these methods and model their reasoning in terms of conceptual and procedural resources[1,2] as well as a description of their framing[3] of the problem and the epistemic games[4,5] they use.

I. BACKGROUND AND TRANSCRIPT

In 2007, students in an intermediate mechanics course using small-group learning materials (the *Intermediate Mechanics Tutorials*[6,7]) took part in group quizzes. In one quiz, students were asked to find the horizontal velocity with respect to time of a beach ball thrown horizontally, assuming a v^2 air resistance and

no appreciable vertical motion during the motion. In one group, two students, Max and Phil (*aliases*), having set up the equation (Eqn. 1) and separated variables to create

$$m\frac{dv}{dt} = -bv^2 \qquad (1)$$

two integrals (Eqn. 2), debate whether to use the "+c" or the "limits" method to integrate the dt side of the separated differential equation.

$$\frac{-m}{b}\int \frac{dv}{v^2} = \int dt \qquad (2)$$

(Earlier, they indefinitely integrated the dv side without adding a constant of integration.)

Max, advocating for "limits," and Phil, advocating for "+c," are explicit about their reasons for choosing each method. Phil's reasoning provides a possible explanation for the failure to find the value of the integration constant observed in earlier studies of students using the "+c" method.

At this point in the discussion, Phil, who has led the group throughout, asserts that the next step in the solution is to indefinitely integrate the right hand (t) side and add an integration constant.

1 *Phil*: so we get negative m over b times negative v
2 to the negative one equals t plus c.
3 *Max*: No, this is where you should be going just
4 plain t, or t-naught to t, or zero to t, because initial
5 time is when you throw it to some time, so just go
6 to t instead of t plus c. <pause> Because then
7 your integration of time would probably be going
8 from zero, when you throw it, to t, some time
9 later. So you don't need c there

In an earlier part of the problem (transcript not given), Max had not objected that limits were not used when performing the v-integral. In the t-integral, he objects to the addition of a constant. His use of language (line 3: "you should be going" and lines 5 and 6: "just go to t") implies that he would prefer using limits of integration based on the physical situation. Phil, however, is not convinced. Although he offers to change the name of the constant to indicate that it represents a time, he is unwilling to assign that constant any particular value.

10 *Phil*: Hmmmmm... or you could do t plus
11 t-naught. You need a constant though. I mean
12 yeah, if you did it

When Max asks why a constant is needed (line 13), Phil reveals his reasoning: without a constant, you don't have a "function" (lines 22–24):

13 *Max*: Why would you need a constant in time?
14 You just go from 0 to t.
15 *Phil*: Because it's <sigh> --
16 *Max*: -- t-naught to t
17 *Phil*: -- that's not a function, well
18 *Max*: well, you could go from t-naught to t, but
19 t-naught is just initially zero when you start
20 throwing the ball.
21 *Phil*: If you, ssssss, no, you can't do that, because
22 then you're not setting up a function. You're
23 setting up a function that's only ok in certain
24 cases, you see what I mean?

From lines 22 to 24, we infer that what Phil is actually referring to is a family of functions. A function "that's only OK in certain cases" is not general enough. It seems that, to Phil, the integration constant must remain as general as possible. In contrast, Max is trying to use information specific to this problem in working through the mathematics.

Later, Phil and Max are still discussing the limits versus integration constant issue. Now, Phil does not object to using limits if they are arbitrary (for example, from t_0 to t) but still does not want to apply a specific boundary condition to the equation. The problem statement does not specifically say that you throw the ball at t = 0, and Phil seems uncomfortable using that common physics convention. Again, he argues that the final equation must be one that "anybody can use" (a family of functions) and not "one that you know" (an equation with a specific boundary condition applied):

25 *Phil*: I just can't imagine integrating without
26 having an extra constant. I guess if you said
27 t-naught is zero, so, I mean, yeah
28 *Max*: but t-naught is the initial time
29 *Phil*: But you don't know that's zero.

30 *Max*: You can just mark it as zero anyway.
31 <pause> for consistency. Why, why do you need
32 some extra time there?
33 *Phil*: Yeah, but then you're not creating an
34 objective function that anyone can use. You're
35 making a function that **you** know, and that's great,
36 but it's not really, I mean,
37 *Max*: Well, no.
38 *Phil*: the math has to cover everything. We'll
39 compromise, we'll say t minus t-naught. Write
40 that. That work?
41 *Max*: That's fine
42 *Phil*: OK

In lines 33 to 35, we again see Phil trying to be as general as possible. Max, though, is looking for consistency between the physics and the math (lines 30 to 32). As we describe below, their compromise (lines 39 to 41) satisfies both their agendas in solving this problem, and they are able to move forward without actually resolving the situation.

II. CONCEPTUAL AND PROCEDURAL RESOURCES

Resources are a model to describe small chunks of student knowledge[1,2]. Although most resources discussed in the literature are conceptual, the definition of resources as pieces of knowledge also leaves room for procedural, factual, and analytical resources.

We describe several procedural resources that we observe students using during the integration described above. In particular, these resources are used after the students have separated variables but before the final solution has been reached. These resources are part of larger activities, *Definite Integration* and *Indefinite Integration*, which describe the algorithmic computation of an integral. We have found, in several years of observation, that a student's past exposure to integration in math and physics classes often has a strong influence on which of these resources is activated. We state explicitly that we use the term "resources" here to describe constructs and procedures that often require and contain sub-steps that we do not make explicit. Instead, our goal is to clarify the many different kinds of procedures that students must carry out when solving integrals in a physics class. Analyses at different grain sizes (*e.g.*, smaller procedural steps) are possible.

Two procedural resources that students might use when integrating a physically meaningful equation are *Find Value* and *Choose Limits*. Each connects the physical meaning of the equation to the mathematical formalism and involves several steps when applied. These resources are frequently used in conjunction with the resource *Extract Boundary Condition*, in which the student uses the problem statement to develop the mathematical form of the boundary conditions.

We have found (though the transcript above does not show it) that the presentation statement of the boundary condition can influence how strongly the *Extract Boundary Conditions* resource is primed and what other resources it gets linked to. In a math-like problem, the boundary condition, if it exists, is already stated in its mathematical form; it is pre-extracted in the form $v(0)=0$ (note the lack of units in this math-like formalism). On the other end of the continuum, a very physics-like problem has no boundary conditions stated, either mathematically or linguistically. In this case the student must reason about the described situation to determine the boundary conditions. An intermediate type problem might state the boundary conditions explicitly in words, leaving the student to do the translation to mathematics.

In this language, we observe Phil and Max using different procedural resources and arguing about the implementation of *Extract Boundary Conditions* and the subsequent of activation of *Find Value* for Phil and *Choose Limits* for Max. From this, we can construct a more complete description of their work, combining the representation of resource graphs [2] with a description of epistemic games in physics [8].

III. DESCRIBING EPISTEMIC GAMES WITH GRAPHS OF PROCEDURAL RESOURCES

Much as resources are a kind of schema, epistemic games are one kind of script [8]. (The former describes conceptually-linked ideas, the latter time-ordered actions.) Epistemic games are sets of rules that are followed when creating new knowledge. They have starting conditions, moves, and an ending condition, known as the epistemic form. When solving a problem, a student takes a particular path through the primed resources of his or her frame. If the links between these resources are robust and often activated in the same order in varying situations, we can call this path an epistemic game. Because the resources primed in a setting depend on other resources activated in that setting, the epistemic games available for play are also heavily personally and contextually dependent; more than one game may be available in a particular situation. Similarly, since resources can be applied in a variety of situations, the same game may be available in different situations. In physics, epistemic games have been described at a fairly large grain size [5].

In our case, both Max and Phil can be seen as playing variations of the game "Mapping Meaning to Mathematics," described by Tuminaro [5]. This game has the moves *Develop story* about physical situation, *Translate quantities* in physical story to mathematical entities, *Relate mathematical identities* in accordance with physical story, *Manipulate symbols*, and *Evaluate story*. Specific details differ for Max and Phil, though.

Each specific move in the "Mapping Meaning to Mathematics" game can be described as a procedural resource. To describe the game as a whole, we build a resource graph out of these procedural resources. In general, at this large grain size, the network is relatively linear and strongly unidirectional. In practice, we find that certain loops exist and are commonly activated as a whole. This phenomenon is even more evident at a smaller grain size, where is it is often impossible to determine the order in which resources are activated. A resource graph allows us to represent this situation.

IV. FACETS OF EPISTEMIC GAMES

Although Phil and Max have identical end conditions at a large grain size (an equation describing the speed of the ball), in a finer grain analysis, they differ. Phil requires an equation that "can cover anything" (line 38), while Max questions the need for an unspecified constant. Thus, Phil and Max can be described as playing different versions of one epistemic game.

We call these games "Finding a Family of Functions" and "Fitting the Physical Situation." Both have similar moves and are facets of the "Making Meaning of Mathematics" epistemic game. The early moves in each game are so similar that it is not until near the end of the problem solution that Phil and Max have a difference of opinion on how to proceed.

Phil is playing the game "Finding a Family of Functions." Several procedural mathematical resources are part of this game, including *Compute Antiderivative* and *Add Constant*. The entrance conditions for this game are a differential equation without boundary conditions. The problem statement is very physics-like; the boundary condition is not stated in words or math. Thus, he frames the activity as one of finding the most general solution to a physics problem. Without an explicit boundary condition statement, Phil does not activate *Extract Boundary Conditions*. He begins to solve the problem using "+c," but through argument with Max does not arrive at an ending point.

In Max's case, *Extract Boundary Conditions* is activated by the physical scenario and by his intention (line 30-32) for "consistency" (seemingly between the mathematics and the physics). He makes explicit use of physics conventions in the broader game of "Mapping Meaning to Mathematics." Although the problem statement does not explicitly state that the initial time is equal to zero, Phil feels that an initial time must be defined, and that zero is a conventional initial time for this type of problem. As a result, Phil can more easily be described as playing the "Mapping Meaning to Mathematics" game, even as his variation is one of "Fitting the Physical Situation."

To better represent the "Making Meaning of Mathematics" game and its two facets of "Finding a Family of Functions" and "Fitting the Physical Situation," we represent the individual procedural resources in a resource graph with directed arrows to indicate the possible allowable moves (see Fig. 1). We note that our resource graph is situation and student dependent. In this case, it is drawn to show the moves of two different students

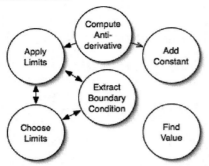

Figure 1: A partial resource graph of two epistemic games. Phil's game is represented by open arrows; Max's by solid

Both Phil and Max begin at *Compute Antiderivative*. Phil's game is, from here, very simple. He needs only *Add Constant* to make a family of functions from the antiderivative. It is clear that these resources are very strongly linked. In lines 25-26, Phil states that he "just can't imagine integrating without having an extra constant". However, although undoubtedly Phil has the resources *Extract Boundary Conditions* and *Find Value*, they are not activated here. In line 29, he states that boundary conditions cannot be used because "you don't know that's zero". To Phil, only the problem writer can determine boundary conditions, and we can infer that if the boundary condition were given in the problem statement, his framing of the problem would be such that *Extract Boundary Condition* and *Find Value* would be activated.

Max's game is rather more complex, and the exact order in which the resources are applied is difficult to determine from the transcript. This is represented in the resource graph by the use of double-ended arrows. Evidence exists for the activation of three resources -- *Apply Limits, Choose Limits, and Extract Boundary Conditions*. Additionally, since Phil does not object when Max integrates both sides of the equation, we can infer that *Compute Antiderivative* is also active.

In several places (lines 3-9, 14, 18), Max uses the language "go" or "going." In these cases, *Apply Limits* is the resource in action. In lines 3-6, Phil gives several possibilities for limits; it is clear that he needs *Choose Limits* to narrow down the options. At the end of that statement (lines 7-9) he uses just one set of limits, zero to t, and describes the reason for his choice.

In lines 18-20, we see a good example of *Extract Boundary Conditions*. Phil states that although you could integrate from t_0 to t, it reasonable to use the common physics convention of calling t_0 "zero," even though this information is not specifically given in the problem. It is worth noting that *Choose Limits* and *Extract Boundary Conditions* are not necessarily linked; a student's

procedure for choosing limits may not involve the boundary conditions.

SUMMARY

Epistemic games can be described as a particular pathway through activated procedural resources. As an example, we look at an interaction between two students as they attempt to solve a first order, separable differential equation. One student prefers the "+c" method, the other the "limits" methods. The openness of this discussion allows a clear look into the reasoning that each student uses.

Using the transcript, we define several procedural resources, show how they can be organized into two facets of a previous described epistemic game, and produce a resource graph which allows visualization of this portion of the epistemic games. By representing two correct mathematical procedures in terms of shared resources, we help clarify the types of thinking in which students engage when learning to apply mathematical reasoning to physics.

ACKNOWLEDGMENTS

We thank Kate McCann for her assistance in gathering the video data.

This material is based upon work supported by the National Science Foundation under grants DUE-0441426, DUE-0442388, and REC-0633951.

REFERENCES

1. D. Hammer, American Journal of Physics 67 (Physics Education Research Supplement), S45 (2000).
2. M.C. Wittmann, Physical Review Special Topics - Physics Education Research 2 (2), 020105 (2006).
3. D. Tannen and C. Wallat, in The Discourse Reader, edited by A. Jaworski and N. Coupland (Routledge, New York, 1987), pp. 346.
4. A. Collins and W. Ferguson, Educational Psychologist 28 (1), 25 (1993).
5. J. Tuminaro and E.F. Redish, Physical Review Special Topics - Physics Education Research, submitted (2006).
6. B.S. Ambrose, American Journal of Physics 72, 453 (2004).
7. M.C. Wittmann and B.S. Ambrose, "Intermediate Mechanics Tutorials," available at http://perlnet.umaine.edu/imt/.
8. J.P. Byrnes, Cognitive Development and Learning in Instructional Contexts, 2nd ed. (Pearson Allyn & Bacon, 2000)

Measuring Student Effort and Engagement in an Introductory Physics Course

Scott Bonham

Department of Physics and Astronomy
Western Kentucky University
1906 College Heights Blvd., Bowling Green, KY 42101-1077

Abstract. Multiple scales reflecting student effort were developed using factor and scale analysis on data from an introductory physics course. This data included interactions with an on-line homework system. One of the scales displays many characteristics of a metric of the individual level of engagement in the course. This scale is shown to be a good predictor of performance on class exams and the Force Concept Inventory (FCI). Furthermore, normalized learning gains on the FCI are well predicted by this scale while pre-instructional FCI scores provide no additional predictive ability, agreeing with observations by Richard Hake. This scale also correlates strongly with epistemological beliefs that learning is related to effort and is the responsibility of the student. The factors that enter into this scale, writing and mastering expert-like problem-solving, are consistent with this being a measure of individual levels of class engagement.

Keywords: On-line homework, student effort, interactive engagement, Just-in-Time-Teaching, normalized FCI gain.
PACS: 01.40.Di, 01.40.Fk, 01.40.gf, 01.50.H-

INTRODUCTION

Student engagement and effort were demonstrated by Hake to be key elements of learning physics.[1] This has been reinforced by other studies, including learning gains due to requiring students to do homework via an on-line system.[2] However, Hake gave only a vague definition of interactive engagement as "minds-on" learning and no quantitative measure has since been offered. Such would be valuable to better understand variation among student learning gains within the same course and why some students may struggle in a course in spite of appearing to put forth significant effort in completing on-line homework assignments and attending class. Clearly engagement is not simply effort. Recent work has suggested that the difference is not entirely due to interactive engagement but to reasoning skills. [3]

Constructing useful measures of student effort and engagement is challenging. Self reporting, e.g. on surveys, has significant drawbacks including subjective self-judgment, influences from student expectations, and providing only a single measurement. An alternate approach is to use the student's on-line homework system itself as a research tool,[4] since the information it collects is quantitative and consistent across all students, reflects actual student behavior, and tracks behavior throughout the entire course. On-line homework systems collect a large amount of information, including when students looked at and submitted assignments, the nature of those responses and general patterns of working, but none of the measures is specifically designed to measure effort and engagement. Therefore any measures of such will need to be constructed and shown to reflect the desired quantity. The focus of the present work is to demonstrate the construction of several measures reflecting student effort in a particular introductory physics class. Data from on-line homework and other class performance measures are analyzed using factor and scale analysis. Finally, it will be shown that one of the constructed scales appears to reflect the level of individual student engagement.

SETTING

The class structure and performance measures will be described in some detail, since the results may be context dependent. The course was an introductory algebra-based physics course taught by the author, taken primarily by non-life science students from the college of science and engineering fulfilling a requirement for their major. University admission is

CP951, *2007 Physics Education Research Conference*, edited by L. Hsu, C. Henderson, and L. McCullough
© 2007 American Institute of Physics 978-0-7354-0465-6/07/$23.00

not highly selective, resulting in a broad range of student abilities and backgrounds. About half report this to be their first physics course, and most do not continue into a second course. As it is the only physics course many will ever take and the course is not taken by physics, engineering or allied health students, the course is intentionally structured such that all students who will put forth sufficient effort will pass the course, at least with a 'D', even if mastery of physics concepts and skills is less than desirable.

The course was taught in an integrated lecture-lab format in a SCALE-UP classroom[5] with five contact hours over four days (M-Th). Course components included guided-discovery computer laboratories similar to RealTime Physics[6], two-page chapter summary exercises, cooperative group problem solving,[7] Just-in-Time-Teaching pre-class questions[8], on-line homework, written homework, and several exams throughout the semester. Students were assigned to heterogeneous groups of three for laboratories and classroom problem solving.

An explicit problem solving approach[7] was taught, modeled, and reinforced through templates used in class, on homework and exams. During in-class problem solving exercises students used problem-solving templates that provided a decreasing amount of scaffolding of the approach over the semester. Selected on-line homework exercises were linked to problem-solving templates in Adobe PDF format for students to use in solving the problems and to turn in for a grade. Grades on these written solution assignments were based on correctness and completeness of the problem-solving steps and not the final answer. Written homework assignments consisting of constructing force, energy and torque diagrams were due the class period following that in which they were introduced. Both types of written homework were graded on simple scales, as illustrated by the distribution of points on the force diagram assignment. Six total points were given: attempting all (some) exercises was worth two (one), arrows completely (partially) correct was worth two (one), and labels completely (partially) correct was worth the final two (one).

The pre-class questions and on-line homework were assigned, collected and graded using the WebAssign homework system.[9] Pre-class assignments due several hours before class consisted primarily of short essay questions while a few involved multiple-choice questions or graphs using GraphPAD.[10] Homework assignments were due twice a week, and contained numerical exercises, graphical response questions,[10] multiple choice, free response questions and symbolic (equation) questions. Approximately half were from the textbook, the remaining exercises were developed by the author or from RealTime Physics.

Three midterm exams and a final exam were given. They all included questions involving definitions, concepts, graphs and diagrams, single and multiple step numerical problems, and a "set up" problem where students used a problem-solving template to set up but not actually solve a challenging problem.

The Force Concept Inventory[11] (FCI) was administered the first and last weeks of the semester. Students completed the Epistemological Beliefs Assessment for Physical Sciences[12] (EBAPS) on-line[13] at the beginning of the semester.

The final grade was calculated as 12% for each midterm exam and 20% for the final, 20% for homework, 20% for in-class assignments (laboratories, group problem solving) and 2% each for chapter summaries and pre-class questions. Most chapter summaries and pre-class questions received full credit, and laboratory and group problem solving scores included significant amounts of participation credit. Partial credit on exams was largely based on the number of correct problem-solving steps present. The class on which this work is based was the fall of 2006, which began with 28 students of which 22 completed the semester. There were a total of 42 pre-class, 25 on-line homework, 10 written solution and 3 diagram assignments collected.

DATA COLLECTION AND ANALYSIS

Data was compiled as follows. Data on student interactions with WebAssign included information on each viewing and submission of an assignment, identifying the student, assignment, time, question and question part, student response and if correct. This data was imported into a Microsoft Access database and various measures were constructed. This included the number of homework and pre-class assignments submitted, the average time before the assignment was due that the assignment was first downloaded and submitted, the average fraction of assignments that was attempted, average fraction of assignments correct on the first and last attempt, average response length (in characters) of pre-class question responses, and other measures. Data in addition to that from the on-line homework included the number of written homework assignments submitted and scores, number of chapter summaries submitted, scores on exams, total score by category for pre-class, in-class, chapter summary and homework assignments, final grades, pre- and post-FCI scores and scores on each of the five EBAPS scales.

The data described above contains about a dozen measures related to student effort and engagement. A factor analysis (Table 1, produced in SPSS with Varimax rotation) shows that they cluster into three factors. Collectively these three factors account for 83% of the variance. A scale analysis for the measures in each factor shows that each cluster forms a good scale, with the value of Cronbach's Alpha being 0.88, 0.92, and 0.75 for Factors 1, 2 and 3, respectively. Cronbach's Alpha is measure of to what degree the different measures agree with each other; a value of 0.7 is acceptable, and 1.0 is the maximum value possible.

TABLE 1: Clustering of measures of activity. Rotated factor analysis was used to obtain these loading factors. Values in bold face are the factor with the greatest loading.

Student activity	Factor		
	1	2	3
Number assignments turned in			
Pre-class assignments	**.89**	.27	.14
Chapter summaries	**.94**	.22	.00
Online homework	**.91**	.17	.05
Paper homework	**.66**	.23	.62
Time before assignment due			
Looked at pre-class	.17	**.92**	-.04
Looked at homework	.41	**.85**	.10
Submitted homework	.39	**.81**	.16
Fraction assignment attempted			
Pre-class	.00	**.81**	.27
Homework	.40	**.77**	.37
Score on diagram assignments	.21	.14	**.78**
Score on written solutions	.01	-.02	**.87**
Length of pre-class responses	-.05	.44	**.76**

Three factors are identified representing different aspects of student effort and engagement. Factor 1 ('Dutifulness') reflects the completeness of students turning in assigned work, and Factor 2 ('Online Discipline') reflects early and complete work on on-line assignments. The interesting thing about Factor 3 ('Engagement') is that average scores on written assignments and length of pre-class question response lengths are not obviously related quantities, suggesting that this scale reflects something more fundamental. This scale also correlates at the $p < 0.05$ level with two EBAPS scales: Nature of Knowing and Learning (0.47) and Source of Ability to Learn (0.57).

Factor 3 also proves to be the most useful measure in explaining student course performance. Stepwise linear regressions were carried out on each of the major grade categories, exams, and FCI scores as functions of these three factors plus pre-instructional FCI scores, the last serving as a measure of initial physics understanding. (See Table 2.) Factor 3 (Engagement) demonstrates a significant predictive ability on eight of the ten measures of student

performance listed, and is the only significant factor on five measures. The last measure, gain on the FCI was calculated following Hake,[1] except for each student individually:

$$FCI\ gain = \frac{FCI_{post} - FCI_{pre}}{100\% - FCI_{pre}} \qquad (1)$$

TABLE 2: Predictive value of the Engagement Factor. Shown are the normalized linear regression coefficients (β) for ten measures of student performance in the course as a function of the three effort scales and pre-instructional FCI scores.

	Factor			Pre
	1	2	3	FCI
Pre-class Total	.39	.09	**.59***	-.13
Chapter Total	**.69***	.16	-.15	-.27
In-class Total	.37	.22	.31	.12
Homework Total	**.82***	-.03	**.22***	-.02
Exam 1	.01	-.15	**.44***	**.58***
Exam 2	.17	-.16	**.61***	.35
Exam 3	-.23	-.40	**.64***	.28
Final Exam	.14	-.28	**.53***	.17
Post FCI	-.12	-.26	**.46***	**.54***
Gain on FCI	-.18	-.29	**.53***	.01

* Significant at the $\alpha=0.05$ level.

Note the very small value of beta for pre-instructional FCI into the FCI gain, which has a p value of nearly 1 (no relation). Note also in Table 3 that the Engagement Factor proves to be a better predictor of FCI gains than any of the individual components it comprises.

TABLE 3: Comparison of linear regression of gain on the FCI by the Engagement Factor and the three measures it is comprised of. The combination of the three is a better predictor than any individually.

Independent variable	R^2	β	p
Engagement Factor	.28	.53	.020
Score on diagram assignments	.25	.50	.030
Score on written solutions	.20	.45	.053
Length pre-class response	.19	.43	.065

RESULTS AND DISCUSSION

Three different scales representing different aspects of student effort and engagement were constructed, and one proved to be productive in predicting many measures of student performance, including gains on the Force Concept Inventory. The statistical evidence arguments that the Engagement Factor reflects student engagement will be summarized, followed by a discussion as to why these particular measures seem to reflect student engagement in the course.

The Engagement Factor is a significant predictor of all but chapter summary and in-class scores. The chapter summary grade depends largely on simply turning in the assignments (reflected in Factor 1) and similarly the overall homework score is most affected by missed assignments. In-class score reflects group activities with significant participation credit, so it largely reflects class attendance. The pre-instructional FCI score is only significant on the first exam and would enter in with a decreasing beta with each subsequent exam, consistent with the author's observation that students with strong physics backgrounds often can 'cruise' through the first unit, but that advantage disappears with subsequent exams. The Engagement Factor is a strong predictor of FCI gains while pre-instructional FCI is not, consistent with Hake's original claim of the utility of the normalized gain. The Engagement Factor also correlates significantly with two scales on the EBAPS, the only one of the three factors and pre-FCI that correlate with any of the EBAPS scales. Furthermore, it correlates significantly with the two EBAPS scales that seem most related to productive student engagement: *Nature of Knowing and Learning* which probes how much they believe learning is a matter of effort compared to innate ability, and *Source of Ability to Learn*, which concerns the role of the student compared to that of teacher/instructional materials. The combination of the three measures into the Engagement Factor is also a better predictor than each alone. The Engagement Factor demonstrates statistical characteristics that would be expected of a quantitative measure of student engagement.

The reasons that these particular measures combine into a scale reflecting student engagement may reflect particular aspects of the course. The diagram assignments were due the day after the diagrams were introduced and students had practiced making such diagrams in class. As the assignments concerned different situations than class examples, assignments scores reflected in part the student's ability to grasp and internalize the principles used in class and apply them in a different situation. The written solutions were typically on more challenging homework exercises following examples worked out in class and group-problem solving of problems for that topic; again this represents in part the student's ability to extract, internalize, and apply principles to slightly different situations. The pre-class questions were graded 'on effort,' which in practice meant full credit for any meaningful response. Therefore, the length of the responses was entirely up to the student, reflecting the degree that students chose to answer questions completely, explain their reasoning, describe their own experiences, ask questions themselves, and otherwise fully engage themselves in answering the questions.

CONCLUSIONS

This work demonstrates the usefulness of on-line homework systems in constructing measures of student behavior. One such measure, composed of written homework scores and length of pre-class responses, not only demonstrates strong predictive ability for class test scores but also explains gains on the FCI completely independent of pre-instructional scores. Thus this measure appears to capture, at least in part, the level of individual student engagement with the material. Further work needs to look at the degree this or similar measures are predictive in different instructional settings.

ACKNOWLEDGMENTS

The author thanks Brian Pittman and colleagues for providing WebAssign data, and Dr. Gordon Scott Bonham for assistance with the statistical analysis.

REFERENCES

1 Richard Hake, American Journal of Physics 66 (1), 64 (1998).
2 K. Kelvin Cheng, Beth Ann Thacker, Richard L. Cardenas et al., American Journal of Physics 72 (11), 1447 (2004).
3 Vincent P. Coletta and Jeffrey A. Phillips, American Journal of Physics 73 (12), 1172 (2005).
4 Scott W. Bonham, Robert J. Beichner, Aaron Titus et al., Journal of Research on Computing in Education 33 (1) (2000).
5 Robert Beichner, Jeff Saul, David Abbott et al., in PER-Based Reform in University Physics, edited by Edward Redish and Patrick Cooney (American Ass. of Physics Teachers, College Park, MD, 2007).
6 David R. Sokoloff, Ronald K. Thorton, and Priscilla W. Laws, RealTime Physics: Active Learning Laboratories. (John Wiley & Sons, New York, 1999).
7 P. Heller, R. Keith, and S. Anderson, American Journal of Physics 60, 627 (1992).
8 Gregor M. Novak, Evelyn T. Patterson, Andrew D. Gavrin et al., Just-in-Time Teaching: Blending Active Learning with Web Technology. (Prentice-Hall, Inc., Upper Saddle River, 1999).
9 WebAssign, WebAssign, (North Carolina State University, 2005), http://webassign.net/info/
10 Scott Bonham, in The Physics Teacher (2007), Vol. In press.
11 David Hestenes, Malcolm Wells, and Gregg Swackhamer, Physics Teacher 30 (3), 141 (1992).
12 Andy Elby, American Journal of Physics 69 (7), S54 (2001).
13 Scott Bonham, Phys. Rev. ST Phys. Educ. Res., submitted (2007).

Voltage is the Most Difficult Subject for Students in Physics by Inquiry's Electric Circuits Module

Carol Bowman and Gordon J. Aubrecht, II

The Ohio State University Marion Campus

Abstract. We report on the investigation of multiple sets of data from an electric circuits Physics by Inquiry course on students' ranking of topic difficulty. Students ranked the difficulty of the preceding class almost every class day and they ranked the difficulty of various course sections on a diagnostic (one diagnostic per section). In the OSU Physics by Inquiry (PbI) class, students—a majority of education undergraduates—work in groups, and are checkpointed as they do experiments in a section. In addition, there is a question of the day at the beginning of almost every class. Here, students are also asked to rank the difficulty, but of the preceding day's classwork. These "difficulty rankings" and student grades (used as a measure of performance) constitute our dataset. We compiled data from four sections of the Spring 2006 and one section of the Spring 2007 Physics by Inquiry electric circuits class. The sections on potential difference appear to be the most difficult.

Keywords: Physics by Inquiry, laboratory survey, physics education research
PACS: 01.40.Di, 01.40.Fk, 01.50.Qb

INTRODUCTION

There have been many descriptions of OSU's Physics by Inquiry (PbI) classes, [1-4] based on University of Washington materials [5,6]. There are three different PbI classes offered—on properties of matter (V.1 [5]), on electric circuits (V. 2 [6]), and on optics and astronomy by sight (V.1 and V.2 [5,6]). Our present study is part of our ongoing formative evaluation of PbI. This paper discusses the topics in the electric circuits class. Our personal experience has long been that voltage is the most confusing topic faced by students. Also, Kim, Bao, and Acar earlier found that PbI student stress levels peak at various points [2], and are especially high in the voltage sections.

There seem to be several reasons for this difficulty. Students learn about current, ammeters, and Kirchhoff's current rule before they study voltage. Potential difference ("voltage" is the term used in the PbI class, so we use it here) is sufficiently different from current that carefully-honed student ideas about current interfere with students' ability to address voltage despite careful text attempts to demonstrate the difference as well as the difference between the respective meters and how they are used. Students at first apply the idea of flow, developed in the model of current, to voltage. For many students, voltages across parallel branches are to be added together as currents are, and these ideas persist despite the measurements made by these same students that show voltages across parallel branches are the same.

PbI instructors have attempted to remedy such difficulties by several avenues of approach. A module introducing ammeters and voltmeters nearly simultaneously has been written with the thought that the sequencing leads to student confusion. A different module addresses student difficulties associated with multiple loops. In our opinion, these additional modules have met with mixed success.

While we were convinced that the voltage sections were the most difficult on the basis of exams and checkpoint conversations with students, we decided to see if we could determine what students think about the relative difficulty of the thirteen or fourteen sections (twelve of which are from the book) addressed in the course. The following sections discuss the diagnostic rankings and the question-of-the-day rankings.

USING DIAGNOSTIC RANKINGS

Before our PbI students work on a section, they fill out a diagnostic sheet that asks questions pertinent to the section content (in a space on the left half of the page). After finishing a section, students rework the diagnostic (in the remaining space on the

CP951, *2007 Physics Education Research Conference*, edited by L. Hsu, C. Henderson, and L. McCullough
© 2007 American Institute of Physics 978-0-7354-0465-6/07/$23.00

right half of the page) on the basis of what they've learned within that section. They get a small number of points for each accepted diagnostic (but no points if the diagnostic has not been accepted, in other words, the points are Heaviside functions). Students are also asked to explain what they have learned in the section and to rate the section's difficulty from least difficult (0) to most difficult (6) as they saw it We also keep track of how many submissions were required for students to have their reworked diagnostic accepted.

We might also look at the diagnostics that few students complete, but it should be kept in mind that these data are skewed. While we urge students to keep up with submitting diagnostics, there is a tendency for students to put off submitting the diagnostics, especially from the later sections, until near the end of the course. The proportion of accepted diagnostics therefore falls steeply toward the last few sections covered in the class..

Some diagnostics also are harder than others as rated by students. There were N = 113 students in four sections of PbI in Spring, 2006 and N = 14 in Spring, 2007. Table 1 shows the difficulty rankings for each section given on diagnostics that were accepted. Sections 8 (voltage), 8a (voltage), and 10 (Ohm's Law) rank highest in difficulty as perceived by students.

Another way to determine how difficult a section is involves determining which sections the fewest students completed. Table 2 shows the results for the total number of diagnostics completed (some classes were not able to get as far as others, but no student in any class turned in the diagnostic for Section 12). Apparently Secs. 8, 8a, and 11 are more challenging than other sections. A similar pattern is seen when we examine the number of groups in which all students completed diagnostics, as seen in Table 3. The dip in completions for Sections 8 and 8a is more pronounced.

We kept track of how many times it took students to get the diagnostic accepted. Many tries should indicate a harder section. Table 4 presents the number of nonaccepted diagnostics by section. It is clear that sections 8 and 8a (on voltage) were less likely to be completed than adjacent sections, even with end-of-course falloff. The falloff mentioned above is apparent here, with the same general trend in both years in Tables 2 through 4 of a drop in completions.

Still another piece of data is the number of tries before acceptance, presented in Table 5. The number of tries is more difficult to compare because different instructors evaluated different groups of diagnostics (except for Marion, where Aubrecht evaluated all diagnostics). By this criterion, Sections 6, 7, 8, and 8a were the most difficult. Diagnostics (6, 7, 8, and 8a) in all classes were all evaluated by the same instructor (Aubrecht) and are consistent. The average number of tries in the four sections was 2.70, while in Marion for all the other sections the average number of tries was 2.11, and in Columbus the average number of tries was 1.15. Aubrecht was apparently more demanding in evaluating the diagnostics. However, comparison shows that these sections (6: parallel and series resistors; 7, voltmeters; 8, Kirchhoff's potential rule; 8a, multiple loops), were clearly more difficult for students independent of the evaluator.

QUESTIONS OF THE DAY

At the beginning of each class (excepting midterm days), a short question relating to the preceding class material is administered. Students also rank the preceding day's material in difficulty (from 0 to 6). The first midterm occurred when most groups were around Section 4 and the new meters section and the second midterm occurred when most groups were in Sections 8 and 8a. Hence, the rankings undercount these sections. Also, a total of four days' questions of the day (and associated rankings) were misplaced for the Columbus sections. Still, a total of 187 group rankings were usable.

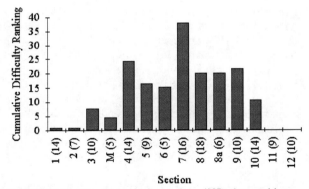

FIGURE 1. Student group section difficulty rankings from questions of the day. The section and its associated number of experiments is shown along the abscissa.

We kept track of each group's progress in order to craft an appropriate question of the day for the next class for the group farthest behind. Therefore, we can bracket the sections covered compared to each group's ranking. Knowing the number of experiments in each section allows us to weight the rankings according to the sections covered the preceding day. This is clearly not the best estimate of time spent and difficulty encountered, but it the best we can have without recording every word and gesture of every group.

TABLE 1. Student ranking of difficulty by section and total ranking (scale: 0, least, to 6, most).

Section Number	1	2	3	4	5	6	7	8	8a	9	10	11
Marion 06	3.00	2.73	3.86	3.50	3.53	4.07	4.09	4.33	4.22	4.67	5.00	0.00
Columbus 9:30	3.48	3.17	3.59	3.83	3.71	4.12	4.18	5.09	4.14	4.23	5.08	4.33
Columbus 1:30	3.15	3.37	3.84	4.27	3.80	4.78	3.86	3.83	4.73	4.05	4.20	
Columbus 5:30	3.55	3.18	4.58	4.71	4.00	4.88	5.13	5.44	4.71	4.40	4.20	
Marion 07	2.57	3.08	3.73	4.30	4.00	3.73	4.38	4.71	5.67	4.60	4.60	
Total Ranking	3.22	3.17	3.93	4.13	3.78	4.30	4.33	4.77	4.50	4.27	4.69	3.88

TABLE 2. Number of diagnostics accepted by section.

Section Number	1	2	3	4	5	6	7	8	8a	9	10	11
Marion 06	17	15	14	13	15	15	8	9	9	10	6	1
Columbus 9:30	36	38	37	35	35	28	22	11	15	29	25	8
Columbus 1:30	33	33	33	29	28	12	16	9	11	25	21	1
Columbus 5:30	26	26	26	24	24	17	17	10	8	16	13	0
Marion 07	14	12	12	10	11	10	8	3	1	4	4	0
Total Accepted	126	124	122	111	113	82	71	42	44	84	69	10

TABLE 3. Number of groups in which all members completed diagnostics by section.

Section Number	1	2	3	4	5	6	7	8	8a	9	10	11
Marion 06	6	5	5	4	5	5	3	2	1	1	1	1
Columbus 9:30	8	10	9	7	7	4	1	0	2	4	3	1
Columbus 1:30	9	9	9	5	3	0	1	0	0	4	4	0
Columbus 5:30	7	7	7	5	4	1	1	1	0	2	2	0
Marion 07	4	2	3	2	1	1	0	0	0	0	0	0
Total	34	33	33	23	20	11	6	3	3	11	10	2

TABLE 4. Number of not-accepted diagnostics by section.

	Students	1	2	3	4	5	6	7	8	8a	9	10	11
Marion 06	16	0	1	2	3	1	1	8	7	7	6	10	15
Columbus 9:30	38	2	0	1	3	3	10	16	27	23	9	13	30
Columbus 1:30	33	0	0	0	4	5	21	17	24	22	8	12	32
Columbus 5:30	26	0	0	0	2	2	9	9	16	18	10	13	0
Marion 07	14	0	2	2	4	3	4	6	11	13	10	10	0
Total	127	3	5	8	20	19	51	63	93	83	52	68	88

TABLE 5. Average number of tries before acceptance.

Section Number	1	2	3	4	5	6	7	8	8a	9	10	11
Marion 06	2.4	2.7	2.9	2.3	1.8	2.7	2.8	2.5	2.0	1.2	1.2	1.0
Columbus 9:30	1.3	1.2	1.2	1.2	1.1	4.0	2.1	2.7	1.9	1.0	1.0	1.0
Columbus 1:30	1.4	1.2	1.1	1.0	1.1	3.1	2.1	2.3	2.3	1.0	1.0	
Columbus 5:30	1.1	1.2	1.4	1.1	1.2	4.3	3.0	2.3	2.4	1.0	1.3	
Marion 07	2.6	2.3	2.0	1.7	1.6	2.1	1.8	2.4	2.0	1.6	1.6	
Total	1.6	1.5	1.5	1.3	1.2	3.4	2.4	2.5	2.1	1.1	1.1	1.0

Figure 1 shows these rankings. Clearly Section 7 is ranked the most challenging, but recall that some sections' rankings, namely, Sections 4, meters, 8 and 8a, were generally missed due to midterm days with no rankings of the preceding day's difficulty (the meters section was used only in 2007). This graph supports a contention that Sections 7, 8, and 8a were seen as especially challenging, in agreement with other evidence presented here.

GRADE CORRELATION

Figure 2 shows the combined averaged data on student rankings from diagnostics in all sections of PbI electric circuits. Figure 3 shows the combined averaged data on student rankings from questions of the day.

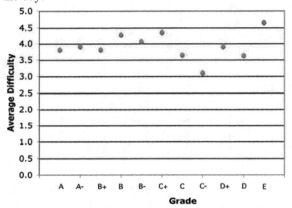

FIGURE 2. The grade distribution and the diagnostic ratings by grade.

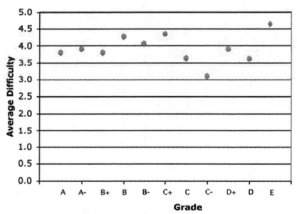

FIGURE 3. The grade distribution and the question of the day ratings by grade. While this graph looks like Fig. 2, it is not.

Note that student rankings of difficulty slowly rise as grade decreases from A to C+, but then appear to fall again to values similar to A students for students with other C and D grades (there is just one E). While the graphs *appear* identical, they refer to different datasets. It is this near identity that convinces us that this is no fluctuation. Perhaps the better students recognized more clearly that they were having difficulties, as the rated difficulty of the sections goes up from A through C. Poorer students may not even have recognized that their difficulty.

CONCLUSIONS

Because Physics by Inquiry is based on research, it does an outstanding job of dealing with many student difficulties in the topics covered. Nevertheless, we found students still had difficulty with certain concepts involving voltage.

Parts of the material related to voltage caused problems in student understanding. Section 8a, for example, was written to deal with difficulties we found students had with voltage in multiple loops. We are currently using another revision, meters, that attempts to help students with these concepts, but, as mentioned above, this has shown mixed success.

This study has shown that indirect formative evaluation such as demonstrated here may be useful in supporting student understanding through course revision. This study does not address whether these difficulties are intrinsic to the concepts independent of the PbI approach or whether the PbI approach itself may have generated some problems.

REFERENCES

1. Lin, Y., Demaree, D., Zou, X., and Aubrecht, G. J., "Student Assessment Of Laboratory In Introductory Physics Courses: A Q-sort Approach," Proceedings of the 2005 Physics Education Research Conference, AIP, Melville, NY, 2006, 101-104.
2. Kim, Y., Bao, L., and Acer, O., "Students' Cognitive Conflict and Conceptual Change in a Physics by Inquiry Class," Proceedings of the 2005 Physics Education Research Conference, AIP, Melville, NY, 2006, 117-120.
3. Aubrecht, G. J., Lin, Y., Demaree, D., Brookes, D., and Zou, X., "Student Perceptions of Physics by Inquiry at Ohio State," Proceedings of the 2005 Physics Education Research Conference, AIP, Melville, NY, 2006, 113-116.
4. Kim, Y. and Bao, L., "Students' Recognition of Cognitive Conflict and Anxiety in PbI Classes," "Students' Cognitive Conflicts and Conceptual Change in PbI Classes," papers presented at the AAPT meeting, Summer 2005 and Kim, Y., Bao, L., and Bocaneala, F., "Developing Instruments for Evaluating Anxiety Caused by Cognitive Conflict," paper presented at the AAPT meeting, Summer 2004.
5. McDermott, L. C., *Physics by Inquiry*, vol. 1, Wiley, New York, 1995.
6. McDermott, L. C., *Physics by Inquiry*, vol. 2, Wiley, New York, 1995.

Reading Time as Evidence for Mental Models in Understanding Physics

David T. Brookes*, José Mestre* and Elizabeth A. L. Stine-Morrow†

*Department of Physics, University of Illinois at Urbana-Champaign, 1110 W. Green St., Urbana, IL, 61801
†Dept. of Educ. Psych., University of Illinois at Urbana-Champaign, 1310 S. Sixth St., Champaign, IL, 61820

Abstract. We present results of a reading study that show the usefulness of probing physics students' cognitive processing by measuring reading time. According to contemporary discourse theory, when people read a text, a network of associated inferences is activated to create a mental model. If the reader encounters an idea in the text that conflicts with existing knowledge, the construction of a coherent mental model is disrupted and reading times are prolonged, as measured using a simple self-paced reading paradigm. We used this effect to study how "non-Newtonian" and "Newtonian" students create mental models of conceptual systems in physics as they read texts related to the ideas of Newton's third law, energy, and momentum. We found significant effects of prior knowledge state on patterns of reading time, suggesting that students attempt to actively integrate physics texts with their existing knowledge.

Keywords: physics education research, reading, situation model, misconceptions, recruitment of prior knowledge
PACS: 01.40.Fk, 01.40.Ha

INTRODUCTION

Remembering what we read is strongly influenced by a) our prior knowledge [1], and b) attentional allocation to construct a coherent representation [2]. Science students often fail to learn from reading a text that contradicts their ideas [3]. This may be because they do not allocate attention to resolve these contradictions. Supporting this idea, there is evidence that students with preconceptions learn better from refutational texts that explicitly acknowledge and then refute those preconceptions [4]. It is generally believed that the way in which a text is processed online (i.e., during reading) will affect what is remembered offline (after reading) [5]. However, we know of only one study [6] that has examined students' online and offline processing of a physics text in a single study. Surprisingly, this study showed a) no difference in online processing between groups of physics students who had different (pre-identified) knowledge states, and b) clear evidence of recruitment of prior knowledge in offline processing, even when the text that had just been read contradicted that prior knowledge.

In the constructivist view, students activate (or "recruit") prior knowledge to interpret their experiences. According to the construction-integration model [7, 8], students learning from text attempt to construct a *situation model*, which is a mental representation of the text, integrating ideas given directly by the text with knowledge-based inferences. For example, if you read the sentence, "all the buildings collapsed except for the mint," you might generate the inference "earthquake." "Earthquake" then becomes part of your mental model

of the situation. In this view, inferences are activated promiscuously as part of a neural network. The network then uses the available information to cycle through a series of iterations and relaxes into a final state where a smaller subset of inferences deemed appropriate to the situation remain activated. What remains activated in the network is dependent on the knowledge base of the reader and the contents of the text that is being read.

Research in narrative understanding has shown that when readers encounter statements that conflict with their situation model, they will take longer to read these statements [9]. For example, readers slow down when information (e.g., Mary orders a hamburger) is introduced into the text that is inconsistent with the situation model introduced earlier in the text (e.g., Mary is a vegetarian) — even if these facts are spatially distant in the text. Thus reading time can be a sensitive probe of a) whether a reader has constructed a situation model, and b) the nature of the situation model that has been constructed. So it is perhaps surprising that an earlier attempt to demonstrate this effect in science learning was unsuccessful.

Our general research questions related to learning in physics are: In light of the null result of [6], do students' prior knowledge states have any noticeable effect on their processing of a physics text when they read it? In other words, if students recruit ideas that conflict with the text that they are reading, do they exhibit prolonged reading time? Additionally, can we use differences in reading time to understand something about the nature of the meaning that students construct? Finally we wish to explore the connections between students' online and offline processing.

CP951, *2007 Physics Education Research Conference*, edited by L. Hsu, C. Henderson, and L. McCullough
© 2007 American Institute of Physics 978-0-7354-0465-6/07/$23.00

METHOD

Participants

We recruited roughly 150 educational psychology undergraduate students from a large Midwestern university for a reading study. Approximately half of this group were eliminated because they did not meet the prior knowledge criteria for participation in phase 2 of our study. The remaining 75 students were invited to participate in phase 2 and 30 students from this group chose to participate. The "misconception" group consisted of 16 students (henceforward designated "non-Newtonian"), while the "non-misconception" group consisted of 14 students (henceforward designated "Newtonian"). Students were assigned to each group based on criteria described in the procedures section below. Newtonian and non-Newtonian subjects did not differ on WAIS vocabulary measures [10] $F(1,26) < 1$.

Materials and Procedures

The experiment was divided into two phases. In the screening phase (phase 1) the 150 participants were divided into three groups based on their answers to four Newton's third law questions from the FCI [11]: Those who could answer all four Newton's third law questions correctly (Newtonian), those students who consistently answered that the bigger or faster moving object exerted more force and/or passive or stationary objects don't exert forces (non-Newtonian), and those students who gave a mixed response of correct and incorrect answers to the four Newton's third law questions.

Subjects who were identified as Newtonian or non-Newtonian in phase 1 (about 75 students) were invited by email to participate in phase 2. Thirty students responded and participated in phase 2. In the reading phase of this study (phase 2), subjects were asked to read (on a computer screen) a collection of 29 passages (four segments each) that accurately described different physical situations involving interactions between two equal or unequal sized objects. (See Table 1 for examples.) Subjects were told in the instructions that all passages they were about to read were true. The passages were read segment by segment, self-paced by the reader, who advanced to the next segment by hitting the space bar. As subjects advanced, the previous segment disappeared. Reading time for each segment was recorded by the computer. The last segment of each passage was designated as the "target" segment. These segments were designed to either agree with or conflict with each subject's prior knowledge. Subjects' reading times on the target segments were used in our analysis. After each passage, subjects were asked to rate their understanding of the ideas in the passage on a 1–5 scale. We refer to this as a "judgment of learning," or JOL. Immediately after subjects had read all 29 passages they were asked to answer a 16-question true/false quiz about a new set of scenarios probing understanding of Newton's third law.

In the passages, the relative size of the two interacting objects was either equal, e.g., "two equal mass sumo wrestlers," or different, e.g., "a bowling ball collides with a bowling pin." All scenarios involved either a collision happening or about to happen, or a situation where the two objects were pushing against each other for an extended period of time. The 29 passages included 6 Newton's third law passages where the two objects were of unequal size, and 3 Newton's third law passages where the two objects were of equal size. In addition to the 9 Newton's third law passages, subjects were given 8 energy passages and 8 momentum passages as well as 4 Newton's second law passages. The energy and momentum passages served both to test certain factors involved in processing of the text and as distracters. The energy and momentum sentences were deliberately chosen to be intuitively obvious to the non-Newtonians and therefore some passages may appear vague to a physics expert. The Newton's second law passages served only as distracters and were not included in any analysis.

Subjects were randomly assigned to one of two conditions; a refutational condition, or a non-refutational condition. Examples of the refutational and non-refutational modes may be found in Table 1. Energy, momentum and Newton's second law passages did not have a refutational mode. Thus all subjects read the same energy, momentum, and Newton's second law distracters.

Hypotheses and Predictions

1. If students recruit prior knowledge and construct a situation model when they read a physics text:
   (a) Non-Newtonian students should read N3uneq target segments significantly slower than Newtonian students.
   (b) Non-Newtonian students should read N3uneq target segments significantly slower than they read N3eq segments.
   (c) Newtonian students should read N3uneq and N3eq target segments without any significant difference in time.
2. We hypothesize that there is a linguistic element to students' models of the idea of force. Namely, non-Newtonian students see force, energy and momentum as properties of motion of an object and roughly equivalent in meaning [12]. In the situation model view, when asked to compare which object

TABLE 1. Reading passages examples. // Denotes each new segment in the reading sequence. The target segments, which were used in our analysis, are indicated in bold.

Description	Example(s)
Newton's third law, unequal sized objects (N3uneq)	(Non-refutational mode): A mosquito is flying towards a bus that is coming the other way. // When the mosquito collides with the bus, // compare the force that the mosquito exerts on the bus with the force that the bus exerts on the mosquito. **// Physics says the mosquito and the bus exert equal forces on each other.** (Refutational mode): A mosquito is flying towards a bus that is coming the other way. // Some people think that when the mosquito collides with the bus, // the force that the mosquito exerts on the bus is less than the force that the bus exerts on the mosquito. **// However, the mosquito and the bus exert equal forces on each other.**
Newton's third law, equal sized objects (N3eq)	Two equally heavy sumo wrestlers are trying to push each other out of the ring. // The two remain locked in battle in the middle, not moving at all. // Consider how much force each sumo wrestler exerts on the other. **// It is true that each wrestler exerts the same amount of force on the other.**
Energy	A stunt man plans to smash through a wall with his monster truck. // As the truck roars towards the wall, // consider the energy of the truck compared to the energy of the wall. **// The energy of the truck is a good deal more than the energy of the wall.**
Momentum	With a hammer, you are hitting a nail into a piece of wood. // Just before the hammer connects with the nail, // think about the momentum of the hammer compared to the momentum of the nail. **// The hammer has much more momentum as compared to the momentum of the nail.**
Newton's second law (N2)	A 200-pound sumo wrestler is pushing against a 300-pound sumo wrestler. // As the bigger wrestler pushes the smaller wrestler out of the ring, // consider how much force each wrestler exerts on the floor. // The bigger sumo wrestler exerts more force on the floor than the smaller wrestler.

exerts more force, students may generate an inference like "bigger object has more force therefore bigger object exerts more force." If students see force, energy and momentum as synonymous, students should generate similar inferences about energy and momentum, e.g., "a bigger object has more energy." We predict that non-Newtonian students will read energy and momentum targets quicker than the N3uneq targets since energy/momentum targets will match their situation model.

3. If there is a simple relationship between online and offline processing:

   (a) We predict that non-Newtonian students who score higher on the quiz will show more evidence of hesitation on target sentences that contradict their naïve ideas about force.

   (b) If the refutational mode leads to more learning among non-Newtonians, we should see more hesitation over N3uneq targets in the refutational mode, and correspondingly higher performance on the quiz.

RESULTS AND ANALYSIS

Online Processing

1. Non-Newtonian students read N3uneq targets significantly slower than the Newtonians did: $\bar{t} = 3.6$ seconds versus $\bar{t} = 2.3$ seconds, $p<0.0001$, two tailed, unequal variance t-test, $n = 30$.

2. Non-Newtonian students read N3uneq targets significantly slower ($\bar{t} = 3.6$ seconds) than N3eq targets ($\bar{t} = 2.4$ seconds): $p < 0.001$, $n = 16$.

3. Newtonians' reading times on the N3uneq ($\bar{t} = 2.3$ s) and N3eq ($\bar{t} = 2.8$ s) targets did not differ significantly, but tended in the opposite direction relative to non-Newtonians, $p= 0.07$, $n = 14$.

4. Non-Newtonian students read N3uneq targets significantly slower ($\bar{t} = 3.6$ seconds) than the equivalent energy ($\bar{t} = 2.6$ seconds) and momentum ($\bar{t} = 2.9$ seconds) targets. $p = 0.0015$ and $p = 0.014$ respectively.

5. Conversely, Newtonians read N3uneq targets significantly faster ($\bar{t} = 2.3$ seconds) than equivalent energy (3.0 seconds) and momentum (3.2 seconds) targets. $p < 0.01$ and $p < 0.001$ respectively.

6. There was no significant difference between non-Newtonians' and Newtonians' reading of the energy and momentum targets ($p>0.15$).

Offline and Online Processing

The Newtonians' average score on the true/false quiz was 15.7 out of 16. On the other hand the average score of the non-Newtonians was 8.7, or 4.7 out of 12 if we subtract out four questions where the objects presented were of equal size. (All non-Newtonians answered these four questions correctly.) Here the statistical null hypothesis is that if students randomly guess true or false they should get 6 out of 12. An average score of 4.7 places this group at $0.1<p<0.2$ on a binomial sampling distribution — their responses are not significantly below random guessing.

In the non-Newtonian group:

1. There is no effect of refutation on reading time for N3uneq targets, nor on quiz performance.

2. Although some students showed they were confused by N3uneq targets (from JOL ratings), there was no correlation between JOL and reading time on N3uneq targets. (Pearson $R = 0.229$, $n = 16$, $p = 0.394$ - two-tailed t test).

3. There is no evidence of correlation between reading time on N3uneq targets and performance on the post-reading true/false quiz (Pearson $R = 0.162$, $n = 16$, $p = 0.548$).

4. There is a significant positive correlation (Pearson $R = 0.568$, $n=16$, $p = 0.022$) between non-Newtonian students' JOL on N3uneq passages and performance on the T/F quiz. In other words, students who rated their understanding higher performed better on the quiz.

CONCLUSIONS

In contrast with [6], we found evidence of online processing when students read a physics text. It is clear from the differential reading times that non-Newtonian students and Newtonian students are generating inferences (and therefore recruiting prior knowledge) and forming a situation model. When non-Newtonian students read the N3uneq targets, they hesitate over those target segments that contradict their inferences and read target segments that agree with their inferences (N3eq) significantly faster. Newtonian students appear to be reading target segments that are consistent with their prior-generated inferences (as evidenced by their quicker reading time).

Our prediction that non-Newtonians would read energy and momentum targets faster than N3uneq targets is confirmed by our data. This is consistent with the idea that students with "non-Newtonian" conceptions of force and motion and Newton's third law have difficulty with the meaning of the term "force," confusing it with "energy" and "momentum" as properties of an object and/or its motion [12].

Interestingly, Newtonians read the Newton's third law targets more quickly than comparable targets about energy and momentum. We propose one possible explanation for this result: Situation model theory suggests that, depending on existing knowledge structures, individuals generate inferences as they read so as to create a globally coherent representation of the situation. In this view, students who have relatively rich knowledge sys-

tems about energy and momentum will activate a wider array of inferences when reading about these concepts For example, a student with sufficient physics knowledge might question "what forms/types of energy are we talking about?" "Momentum is a vector," "the passage does not tell me which object is moving faster, therefore I'm not sure which object has more momentum." The greater number of inferences generated means that it will take the neural network longer to relax into a single state, hence the longer reading times amongst the Newtonians.

There is evidence of online processing (differential reading times) but there is no correlation between reading time and quiz performance. Our data clearly show that, at a perceptual level, non-Newtonian students are surprised by statements that contradict their situation model. Further evidence about how ideas in the text are integrated into students' knowledge is inconclusive, and only suggestive. The correlation between JOL and quiz performance suggests that metacognition may be playing a role. Confusion arising from the text is insufficient for any learning to take place. Students also make conscious decisions to either accept or reject the ideas that they've just read, based on possible factors such as their epistemological beliefs, the manner in which they have framed the experimental situation, and so on. In future research we hope to explore the interaction of students' metacognition with their online and offline processing of the text they are reading.

REFERENCES

1. F. C. Bartlett, *Remembering: A study in experimental and social psychology*, Cambridge University Press, Cambridge, England, 1932.
2. E. A. L. Stine-Morrow, L. M. S. Miller, and C. Hertzog, *Psychological Bulletin* **132**, 582 – 606 (2006).
3. D. E. Alvermann, L. C. Smith, and J. E. Readence, *Reading Research Quarterly* **20**, 420 – 436 (1985).
4. B. J. Guzzetti, T. E. Snyder, G. V. Glass, and W. S. Gamas, *Reading Research Quarterly* **28**, 116 – 159 (1993).
5. W. Kintsch, *Cognition and Instruction* **3**, 87 – 108 (1986).
6. P. Kendeou, and P. van den Broek, *Journal of Educational Psychology* **97**, 235 – 245 (2005).
7. W. Kintsch, *Psychological Review* **95**, 163 – 182 (1988).
8. W. Kintsch, *Comprehension: A Paradigm for Cognition*, Cambridge University Press, Cambridge, England, 1998.
9. J. E. Albrecht, and E. J. O'Brien, *Journal of Experimental Psychology: Learning, Memory, and Cognition* **19**, 1061 – 1070 (1993).
10. D. Wechsler, *Wechsler Adult Intelligence Scale — Revised*, Psychological Corporation, San Antonio, TX, 1981, 7th edn.
11. D. Hestenes, M. Wells, and G. Swackhamer, *The Physics Teacher* **30**, 141 – 158 (1992).
12. D. T. Brookes, and E. Etkina, *Cognition and Instruction* (in review).

The Dynamics of Students' Behaviors and Reasoning during Collaborative Physics Tutorial Sessions

Luke D. Conlin[*], Ayush Gupta[*], Rachel E. Scherr[*], and David Hammer[†]

[*]Department of Physics, University of Maryland, College Park, MD 20742
[†]Department of Curriculum & Instruction and Department of Physics,
University of Maryland, College Park, MD 20742

Abstract. We investigate the dynamics of student behaviors (posture, gesture, vocal register, visual focus) and the substance of their reasoning during collaborative work on inquiry-based physics tutorials. Scherr has characterized student activity during tutorials as observable clusters of behaviors separated by sharp transitions, and has argued that these behavioral modes reflect students' epistemological framing of what they are doing, i.e., their sense of what is taking place with respect to knowledge. We analyze students' verbal reasoning during several tutorial sessions using the framework of Russ, and find a strong correlation between certain behavioral modes and the scientific quality of students' explanations. We suggest that this is due to a dynamic coupling of how students behave, how they frame an activity, and how they reason during that activity. This analysis supports the earlier claims of a dynamic between behavior and epistemology. We discuss implications for research and instruction.

Keywords: Mechanistic reasoning, inquiry, behavior, explanation, frames, framing, epistemology.
PACS: 01.40.Fk

INTRODUCTION

In an analysis of students collaborating on inquiry-based physics tutorials, Scherr [1] indicated preliminary evidence of a connection between students' behaviors and the level of their scientific reasoning. Students' behaviors and reasoning during an activity are influenced by the set of expectations, i.e. 'frame,' they associate with the activity. The tutorials are meant to provide students with opportunities to practice their inquiry and reasoning skills. But is this how the students frame these activities? In the present paper we report more episodes of behavior and reasoning during tutorials, provide strong evidence for the connections between behavior, framing and reasoning, and offer reasons for such connections.

MECHANISTIC REASONING

One of the goals of inquiry-based science teaching is to give students opportunities to learn via authentic scientific practices. Constructing scientific explanations of phenomena is a central part of authentic science, but what constitutes a scientific explanation? Based on philosophy of science literature we may adopt a tentative definition of a scientific explanation: an argument that lays out the causal mechanisms by which the phenomenon occurs, given the laws and initial conditions [2]. If constructing mechanistic explanations is an essential feature of doing science, then inquiry-based science instruction should give students opportunities to develop mechanistic reasoning skills. Indeed, inquiry in science may be described as "the pursuit of coherent mechanistic accounts of phenomena" [3].

In the present work, we use the coding scheme developed by Russ [4] to recognize and analyze mechanistic reasoning in students' verbal responses during physics tutorials. Mechanistic reasoning about a phenomenon involves several elements that Russ roughly organized into a hierarchy of increasing quality of evidence: describing target phenomenon, identifying set up conditions, identifying entities, identifying actions, identifying properties of entities, identifying the organization of entities, and "chaining." Chaining is the most complete of these elements, and involves linking several of the elements together, either to make a prediction or to reason about how things must have been in the past. A wide range of conversations produce lower level codes; mechanistic reasoning is best evidenced by chaining.

CP951, *2007 Physics Education Research Conference*, edited by L. Hsu, C. Henderson, and L. McCullough
© 2007 American Institute of Physics 978-0-7354-0465-6/07/$23.00

An example of chaining occurred when one group was discussing why rubbing two hands together does not produce sparks. The group had identified moisture as an entity relevant to explaining this, but could not explain further what moisture had to do with it. John [5] finally chained together an explanation based on the entity (moisture) and its properties (it is a conductor). He argued, "It's a conductor so like it's not going to let charge build on your hands because moisture's a conductor so it's like going to dissipate off into the atmosphere..."

Russ's framework allows a systematic means of assessing the substance of students' reasoning with respect to causal mechanisms, whether or not those mechanisms correspond to the "correct" account. In this paper, we examine the mechanistic reasoning of physics students during the introductory physics tutorials, and note how it is influenced by the students' framing of the tutorial activity.

FRAMING AND BEHAVIOR

A 'frame' is the set of expectations each participant brings into a situation that corresponds with their sense of "what is going on here" [6]. For example, students in tutorial may frame the activity as a 'complete the worksheet' task, or as a 'make sense of physics' task. What counts as an appropriate explanation depends on the experience and expectations of those providing and evaluating the explanations [7]. Therefore, we should expect that student explanations of phenomena should be influenced by how they frame the activity.

Although we cannot directly observe a set of expectations, there are linguistic and behavioral cues that can inform us how a student is framing an activity. Scherr used vocal register, linguistic cues, body language, and gesture to identify four locally stable clusters of behaviors, which she calls "behavior modes," among groups of physics students working on tutorials. She initially coded the student groups for changes in behavior, using theory-neutral colors, and then used these modes as indicators of how students frame the activity. For example, the "blue" behavioral mode, in which students have their eyes on their worksheets and speak in soft voices, indicates students framing the activity as "completing the worksheet." The modes and frames are summarized in Table 1. Each behavior mode is easily observable, stable for time periods ranging from seconds to minutes, delineated by sharp transitions, and largely consistent for the entire group of students working together. For the current investigation, we use this methodology to characterize several locally coherent behavior modes,

and examine the nature of explicit reasoning during these modes.

TABLE 1. Behavior Mode Coding Scheme

Color Code	Behaviors	Frame
Green	Sitting up, eye contact with peers, subdued gestures, lower vocal register	Discussion
Blue	Hunched over, eyes on worksheet, low vocal register, writing	Worksheet
Yellow	Fidgeting, laughing, looking away, touching face/hair	Socializing
Red	Sitting up, eyes on TA, subdued gestures, lower vocal register	Receptive to TA

METHODOLOGY

This study involved 24 college students enrolled in an introductory algebra-based physics course at the University of Maryland. We transcribed 20-minute video clips of 4-student groups working during 6 tutorial sessions. The tutorials consist of 45-minute guided-inquiry sessions delivered via worksheet and attended by teaching assistants. The videos were first coded for behavior without a transcript. The behavior modes are coded somewhat holistically by considering the presence of behaviors in Table 1. Two independent coders agreed on 90% of behavior codes pre-discussion, to a precision of 5 seconds. Then the sessions were transcribed and coded for (a) instances of some level of mechanistic reasoning, and (b) instances of chaining. Two independent coders agreed on 87% of mechanistic reasoning and chaining codes before discussion. The codes for behavior, mechanistic reasoning, and chaining were then matched for each 5 seconds of the tutorial session. In the next two sections, we describe key findings from this data and provide a description of the dynamic between behavior, framing, and reasoning.

DATA AND DISCUSSION

The percentages of mechanistic reasoning codes and of chaining specifically within each behavior mode are reported in Table 3. The pure blue, green, red, and yellow behavior modes account for nearly all of the time spent in the tutorials (86%), while the remaining time was spent in 'mixed modes.' Since the behavior modes practically span the time spent during tutorial, we will refer to the frame rather than the color code scheme for the remainder of the paper.

TABLE 2. Mechanistic Reasoning within Frames (summary for all 6 video clips)

Frame	Color Code	% Time	% Mechanistic Reasoning	% Chaining
Animated Discussion	Green	25%	53%	81%
Completing the Worksheet	Blue	32%	18%	6%
Receptive to TA	Red	24%	15%	4%
Socializing with Peers	Yellow	5%	1%	0%
Other	Mixed	14%	12%	8%

Notably, the majority of mechanistic reasoning occurred during the "animated discussion" frame (53%). A significant amount of mechanistic reasoning also occurred during the "completing the worksheet" frame (18%), and the "receptive to TA" frame (15%).

The percentage of coding in any behavior mode is of course partly a result of how much students speak; students say more in the discussion frame, so there should be a higher percentage of codes by any criteria. As such, we found that the number of mechanistic reasoning codes was proportional to the number of statements made within each mode. This was not true of chaining, which leads to the finding of importance here. The number of codes for chaining in particular is disproportionately high: nearly all of the chaining occurred during the discussion frame (81%), which shows that the nature of student reasoning is different during this frame.

There are several plausible reasons for the high percentage of chaining happening during the discussion frame. Similar to what was observed by Tannen and Wallet [8], one reason is that the discussion frame was often precipitated by conflicting lines of reasoning amongst group members, who would then need to chain arguments together in order to convince each other of their explanation. For example, one group was in the worksheet frame during a tutorial involving the application of Newton's Third Law to a collision between a massive truck and a less massive car. Eric held that the forces between them are equal. Maya disagreed, saying "I don't think there's any way that you can explain to me how a massive truck is going to have the same force…" after which the group abruptly shifts into the discussion frame as they try to convince each other of their respective points of view.

While students' framing of the tutorial may influence their reasoning, it also seems the opposite may occur; the nature of explanations may shift the framing. Students in the worksheet frame tend to make comments coded at low-levels of mechanistic reasoning as they negotiate the target phenomenon to be explained and the entities involved. When a 'critical mass' of lower mechanistic reasoning codes occur, the group will transition into the discussion frame as they collectively chain together how these entities act to bring about the phenomenon. Once a sufficiently chained explanation is agreed upon, the group will shift back to the worksheet frame in order to move on with the tutorial.

The effect of reasoning on the evolution of the group's frame can be seen in graphs such as Figure 1, where a single video clip is coded from start to finish. The arrows indicate chaining, which typically occurs near transitions between worksheet and discussion frames. For this particular case, six of the nine instances of chaining occurred on the brink of such a transition.

COUPLED DYNAMICS OF BEHAVIOR AND REASONING

Since the tutorial involves both group discussion and a worksheet, there seems to be an ever-present tension between these two ways of framing the activity. Out of the individual framing of each student emerges an overarching group dynamic that evolves over the course of the tutorial. This is evident in that all four students occupy the same behavior modes for 86% of the time.

Based on the behaviors, we have seen that the overall framing of the activity varies between tutorials and between groups. For example, Table 3 compares the framing for two groups working on different tutorials. It seems that Group 1 has more of a tendency to frame the tutorial as a discussion, while Group 2 tends to frame it more as a worksheet-completion activity. More research needs to be done to explore what influences groups to frame the tutorial in a certain way.

TABLE 3. Time Spent in Frames. This compares two groups working on different tutorials.

Frame	Group 1	Group 2
Discussion	47%	25%
Worksheet	20%	45%
Socializing	4%	14%
Receptive to TA	17%	13%
Other	12%	3%

Each group transitions in their behavior and framing together, and these transitions are generally quite abrupt. There are many ways that the group's framing may shift. Sometimes the frames switch due

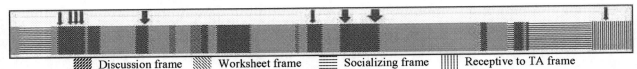

Discussion frame   Worksheet frame   Socializing frame   Receptive to TA frame

FIGURE 1. Tutorial Frame & Reasoning Evolution. This represents a group during a 20-minute clip from a single tutorial on Work & Energy, with time running left to right.

to an explicit frame negotiation from one or more of the members. For example, a group working on a Newton's third law tutorial had gotten off task. They were in the socializing frame, joking and talking about things that were off-topic. After a few minutes, Eric made a subtle appeal to bring the group's focus back to the worksheet, saying, "Alright, alright, alright..." while hunching over, putting his pencil near his worksheet and starting to read. The group soon transitioned into the worksheet frame.

Another reason for each group's shifts in behaviors and framing is due to contrasting student epistemologies. During their discussion about Newton's third law during a car-truck collision, Macy drew on her personal experience: "But I'm trying to think, like, okay for example I was, when I was in a car accident..." which sparks the students into a brief period of discussion. In the end Macy is not satisfied with the group's explanation that the forces are equal: "I don't know it just doesn't seem right to me." The discussion mode ends abruptly when Eric proclaims his differing epistemological stance, in which he appeals to authority: "I don't try to argue with the laws of physics. I just trust that they work."

IMPLICATIONS AND FURTHER RESEARCH

The results here support Scherr's analysis of behavioral modes as indicative of epistemological framing. In addition, we find that the substance of student reasoning shows different patterns during different behavioral modes. Our first interest here is in developing better models of student reasoning in tutorials, but we note one instructional implication: Since behavior modes are so distinct and stable, an instructor can recognize them in real-time and act on them accordingly. For example the discussion frame is more associated with mechanistic reasoning than others, so an instructor may treat it as one sign of a well-functioning group.

The causes of frame transitions need to be examined in more detail. Specifically, the connection between chaining and frame transitions as well as the effects of the teaching assistants on the behavior modes and student reasoning should be further studied. It may also be helpful to examine students' written

work in tutorial, to more accurately assess the nature of reasoning during the worksheet frame.

SUMMARY

We have shown that there is a strong connection between how students behave during a tutorial activity and the substance of their verbal reasoning. Specifically, our evidence suggests that the most advanced mechanistic reasoning occurs while students are engaged in animated discussions. We have given several reasons for frame and behavior mode changes, including contrasting lines of reasoning, contrasting epistemologies, and the dynamic of reasoning throughout the tutorial.

ACKNOWLEDGMENTS

This work supported in part by NSF grant REC 0440113. Further thanks go to Joe Redish, Rosemary Russ, Tom Bing, Renee-Michelle Goertzen, Brian Frank, Matty Lau, Andy Elby, and the rest of the Physics Education Research Group at the University of Maryland for their helpful comments and discussions.

REFERENCES

1. R.E. Scherr, "How Do Students Frame Collaborative Active Learning Activities?" NARST Annual Meeting, April 2006, San Francisco, CA.
2. P. Railton, "A Deductive-Nomological Model of Probabilistic Explanation" in *Philosophy of Science: The Central Issues,* edited by M. Curd and J.A. Cover, NY: W.W. Norton & Co., Inc., 1998, pp. 746-765.
3. D. Hammer, R. Russ, J. Mikeska, and R.E. Scherr, in *Establishing a Consensus Agenda for K-12 Science Inquiry,* edited by R. Duschl and R. Grandy (Sense Publishers, Rotterdam, NL, in press).
4. R. Russ, "A Framework for Recognizing Mechanistic Reasoning in Student Scientific Inquiry," Ph.D. Thesis, University of Maryland, 2006.
5. All student names in this paper are pseudonyms.
6. D. Hammer, A. Elby, R.E. Scherr, and E.F. Redish, in *Transfer of Learning from a Modern Multidisciplinary Perspective,* edited by J. Mestre (Information Age Publishing, Greenwich, CT, 2005), pp. 89-120.
7. J.K. Gilbert, C. Boulter, and M. Rutherford, *Int. J. Sci. Ed.* **20** 83-97 (1998).
8. D. Tannen and C. Wallet, *Soc. Psych. Qrtly.* **50**, 205-215 (1987)

Hands-On and Minds-On Modeling Activities to Improve Students' Conceptions of Microscopic Friction

Edgar G. Corpuz[*] and N. Sanjay Rebello

*Physics & Geology Department, University of Texas-Pan American, 1201 W. University Dr., Edinburg, TX 78539
Department of Physics, 116 Cardwell Hall, Kansas State University, Manhattan, KS 66506-2601

Abstract. In this paper we discuss the development and validation of hands-on and minds-on modeling activities geared towards improving students' understanding of microscopic friction. We will also present our investigation on the relative effectiveness of the use of the developed instructional material with two lecture formats - traditional and videotaped lectures. Results imply that through a series of carefully designed hands-on and minds-on modeling activities, it is possible to facilitate the refinement of students' ideas of microscopic friction.

Keywords: friction, microscopic, students' conceptions, physics education research
PACS: 68.35, 01.40Fk

INTRODUCTION

We are currently at the verge of several breakthroughs in nanoscience and technology and we need to prepare our citizenry to be scientifically literate about the microscopic world. An urgent need exists to fill the lacunae of research in the understanding and learning of phenomena at the nanoscale level by all students regardless of their future academic goals. Friction provides a very good context for making students aware of the disparity between the macroscopic and microscopic world.

Our previous research [1] showed that students' mental models of friction at the atomic level are significantly influenced by their macroscopic ideas. For most students, friction is due to meshing of bumps and valleys and rubbing of atoms. The aim of this research is to develop and validate instructional material that facilitates refinement of students' ideas of microscopic friction building on their prior knowledge and experiences. In this paper we present a description of the instructional material developed and its effectiveness in improving students' conceptions about microscopic friction.

The research questions we seek to address are:

- To what extent are the developed materials effective in improving students' conceptions about microscopic friction?
- How does the effectiveness of the developed materials compare with traditional instruction in improving students' conceptions about microscopic friction?

THEORETICAL FRAMEWORK

We adapt the Vygotskian social constructivist view that learning occurs within a Zone of Proximal Development (ZPD) facilitated by interactions with more capable individuals through scaffolding [2]. We utilized several scaffolding activities, including conceptual change strategies [3], to enable students to refine and extend their models of microscopic friction.

METHODOLOGY

Development of Instructional Module

Teaching interviews [4] were conducted in order to investigate how different modeling activities influence students' conceptual development. The developed instructional module consisted of the hands-on and minds-on activities that were found to be helpful in activating appropriate associations among students during the teaching interviews. We also included questions and hints which scaffolded the construction of productive associations.

In terms of the sequencing of the activities, Karplus' Learning Cycle [5] was adopted. Students were engaged in exploration, concept construction and application activities. The goal of the exploration was to invoke student prior knowledge about friction. The exploration activities included the dragging of a

CP951, 2007 Physics Education Research Conference, edited by L. Hsu, C. Henderson, and L. McCullough
© 2007 American Institute of Physics 978-0-7354-0465-6/07/$23.00

wooden block across a wooden and sandpaper surface and the feeling and sketching of the different surfaces at the atomic level. In this modeling cycle, students' prior ideas about friction were activated.

In the concept construction, students were explicitly required to represent their model using multiple representations. They were asked to sketch a graph of friction vs. surface roughness and talk about what happens to the friction force in different situations. In the application phase, students were given activities or situations where they apply the concepts that they have constructed. This particular application activity involved metal blocks with a smooth surface and other relatively rough surfaces. Here students were asked to make their predictions in which case (smooth on smooth or smooth on rough) they would observe more friction. The application activity produced cognitive conflict since students' predictions differed from their observations.

To resolve this cognitive conflict, students proceeded to the second cycle of the model building process. In the second cycle, students completed the papers and transparency activity for their exploration. In this activity students took a sheet of paper and dragged it across a transparency that had been rubbed with fur. They observed that the paper tends to stick to the transparency and there was resistance to the motion of the paper on the transparency. Later they took the same piece of paper, crumpled it, straightened it out again and dragged it across the same transparency. In this case the paper did not stick and there was no resistance to the motion of the paper on the transparency.

In the concept construction phase, students were asked to explain their observations in the paper and transparency exploration activity. After they explained their observations in terms of the "real" area of contact they then revisited their previous graph and modified it based on their new experiences with the paper and transparency exploration. They were also asked to reflect on their earlier experiences with the metal blocks and resolve their cognitive conflict. Thus, at the end of the concept construction phase students emerged with a model that accounts for friction in terms of the 'real' area of contact between atoms. The model explains their observations with the metal blocks and the relationship between friction and surface roughness both in the microscopic as well as the macroscopic domains. Students also realize the role of the real area of contact and that friction at the atomic level increases with increasing smoothness. They also understand the role of electrical interactions when talking about friction at the atomic level.

In the application activity students predicted the difference in friction in the two cases: block on its broad side vs. block on its narrow side. The

application activity provides a context in which students apply their model that identifies the role of 'real' contact area in a macroscopic context. It allows them to examine how this notion of 'real' area of contact can yield results consistent with the macroscopic result that the force of friction is independent of the area of contact, thus reinforcing the connection between microscopic processes and their macroscopic manifestations.

Validation of Instructional Module

Subject area experts were involved in the content-validation of the materials. The first version of the instructional module was shown to two experts whose research specialization is in surface phenomena. Moreover, these experts had been teaching introductory physics courses for at least six semesters, so they were knowledgeable about the background of targeted students. They examined the instructional module to ensure that the content of the module was scientifically accurate, valid and relevant to the targeted students. The experts' suggestions were minimal, focusing on formatting and inclusion of more pictures. The instructional material was then revised based on the experts' feedback.

Several groups of students were involved in the validation of the developed module. The students typically worked in groups of two or three as they completed the activities. The researcher observed and made field notes of each session. Each session was likewise videotaped with the IRB consent from the students. Post-activity interviews were conducted in order to cross validate the researcher's observations, get students' feedback regarding students' difficulties and confusions and bring forth other issues of concern regarding the implementation of the developed instructional materials.

In addition to completing the post-activity interviews, students were also asked to complete a Likert-scale questionnaire that pertains to the content, appeal, design and difficulty of the developed learning instructional materials. The pilot version of the material was then revised based on the insights gained by the researcher through the observations and the feedback of students during the post-activity interview.

Qualitative Evaluation

Individual as well as groups of two or three students used the developed instructional activities. We kept track of students' conceptual progression by incorporating open-ended questions in the module for them to answer. These reflective open-ended questions were embedded in the module at appropriate points to provide the learner as well as the instructor-researcher feedback about student learning.

In addition to the reflective questions, stopping points were interjected into the module to encourage students to discuss their predictions, observations and thoughts with other members in their group and the instructor. Students were asked to discuss what they were doing in the activities and how these activities influenced their ideas.

Quantitative Evaluation

In addition to the qualitative assessment of student learning we also developed an instrument for quantitative assessment of student learning. To develop the test questions we first identified the learning goals -- target ideas that we wanted students to learn. In ensuring content-related validity of the test, a table of specifications was prepared initially in order to make sure that each target idea that we wanted students to learn was addressed in the test with at least one test item. The table of specifications was shown to two experts in order to cross check whether particular items measured the target ideas. Questions with corresponding distracters were then constructed. The distracters used in each of the items were based on ideas students brought out in the group and individual clinical interviews and teaching interviews conducted in the first two phases of the research. The number of items constructed was constrained by the fact that the test intends only to measure students' understanding on a single topic (microscopic friction). The final version of the test had 10 multiple-choice questions.

The Kuder-Richardson reliability coefficient (KR-20) was 0.67, indicating that the test items were homogeneous, i.e., the test measures the same characteristic of the people taking it. The test is also reliable, i.e., the individual items were producing similar patterns of response in different people. Moreover, it is safe to say that the test is a valid measure of the understanding of microscopic friction in a way that is consistent with the target ideas established for the students.

Gauging the Effectiveness of the Developed Materials

In investigating the relative effectiveness of the developed instructional materials in improving students' conceptions via traditional assessment format, a pre-test post-test control group design was employed. The use of hands-on and minds-on modeling activities were compared with lecture in two formats – traditional lecture and videotaped lecture on microscopic friction. The same ideas that we envisioned students to construct while they did the hands-on and minds-on activities were presented by the instructor in lecture. The time on task for the different groups was almost equal, each being about an hour long. Moreover, the same sets of activities were performed by the instructor in the two lecture formats. The videotaped lecture was conducted by the same instructor who did the traditional lecture.

RESULTS AND DISCUSSION

All of the students participating in this study were enrolled in a conceptual physics course for elementary education majors. There was no statistically significant difference in the performance on class tests and exams between students in the two control groups – videotaped (N = 24), live lecture (N = 56) and the experimental group – activity (N = 66).

Figure 1 shows a comparison between the pre-test and post-test scores of students in the three groups. The three groups also performed very similarly on the pre-test, however their performance on the post-test was markedly different. A two-tailed t-test comparing the pre-test and post-test scores for the activity group indicated a statistically significant difference ($p < 1\times10^{-12}$) A similar t-test for the lecture and video group and lecture group showed a statistically significant difference, but the differences were not as significant as the activity group.

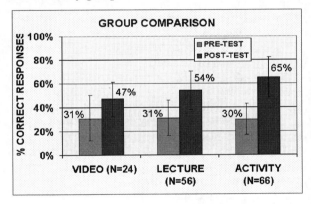

FIGURE 1. Pre-test and post-test scores of the Experimental and Control groups.

In comparing the three groups, we find that the mean pre-test scores are almost all equal to about 30%. The post-test scores, however, are significantly different. A two-tailed t-test comparison between the video and activity groups indicated a statistically significant difference at the $p < 1\times10^{-5}$ level of significance. A similar two-tailed comparison between the lecture and activity groups indicated a smaller albeit statistically significant difference at the $p < 5\times10^{-4}$ level of significance. Thus, the activity appears to be more effective than either the lecture or the videotaped lecture in improving student conceptions of friction as measured by the post-test. Figure 2 indicates the percentage of students in each group that answered each question correctly.

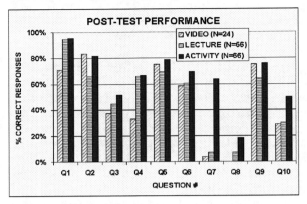

FIGURE 2. Percent of students getting each question correct on the post-test for each group.

Overall, the activity group outperformed the lecture or the video group on most of the questions. The activity group showed the highest percentage of correct responses for question 1. This question assesses the idea that when two surfaces become very smooth the friction force increases because of the increase in the number of atoms that would be electrically interacting with one another. This was one of the important target ideas of the instructional materials. Therefore, it appears that the activity was successful in helping students understand this target idea, as measured by this question.

The greatest disparity between the activity and the lecture or video groups is seen in question 7. This question asks students to choose a graph that depicts the relationship between friction and roughness in the microscopic domain. It appears that students who performed the activity and drew the graph for themselves were better able to do so than those who merely saw the instructor draw and explain the graphs.

In Fig. 2, we see that the question with the lowest percentage of students in all of the three groups answering correctly is question 8. This question asks students to choose an explanation of why it is equally easy to drag a block of wood on its wide side or its narrow side. The correct explanation is that in both cases – wide or narrow side being dragged – the microscopic area of contact is the same. Most students selected the response that states that force of friction does not depend on contact area. This statement, which is commonly found in most textbooks, is used to explain macroscopic friction. Thus, it appears that when a question demands thinking microscopically students tend to revert to commonly held ideas.

The other two low-performing questions (below 60% correct) for the activity group were questions 3 and 10. Question 3 asks students to predict what happens to the force of friction between two surfaces in a weightless environment. Clearly, students had not performed this activity, so when asked they tended to rely on the macroscopic view of friction. Question 10 asks them to select factors that affect friction at an atomic level. Almost all students who answered this question incorrectly included not just microscopic factors (area of contact and electrical interactions) but also macroscopic factors.

CONCLUSIONS

Based on our results, the developed hands-on and minds-on learning materials appear to be effective in enabling students to learn the target ideas as measured by our test. These ideas include the electrical origin of friction and how it varies with roughness in the microscopic domain. Students appear to have difficulty, however, with other ideas pertaining to the role of contact area and the factors affecting friction in the microscopic or macroscopic domains.

The hands-on and minds-on learning materials also appear to be superior to direct instruction either by lecture or videotaped lecture. This latter result aligns with evidence of the superiority of interactive engagement over traditional methods as cited elsewhere [6].

ACKNOWLEDGMENT

This work is supported in part by the National Science Foundation under grant REC-0133621. Opinions, expressed are those of the authors and not necessarily those of the Foundation.

REFERENCES

1. E.G. Corpuz and N.S. Rebello in Annual Meeting of the National Association for Research in Science Teaching., Dallas, TX: NARST (2005).
2. L.S. Vygotsky, *Mind in Society: The Development of Higher Psychological Processes*, Cambridge: Harvard University Press (1978)..
3. B. Posner, K. Strike, P. Hewson & W. Gertzog, *Science Education* **66**, 211-227 (1982).
4. P.V. Engelhardt, et al. "The Teaching Experiment - What it is and what it isn't," in *Physics Education Research Conference*, Madison, WI (2003).
5. R.J. Karplus, *Journal for Research in Science Teaching* **12** 213-218 (1974).
6. R.R. Hake, R.R., *American Journal of Physics,* **66**(1) 64-74 (1998).

Modeling Success: Building Community for Reform

Melissa Dancy[*], Eric Brewe[§] and Charles Henderson[†]

[*]Department of Physics, University of North Carolina at Charlotte, Charlotte NC, 28223
[§]Department of Curriculum and Instruction, Florida International University, Miami FL, 33199
[†]Department of Physics, Western Michigan University, Kalamazoo MI, 49008

Abstract. Dissemination of the Modeling Method of physics instruction to high school teachers has been considerably more successful than typical dissemination efforts in science education. In order to learn from this success, we interviewed five leaders of the Modeling Project. Based on these interviews, along with other information about the project, our preliminary findings suggest that the Modeling Project differed from standard dissemination efforts in several key ways that are likely related to its success. Specifically, the Modeling Project placed primary importance on the building of community. This emphasis on building community manifested itself in nonstandard community-based curriculum development and dissemination activities.

Keywords: Dissemination, Educational Change, High School
PACS: 01.40.Fk, 01.40.J-

INTRODUCTION

The education research community has produced a substantial body of research demonstrating the weaknesses of traditional instructional approaches and has developed a large number of proven alternative methodologies and curricular packages which are widely available [1]. In addition, considerable time, effort, and money have gone into disseminating both the research findings as well as curricula. Despite all these efforts, available evidence indicates that these products are at best only weakly integrated into mainstream physics teaching [2]. Although there is still work to be done, the research base is strong and alternative curricula are widely available. Therefore, for those interested in large-scale reform, attention needs to be given to issues of dissemination. Why does dissemination so often fail and how can we improve the outcomes of our dissemination efforts?

One way to approach this problem is to look at dissemination efforts that have achieved some measure of success. Within physics education, one approach that has impacted large numbers of physics classrooms is Arizona State University's Modeling Instruction in High School Physics Project [3]. Why has the Modeling project been more successful than many other reform projects? What can we learn from the dissemination of Modeling Instruction that will help improve the impact of other reforms? In an effort to answer these questions we conducted open-ended

telephone interviews with five leaders of the Modeling Project. We asked them about the history, goals, philosophies, successes, failures, and evolution of the project. Although analysis of these interviews is ongoing, in this paper we report some preliminary findings.

HISTORY AND SUCCESSES OF THE MODELING PROJECT

Modeling, as a theory-based instructional technique, was conceived between 1980-85 with the goal of making instruction in physics reflect the activities of practicing physicists. The instructional techniques were established by Malcom Wells in his classes from 1985 to 1993 and were formalized by David Hestenes [4]. These instructional techniques formed the foundational curriculum for the Modeling Workshop Project. Evidence for the success of the Modeling approach to physics came from the development of the Force Concept Inventory and led to a pilot workshop project. Modeling gained support from the National Science Foundation during 1990 and held the first Modeling Workshop for high school physics teaching during 1991. Malcom Wells led the initial workshop, which lasted six weeks and included teachers from Arizona. During the summer of 1992, the project expanded to two workshops and included a follow-up workshop for the initial participants. The major support for the project came from a national

CP951, *2007 Physics Education Research Conference*, edited by L. Hsu, C. Henderson, and L. McCullough
© 2007 American Institute of Physics 978-0-7354-0465-6/07/$23.00

dissemination grant in 1994 and expanded the workshops to initially include sites in Arizona and Chicago.

As the number of workshops grew, so did the scope and emphasis. The group of initial workshop participants in 1991 made a commitment to participate in three successive years. The initial workshop focused on Modeling in mechanics, using technology to support teaching and help manage classroom discourse. In subsequent years, the participants began to develop curricular materials that covered electricity and magnetism as well as refining materials for the mechanics curriculum. Many of the participants in the initial workshops went on to become workshop leaders in subsequent years. The leadership team understood that growing the project would require additional leaders and vetted participants to begin leading workshops on their own.

The training of workshop leaders through the national dissemination grant expanded the reach of the project. In 1998, an initial leadership workshop helped train new workshop leaders, and initiated an on-going effort to develop new curriculum, including Modeling Chemistry and Modeling for Physical Science. In addition to expanding the scope of the workshops, leaders were encouraged to work with faculty at local universities to submit proposals to hold local Modeling workshops. Jane Jackson was essential in coordinating local high school physics teachers with faculty, and providing templates for proposals.

Modeling workshops helped teachers gain a better understanding of the content of physics by acting as both participant and teacher in the physics curriculum. Additionally, participating teachers saw significant improvements in conceptual understanding in their students as measured by the Force Concept Inventory. The Modeling Workshop Project Final Report 1994-2000 tracked the FCI scores for students of teachers in Modeling Workshops, the scores are reported in Table 1 [5].

Table 1. Posttest FCI scores for students of Modeling Workshop Participants.

Pre-Workshop	Post-Workshop	Post Follow-up Workshop
42%	53%	69%

By training workshop leaders and encouraging them to submit proposals for local workshops, the leadership team allowed local leaders to adapt the workshops to their particular circumstances and needs. This gave leaders a sense of ownership and encouraged strong allegiance to the workshop project. As a result, at this point over 1500 teachers have

participated in a Modeling Workshop, which represents over 8% of all physics teachers nationally. The numbers are even greater in Arizona, where over 70% of teachers in the state have participated in Modeling Workshops [6]. In light of the conceptual gains for both teacher and student, and the broad participation in the workshops, it is not surprising that in 2002 the U.S. Department of Education cited the Modeling Workshop Project as one of two outstanding educational reform efforts nationally [7].

DATA COLLECTION AND ANALYSIS

Open-ended, semi-structured telephone interviews were conducted with five leaders of the Modeling Project: Larry Dukerich, David Hestenes, Jane Jackson, Colleen Megowan, and Greg Swackhamer. All interviews were conducted by Eric Brewe and then transcribed for analysis. Although a detailed analysis is ongoing, the preliminary analysis reported here is based on identification of dissemination-related patterns and themes in the interview transcripts. All three authors engaged in this process.

WHAT MAKES MODELING DIFFERENT?

In many respects the Modeling Project is similar to other reform efforts. It is based on a good idea, backed up by research which indicates that the modeling approach helps students deepen their conceptual understanding, improve their problem solving skills, and develop more positive attitudes toward science. Additionally, the modeling project faced many of the barriers so common to dissemination in general. Our interviewees indicated that the Modeling Project was significantly impeded by lack of sufficient funding, resistant teachers, and either a lack of administrative support for the reforms or, in some cases, outright resistance on the part of administrators and policy makers. However, the Modeling Project was able to overcome these barriers.

Although every dissemination approach is unique, there are many aspects that are common. For the purposes of this paper, when we speak of the standard dissemination approach we refer to an approach similar to the model of reform advocated by the NSF CCLI program [8]. In other words, an approach which begins with a good idea, demonstrates the effectiveness of the idea through peer-reviewed research, uses research to develop the idea as well as supporting materials such as curriculum and teachers guides, then disseminates the idea nationally through talks, papers, books, and workshops aimed at potential adopters. In this model of dissemination ideas and/or curriculum are developed by educational researchers and then distributed to teachers with some sort of training or guidance in their use.

This method of reform assumes that education researchers are in the best position to develop ideas and curriculum and that the most effective way of dissemination is to give teachers ready-to-use materials that they can implement more or less as is in their own classrooms. There are a lot of reasons why this method of reform makes sense. Teachers don't have the time or resources to continuously test and refine curriculum to the extent that a researcher can. In addition, once someone has put considerable time, effort, and money into working out the most effective method or curriculum, this project should be shared and used by others so they do not have to reinvent the wheel themselves. The basic idea is to develop the best possible project then share with others so they can use it too. In theory it is a great idea, but in practice, it has been limited in its impact.

The Modeling Project Focused on Building Community

Modeling differs from this standard approach. Standard curriculum development and dissemination is focused on a specific idea, methodology or curriculum. Although the Modeling Project appears on the surface to be focused on a specific set of "Modeling" curricular materials and strategies, the interviewee's descriptions reveal that Modeling Project was actually largely focused on the building of a community of "Modelers". Instead of developing and disseminating a product, the project built a community around the idea of teaching through scientific models. Throughout the interviews we found an overarching theme of community building exemplified in quotes like the following: *"Here is an important thing about the modeling program, it is the building of the community and the treatment of the dignity of the members of the community that is so important…"*

A great deal of deliberate effort went into building relationships and in developing communication avenues between modelers. For example, an effort was made to bring the same potential modelers back to Arizona State University each year so that they would get to know each other and feel comfortable turning to each other for support. Additionally, a listserv was created specifically for modelers to communicate with each other and to have support for their attempts to reform their own classrooms. As of this writing that listserv has over 1500 members. Even within the original leadership of the modeling project, our interviews indicate that each person brought something unique to the project and that the project

was not the result of any one (or even two) people, but rather the result of the unique contributions of everyone. Because the project so successfully built a community of modelers, it appears that even if all the initial instigators of the project were to discontinue their involvement, the project itself would go on. Ownership of the project does not fall with any one person or small group, it is shared by all and newcomers are welcoming and encouraged. Recently, the modeling community has established The American Modeling Teachers Association as a means to support the community of modeling teachers. [9] The establishment of this association was done independent of the original leaders of the Modeling Project, even to the extent that the initial listserv post to the modeling community announcing the association included the statement *"Both Dave Hestenes and Jane Jackson are aware and supportive of the organization; they are lifetime honorary voting members."* The modeling community can and does function independently of any particular individuals.

This theme of community building manifested itself in both the curriculum development and dissemination activities used by the modeling program. Each is discussed below.

Community-Based Curriculum Development

Although some curriculum was developed by the original implementers of the project they did not attempt to distribute the curriculum as is. Instead they deliberately encouraged workshop participants to work on refining existing curricular materials as well as developing their own new materials. The workshop participants were then encouraged to share their work with others to serve as a basis for future curriculum development. The leaders of the Modeling Project did not ask the workshop participants to develop curriculum because the curriculum needed to be developed (the leaders were capable of developing high quality curriculum on their own), rather the leaders asked the participants to develop curriculum because it was recognized to be an effective way of giving ownership to the participants. As one interviewee said *"A major part of [the success] is this social aspect, and one of them is raising the level of importance of the teacher. Making it so that it is their theory, getting them involved in developing the materials, and all of that stuff."* It appears that the intended effect was achieved. By encouraging potential adopters to become involved in the development process, teachers came to feel a sense of ownership in the project which appears to have increased their interest and commitment toward adoption. Additionally, the Modeling Project itself benefited by having a collection of curriculum inspired by the ideas of many, a curricula which is therefore likely to be better than what any one person alone would have produced. This community-based approach is still ongoing in the Modeling community with no plans to cease this community-based development.

Community-Based Dissemination

In the standard dissemination approach there are usually a few leaders who take nearly all the responsibility for writing articles, giving talks, and most importantly, running workshops. The Modeling Project diverged from this model of dissemination. Rather than targeting potential adopters in the first workshops the initial leaders *were really trying to take teachers that were not of variable quality, we were trying to take people we thought were the best teachers that we thought could lead workshops elsewhere or that could serve as role models for other people to follow.* The Modeling Project has always encouraged others to join in the dissemination effort, to assume leadership positions, and, just as with the curriculum development, to build ownership in the project. As one interviewee said, *"I became absolutely convinced right then, in the first workshop, that the workshop should not be taught by faculty and it must be taught by teachers and this is one of the crucial things that has made modeling work. It has to be taught by the practitioners themselves who have a great stake in it."* It appears that by encouraging and supporting others to take initiative in the dissemination efforts, the modeling project has benefited from both a large number of people to help with the dissemination and also from having a very enthusiastic group of disseminators who feel a personal connection to the project.

Another feature of the community-based dissemination is the availability of the Modeling materials in editable Microsoft Word format. As one interviewee described this decision, *"[We decided to make the curriculum available] in Word so it is editable and teachers could choose to take it and change it to suit their styles rather than change their styles to suit the materials."* This is, of course, contrary to some dissemination efforts that do not think it is appropriate for teachers to make modifications to the curricular materials.

CONCLUSION

The standard approach to dissemination is hierarchical, with a small group of individuals developing a idea or curriculum to be disseminated and then providing a product to adopters, while retaining ownership of the product and its dissemination. It is an approach that can be characterized as top-down. The Modeling Project did not follow this approach. Their approach was more grassroots and focused on the building of community and the giving away of ownership. It appears that the grassroots approach offered numerous advantages. By giving ownership to the adopters, in addition to curriculum and ideas, the project has developed and has benefited from a very large (and still growing) group of highly enthusiastic implementers/leaders. The project has taken on a life of its own, outside the initial instigators of the project, which has positioned it to survive and grow well into the future.

ACKNOWLEDGMENTS

We would like to thank the five interviewees for participating in this project.

REFERENCES

1. L. C. McDermott and E. F. Redish, "Resource letter: PER-1: Physics education research," American Journal of Physics **67**, 755-767 (1999).
2. J. Handelsman, D. Ebert-May, R. Beichner, P. Bruns, A. Chang, R. DeHaan, J. Gentile, S. Lauffer, J. Stewart, S. M. Tilghman, and W. B. Wood, "Education: Scientific teaching," Science **304**, 521-522 (2004).
3. Detailed information about modeling instruction and its effectiveness can be found online at: http://modeling.asu.edu/.
4. D. Hestenes, "Modeling games in the Newtonian world," American Journal of Physics **60**, 732-748 (1992).
5. "Modeling Instruction in High School Physics Final Report," Report to National Science Foundation, (2000).
6. J. Jackson, private communication.
7. U.S. Department of Education, Expert Panel Report, December 2002 excerpted and accessed from http://modeling.asu.edu/modeling/Expert_Panel_Mod_Instr.pdf on July 7, 2007.
8. National Science Foundation, *Course, curriculum, and laboratory improvement (CCLI): A solicitation of the division of undergraduate education (DUE)* (National Science Foundation, Arlington, VA, 2005).
9. More information about the American Modeling Teachers Association can be found online at: http://www.modelingteachers.org/

Measuring the Effect of Written Feedback on Writing

Dedra Demaree

Department of Physics, College of the Holy Cross, Worcester, Massachusetts, 01610

Abstract.
   Members of the Physics and English departments at The Ohio State University (OSU) and Rochester Institute of Technology are involved in an ongoing study addressing issues related to writing activities in the physics classroom. Students in the physics 103 and 104 course sequence at OSU "The World of Energy" view weekly videos then turn in summaries as part of their homework grade. These summaries are given one point if turned in; they are not graded for the quality of their content. In winter quarter, 2006, some students were given substantial feedback on these summaries with comments aimed to improve their writing. Feedback-induced improvement in their video summaries is demonstrated in this paper.

Keywords: writing to learn
PACS: 01.40.Fk

INTRODUCTION

Writing within content areas is often promoted as a way to both improve specific writing skills and as a way for students to learn content [1, 2]. Members of the Physics and English Departments at both The Ohio State University (OSU) and Rochester Institute of Technology (RIT) have been studying writing to learn within physics to find ways to quantify claims in the literature and measure any effects of writing on learning [3, 4].

Writing is sometimes used in content areas as a formative assessment tool since it provides the instructor opportunity to give specific feedback to student ideas [5, 6]. In order for formative assessment to be effective, the learner must realize there is a gap between their present state and a desire goal, and take steps to close that gap [7]. Students must be able to either self-assess their gap, or instructors must be able to communicate to the students about how they are doing. In [7], Black and Wiliam warn that "there are complex links between the way in which the message is received... and the learning activity which may or may not follow."

Traditionally, many writing to learn studies have been qualitative, and such studies seldom try to *quantify* the claimed benefits of writing [8]. In addition, there is sometimes little information given on the specific nature of the feedback, or more importantly, how the quality of the student writing was measured. It seems that at a minimum, if student learning is taking place based on feedback there should be some quantitative, measurable effect of that feedback. This paper uses Physics Education Research techniques to conduct a small study to look for a quantitative effect of feedback on student writing without external motivators, namely without giving students additional points toward their grade.

IMPLEMENTATION DETAILS

In the winter quarter of 2006, a study was conducted in order to see if student writing could be improved based solely on feedback given in the form of written comments on students' papers. Students in Physics 104 ("The World of Energy") at The Ohio State University (OSU) watch eight weekly videos on various topics related to energy issues, from pollution in Los Angeles to cleanup efforts at Chernobyl. After each video, students write at least a two paragraph summary of the contents. Students in this course are graded in a typical fashion based on homeworks, two midterm exams, and a final exam. The video summaries count for 1 point each out of a total of 140 course points. Students receive this point if they submit the video summary. The content of the video summary is essentially ungraded.

The Physics 104 course has been taught in some form for over 30 years at OSU. From the start, this was a studio-based course, with ample hands-on explorations. The videos are aimed at helping students see the connection between the technology discussed in the course and daily life. The video summaries primarily serve as assurance that the students watched the videos.

This course provides an ideal situation for this study because there is a situation where one set of students can be given additional feedback without disadvantaging the students who do not receive feedback. Since all students receive one point for submitting the summary regardless of the quality, giving feedback will not have any affect on student grades. There is also no concern if students who receive feedback learn more than the other students, because neither the content of the video summaries or their writing abilities are tested in the course.

There are three instructors for this course: A, B and C. Two of them teach one section each, while the third

CP951, *2007 Physics Education Research Conference*, edited by L. Hsu, C. Henderson, and L. McCullough
© 2007 American Institute of Physics 978-0-7354-0465-6/07/$23.00

teaches three sections. One section is taught on Saturday and contains a large percentage of graduate students (in-service teachers and pre-education majors) who have additional assignments. Instructor A typically does not give any feedback to the students when he grades; call his class N1. Instructor B does not give extensive feedback though she feels it is important to give positive encouragement and often wrote things such as "very nice" or "I agree" on students papers; call her class N2.

Instructor C (with three sections) typically only writes "okay," "good," or "very good" to indicate his level of satisfaction with the student work. In two of his sections I collected the summaries and gave extensive feedback on their writing; call them C1 and C2 (C2 is the above mentioned Saturday section). When he received the commented papers from me he also read them and wrote one of his usual three comments before marking down their point. In his third section, only his normal comments were given; call that section N3.

Sections N1, N2 and N3 are therefore the control sections, where no additional feedback is given, while C1 and C2 were given special treatment. These details are summarized in Table 1. I was not involved with the course in any way except to provide the additional summary feedback.

TABLE 1. This table shows the five sections involved in this research study.

Instructor	A	B	C
Type of Instructor Feedback	None	Minimal but positive	Minimal with some indication of satisfaction
No Additional Feedback (Control)	N1	N2	N3
Additional Feedback			C1 and C2

The students were not given any specific instructions from the course instructors about how to write good summaries. The instructions to the students were to "write at least two paragraphs summarizing the video and what you see as the major issues presented by the video." The students were also given a list of questions specific to each video as a guide to help them take notes.

At the start of this study, a collaborator from the English Department gave me extensive information on what would normally be expected from a video summary, as well as what kind of feedback would be appropriate to give. These ideas were used to design both the feedback I gave and the grading rubrics for the purpose of this study. In general, a summary should have a clear order and there should be some indication as to why the information is important. The organization of the summary should reflect what is being summarized, for instance if the video

had two distinct parts the summary should too. The ideas in the summary should be related to each other and related to the overall point of the video and not just given in a list. The summary should reflect the balance of importance of points: more time should be spent on main ideas, and less should be spent on supporting stories.

DATA OBTAINED

During the course of the quarter, I collected summaries from sections C1 and C2 from the instructor the day they were handed in, and I promptly returned them to the instructor with my written feedback directly on the student papers. For all sections, after the instructors had a chance to look at the summaries and write their own comments, I collected them and made copies for my records. Students were given one point for submitting the summaries regardless of the quality of the content.

After the quarter was completed and the students' grades were recorded, I used the copies of the summaries to grade them in detail based on the rubrics. From here on out, reference to the summary grade refers to this post-quarter grading for research purposes, and does not reflect the single point the students received during the quarter toward their course grade. I also recorded the number of instructor comments, if they were positive or negative, and the nature of comments I gave to each summary.

TABLE 2. This table shows the details of the five grading rubrics.

Rubric	Explanation
Balance	Does the emphasis placed on ideas in the summary reflect their importance?
Flow	How readable is the summary, and are the ideas related or just given as a list?
Content	Is the content correct, is terminology defined and are all important points included?

In line with the above information about what makes a good video summary, I graded the summaries based on three main parameters: Balance, Flow, and Content. The explanation for each of these is given in Table 2. For each of these parameters students received a grade ranging from 0 to 3 points, similar to the grading rubrics developed for the *ISLE* classroom by the Rutgers Physics and Astronomy Education Group [9]. 0 indicates that there is no attempt made by the student toward this parameter. 1 indicates that there is an attempt but it is barely perceivable. 2 indicates that the student made a decent attempt toward that parameter. 3 indicates that the student did a good job with that parameter. When I gave the written feedback to the students I mainly gave comments asking how ideas were related, what was the importance of certain statements, or I asked for an explanation if something the student stated was not clear.

In addition, I also gave a voluntary feedback survey to the students in the C1 and C2 sections, asking them about the additional written feedback. Students were asked if they read the additional feedback, what they thought of it, if they felt it impacted how they wrote the video summaries, and were asked to comment generally on the usefulness of the videos and the video summaries. Since the survey was voluntary, I did not get many returned, but the range of student opinions is presented here.

RESULTS

Although there were roughly 120 students in the five sections of Physics 104, only 41 students were used in this study. 21 students were in the C groups and 20 were in the N groups. Almost none of the students in section N1 signed consent forms, and some students from other sections declined to participate as well. In addition, the graduate students were removed from the study since they did not have a comparable control group. Since the video summaries were only worth one point each, students from both the C and N groups who completed less than six of the eight summaries were also excluded from the study. When ranked by final course grade, students from the C groups are very equally mixed among students from the N groups, indicating that the two groups were similar in overall course content knowledge. Additional analysis described below shows that any effects of having a non-ideal selections of students for this study did not effect the results.

The remaining 41 students were given a grade for each video summary based on the three parameters as discussed above. These students were then separated into groups based on if they received the additional feedback, on their final course grade, and on their video summary grades. Summary grades were also analyzed based on three methods. First, I looked at the individual summary grades for each week, second, I looked at the difference between each student's initial and final summary grade, and third, I did a linear fit for the eight summaries of each student to see if students had consistent improvement in their video summaries throughout the quarter.

The strongest indicator would be seeing a consistent significantly higher summary grade for the later weeks for the C1 and C2 groups (with respect to the control groups, N1, N2 and N3). However, this would be difficult to achieve with such small numbers of students. Indeed, the students in the C groups have lower (5.7/9 compared to 6.5/9, not significantly with $\rho = 0.1$) summary grades the first week, and slightly higher summary grades (7.2/9 compared to 6.9/9) in the last week. It could be that the students in the N groups were initially

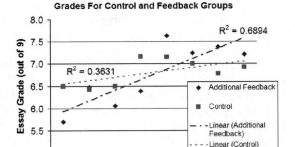

FIGURE 1. This figure shows the average grades for the Control groups and the Feedback groups, as well as a linear fit showing the higher slope for the students who received additional feedback.

better at writing the video summaries, and that this is masking any ability to see significantly higher grades in the C group at the end of the quarter.

In order to remove any initial writing level, I also looked at the students' final summary grade minus their initial summary grade. Students in the C groups had an average improvement of 1.1 point out of 9, while students in the N groups had an average decrease of 2 points out of 9. This significant ($\rho = 0.003$) result is sufficient to show that students in the C groups improved their video summaries throughout the quarter more than those in the N groups.

Another interesting way to look at the data is to look for signs of improvement throughout the quarter (not just comparing the final and initial summaries.) A positive slope as found from a linear fit to any individual student's summary grades indicates a net improvement. Linear fits may not be strong as grades can fluctuate based factors such as the video content and how much time the student devotes to a particular assignment. The average slope for students in the C groups is 0.24, indicating that over the course of eight summaries their grade would improve by nearly 2 points out of 9. However, for the N groups, the average slope is only 0.06, indicating that their grade would improve by only half a point out of 9. This result is statistically significant with $\rho = 0.005$). The average Correlation Coefficient for the linear fits is only $R^2 = 0.310$ for the C groups, and $R^2 = 0.180$ for the N groups. This reflects the difficulty of making a good fit to this type of data. However, when a linear fit is done for the average scores for all students in each group, a clear difference in the slopes can bee seen (see Fig. 1), indicating that the improvement in the C group summaries was consistent throughout the quarter.

It is possible that only a certain subset of students

(for example, those with better grades, or those working with a particular instructor) would show improvement on their summaries regardless of feedback. I therefore also ranked students based on their final course grade. There was no statistical difference in the calculated slopes for students who did better in the course than others ($\rho = 0.38$). Students who had higher course grades did tend to have better video summary grades, but those grades did not change any more through the quarter than the lower scoring students. Similarly, students who started with high video summary grades were neither more nor less likely to improve than those who started with low grades ($\rho = 0.91$). These results indicate that the observed improvements in essay grades are dominantly based on the additional feedback given to students in the C group.

Students in the C groups improved more than the N groups in each individual parameter. However, the results for the individual parameters are not statistically significant. Students from the C groups improved most in the Flow and Content parameters (with an average of 0.5 out of 3, and 0.7 out of 3 point improvement through the quarter respectively). Effectively no improvement was observed in the N groups for these (with an average of -0.1 and 0.2 point improvement through the quarter respectively).

Out of the eight summaries, students in the C groups received on average 4.8 positive comments, or approximately one positive comment for every two summaries turned in. Those in the N groups received on average 6.5, so on nearly two thirds of their summaries. Recall that instructor B made a point to give positive feedback to the students. This shows that the simple presence of positive feedback does not explain the improvement in student writing; it is the specific feedback on how to write that made a difference. Students in the C group received on average 30 comments over the course of their eight summaries, while those in the N group received on average only 9.3. Students in both groups rarely received negative feedback, with only 0.5 comments for the C groups and 0.1 for the N groups (1 in 10 students).

Of the thirteen C group students who returned the voluntary surveys, nine of them said they always read the additional feedback, two sometimes did, and three never did. About three fifths of the students said the feedback was either clear, helpful, or good. Despite the lack of explicit negative feedback, about two fifths of them said it was annoying and/or unclear. Their reason was mainly that they were being criticized when they knew they would get full points anyway. Slightly less than half of the students thought the feedback impacted the way they wrote their summaries. Integrating the feedback with instruction about how to write the summaries would probably make the experience more positive for the students. The students overall enjoyed the videos but thought the summaries were not very useful for the course.

CONCLUSIONS

Students in this study who received additional feedback on their video summaries showed more improvement in the quality of their writing over the course of the quarter than those who did not receive additional feedback. They had significantly higher differences between their final summary grade and their initial summary grade, as well as significantly higher slopes as found by taking a linear fit of their summary grades through the quarter. These improvements were observed despite the fact that there was no motivator for students to improve their writing; they received full points toward their course grade regardless of the quality of their summaries.

Writing activities can provide the opportunity for formative assessment in the physics classroom, provided students reflect on the feedback provided to them. This study provides evidence that written feedback does impact student writing. This indicates that students did reflect on the feedback. This study also demonstrates one possible method for quantifying the quality of student writing, in lines with our desire to quantify writing to learn in physics.

ACKNOWLEDGMENTS

The author would like to thank Mary Wildermuth, James Bihari, and Bill Davis for their advice and for facilitating the collection of the summaries. The author would also like to thank Bernard Mulligan for his suggestion that I use this course in my studies, and for his useful discussion on writing in physics. The author would also like to thank Cat Gubernatis for her discussions about how to give feedback to the video summaries.

REFERENCES

1. C. W. Keys, *Science Education*, **83**, 115–130 (1999).
2. P. D. Klein, *Educational Psychology Review*, **11**, 203–270 (1999).
3. D. Demaree, C. Gubernatis, J. Hanzlik, S. Franklin, L. Hermsen, and G. Aubrecht., *PERC proceedings* (2006).
4. S. Franklin, and L. Hermsen, *In preparation for Educational Researcher* (2006).
5. T. Hein, *Eur. J. Phys.*, **20**, 137–141 (1999).
6. E. Etkina, and K. A. Harper, *Journal of College Science Teaching*, **31** (2002).
7. P. Black, and D. Wiliam, *Assessment in Education: Principles, Policy and Practice*, **5** (1998).
8. J. Ackerman, *Written Communication*, **10**, 334–370 (1993).
9. E. Etkina, A. V. Heuvelen, D. Brookes, and D. Mills, *The Physics Teacher* (2002).

From FCI To CSEM To Lawson Test: A Report On Data Collected At A Community College

Karim Diff and Nacira Tache

Department of Natural Sciences, Santa Fe Community College, Gainesville, FL 32606 USA

Abstract. As part of an ongoing assessment of our introductory physics courses, we have administered the Force Concept Inventory (FCI) and the Conceptual Survey of Electricity and Magnetism (CSEM) in the three different levels of physics courses offered at Santa Fe Community College: Applied physics, algebra-based physics and calculus-based physics. We present data collected this past year, including an analysis of the correlations between normalized FCI and CSEM gains for the past four years. In addition, we report results obtained this past year in a study of correlations between the Lawson classroom test of scientific reasoning and gains on the FCI and CSEM.

Keywords: Lawson Test, FCI, CSEM, scientific reasoning.
PACS: 01.40.Fk, 01.40.Ha

INTRODUCTION

Assessment instruments such as the Force Concept Inventory (FCI) [1] or the Conceptual Survey of Electricity and Magnetism (CSEM) [2] are valuable tools to measure instructional effectiveness in introductory physics courses. Recent articles [3,4,5] have attempted to determine whether some "hidden variables" such as scientific reasoning abilities as measured by the Lawson Test[6] or mathematical skills may correlate with normalized gains on diagnostic tests. To investigate that idea, as part of an ongoing assessment of our courses, we gave the FCI, the CSEM, and the Lawson test to three different student populations (applied physics, algebra-based physics, calculus-based physics). Preliminary results show positive, significant correlations between scores on the Lawson test and normalized gains on the FCI and CSEM. We also identified a cohort of students who took both the FCI and the CSEM but found no significant correlation for this case.

FCI-LAWSON CORRELATIONS

Following a suggestion made by Coletta and Phillips [3] we examined the relationship between normalized gains on the FCI and scores on the Lawson test for a group of students taking their first semester physics course at Santa Fe Community College in the Fall 0f 2006. The students were split among three

different physics courses: Applied physics (which requires Intermediate Algebra as its pre-requisite) and the usual algebra-based and calculus-based courses. Our aggregate results are shown in figure 1 below. A small ($r = 0.36$), significant ($p = 0.00007$, at the usual 95% confidence interval) correlation is found.

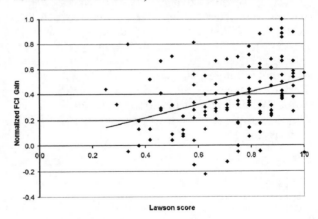

FIGURE 1. Normalized FCI gain versus Lawson score. Number of students: N = 116. Correlation: r = 0.36. Significance level p = 0.00007. Slope s = 0.50.

We also examined the relationship between pre-instruction scores on the FCI and the normalized gains. In his extensive survey published in 1989, Hake [7] found no correlation between class-averaged pre-instruction scores and normalized gains. Our results, shown in figure 2, indicate a positive correlation,

CP951, *2007 Physics Education Research Conference*, edited by L. Hsu, C. Henderson, and L. McCullough
© 2007 American Institute of Physics 978-0-7354-0465-6/07/$23.00

$r = 0.37$, with a significance level $p = 0.0004$, for single-student normalized gains. Coletta and Phillips suggest some explanations for this discrepancy, but we need further analysis of our data to make a determination.

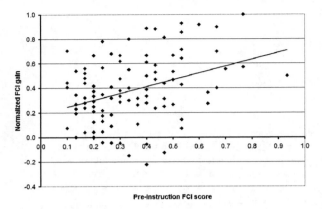

FIGURE 2. Normalized FCI gain versus pre-instruction score. Number of students: N = 116. Correlation: $r = 0.37$. Significance level $p = 0.0004$. Slope $s = 0.56$.

CSEM-LAWSON CORRELATIONS

During the spring and summer 2007 semesters we examined the relationship between scores on the CSEM and Lawson test. Although this was not part of our study, we took advantage of the fact that many students were returning from the previous semester to conduct an ad-hoc pre-post analysis of their Lawson test scores and found no significant changes. A more thorough investigation of this aspect will be the subject of a future publication. In figure 3 we show normalized gains on the CSEM versus Lawson test scores. We find another positive correlation, $r = 0.35$, with a significance level $p = 0.0004$ for N=100 students tested.

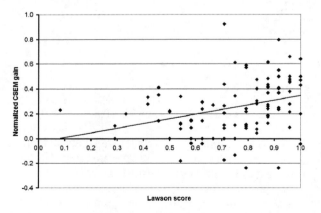

FIGURE 3. Normalized CSEM gain versus Lawson score. Number of students: N = 100. Correlation: $r = 0.35$. Significance level $p = 0.0004$. Slope $s = 0.38$.

It is interesting to note that in this case no correlation is found between normalized CSEM gains and pre-instruction CSEM scores as shown in figure 4. As discussed by Maloney et al. [2] students are rarely familiar with the more abstract concepts, language and phenomena of electricity and magnetism (as opposed to mechanical concepts studied in first semester physics.) The correlation between scores on the Lawson test and the CSEM normalized gain opens an interesting possibility concerning the predictive power of the Lawson test.

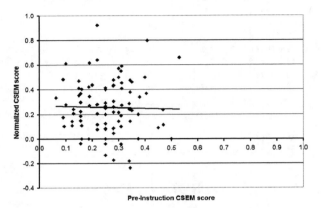

FIGURE 4. Normalized CSEM gain versus pre-instruction CSEM score. Number of students: N = 100. Correlation: $r = -0.02$. Significance level $p = 0.8$. Slope $s = -0.05$.

FCI-CSEM CORRELATIONS

While casual observation indicates that students struggle more with second-semester physics than with first-semester physics, we wanted to know whether students who had large normalized gains on the FCI would also have large normalized gains on the CSEM. Normalized FCI gains ranging from 0.2, for traditional classes, to 0.70 for highly interactive classes have been reported extensively[7]. For CSEM gains, the national average is 0.25[8]. Using data collected over the past four years we were able to identify a cohort of students (N= 54) in our calculus-based physics who completed the pre- and post-instruction assessments both times. Their average FCI gain was 0.40 and their CSEM average gain was 0.41. However, as shown in the figure below, we obtained a small ($r = 0.21$) but not significant ($p = 0.12$) correlation. To obtain reasonable sample sizes, given the small size of our classes, we had to collect data over multiple years. This limits our ability to interpret the data as classroom conditions may change significantly over such a time span.

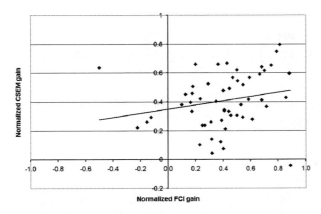

FIGURE 5. Normalized CSEM gain versus normalized FCI gain. Number of students: N = 100. Correlation: *r = 0.21*. Significance level *p = 0.12*. Slope *s = 0.15*

One of the striking results of our analysis is the wide discrepancy that exists across our three different types of courses. Table 1 below breaks down the normalized gains and Lawson averages by class type. Cognitive factors, as measured by the Lawson test may offer some clues about this discrepancy, but there could be other factors at work as well. Mathematical skills vary widely across the three classes and could interfere with conceptual learning as discussed by Meltzer [4]. The effectiveness of the classroom techniques used may also be a factor. While students in the algebra and calculus-based courses have been very receptive of the various modifications made to the courses, with an emphasis on interactive engagement, students in the applied physics courses

have been less willing to embrace departures from the classical lecture model. We are currently implementing a more hands-on curriculum [9] that we hope will result in greater learning gains for our students in applied physics.

CONCLUSION

We have investigated the relationship between student scores on the Lawson Classroom Test of Scientific Reasoning and their normalized gains on the FCI and the CSEM. We find small, but significant correlations between Lawson scores and gains on both the FCI and the CSEM. If confirmed by further analyses, these correlations could help instructors identify at-risk students and establish effective intervention programs. This could also contribute to a better understanding of factors, other than instruction format, that contribute to significant gains on standardized assessments. On the FCI scores we also find a correlation between pre-instruction scores and normalized gains. Such a correlation does not exist, in our data, between pre-instruction and normalized gains for the CSEM. We also do not find a significant correlation between FCI normalized gains and CSEM normalized gains for a cohort of students in calculus-based physics. Another finding of our analysis is the spread of learning gains between students in our applied physics courses and their counterparts in algebra-based and calculus-based physics.

TABLE 1. FCI, CSEM, and Lawson test by class type.

Course	Normalized FCI gain	Normalized CSEM gain	Lawson score (%)
Applied Physics	0.23	0.15	69.7
Algebra-based Physics	0.38	0.32	70.5
Calculus-based Physics	0.48	0.29	78.8
All students	0.38	0.25	73.6

REFERENCES

1. D. Hestenes, M. Wells, and G. Swackhamer, *Phys. Teach.* **30** (3),141-158 (1992). The revised version of 1995 was used for this study. The test is available online at http://modeling.asu.edu/R&E/Research.html to authorized educators.
2. D. P. Maloney, T. L. O'Kuma, C. J. Hieggelke, and A. Van Heuvelen *Phys. Educ. Res., Am. J. Phys.Suppl.* **69** (7) S12-S-23 (2001).
3. V. P. Coletta and J. A. Phillips, *Am. J. Phys.* **73** (12) 1172-1182 (2005).
4. D. E. Meltzer, *Am. J. Phys.* **70**(12) 1259-1268 (2002).
5. V. P. Coletta, J. A. Phillips, and J. J. Steinert *Phys. Rev. ST Phys. Educ. Res.* **3**,010106 (2007)

6. A.E. Lawson, *J. Res. Sci. Teach.* **15** (1), 11-24 (1978)
7. R. R. Hake, *Am. J. Phys.* **66** (1) 64-74 (1998). Additional details concerning this study can be found online http://www.physics.indiana.edu/%7Esdi/IEM-2b.pdf
8. "Creating a New Physics Education Learning Environment."http://www.wcer.wisc.edu/archive/Cl1/ilt/case/joliet/summative_outcomes.htm
9. ICP21: Introductory College Physics/21st century. http://www.ztek.com/

Spending Time On Design: Does It Hurt Physics Learning?

Eugenia Etkina, Alan Van Heuvelen, Anna Karelina, Maria Ruibal-Villasenor, and David Rosengrant

Graduate School of Education, Rutgers University, New Brunswick, NJ 08901

Abstract. This paper is the first in a series of three describing a controlled study "Transfer of scientific abilities". The study was conducted in a large enrollment student introductory physics course taught via Investigative Science Learning Environment. Its goal was to find whether designing their own experiments in labs affects students' approaches to experimental problem solving in new areas of physics and in biology, and their learning of physics concepts. The theoretical framework for the design of the study was based on transfer theories such as "preparation for future learning", "actor-oriented transfer", "transfer of situated learning" and "coordination classes". In this paper we describe the design of the study and present data concerning the performance of experimental and control groups on multiple-choice and open-ended exam questions and on the lab exams that assess student understanding of the physics and the reasoning processes used in the lab experiments. We found that the experimental group outperformed the control on lab-based and traditional exams and the difference increased as the year progressed. The project was supported by NSF grant DRL 0241078.

Keywords: Reformed instruction, design labs.
PACS: 01.40Fk; 0140.gb; 01.50Qb.

INTRODUCTION

This manuscript is the first of three papers that describe an experimental design study in an introductory course whose goal was to investigate the effects of design labs on student learning of physics and their acquisition and transfer of scientific abilities.

Scientific abilities are approaches and procedures that scientists use when engaging in the construction of knowledge or in solving complex problems [1]. In this study all students enrolled in the course attended the same large room meetings and recitations that followed the *ISLE* curriculum [2]. However, they were randomly split into two groups in the labs. The experimental group designed their own experiments. Their scaffolding included questions concerning the scientific abilities needed for the *ISLE* design labs [3] and on self-assessment rubrics [1]. In the control group students performed the same experiments but with the design provided in the write-up and supported by conceptual questions that helped students work through the physics. Throughout the semester the groups were compared on their physics learning. At the end all students performed two lab transfer tasks: one to design an experiment to investigate a physics problem in new area of physics and the other an experiment in biology. This paper describes the part of the study related to the comparison of the groups on the paper-and-pencil exam problems and lab-based problems. The research question in this paper is: if students in the labs focus on designing their own experiments without having the "right answer" and on the elements of the scientific investigation instead of on solving physics problems, do they learn less physics than those who have a good experimental design provided for them and more opportunities to engage in physics problem solving? Two other submitted papers describe the aspects of the project related to transfer in the physics and biology content.

MOTIVATION

There are two big motivations for this study: (1) Recent reports concerning science and engineering education encourage student acquisition of conceptual and quantitative understanding of physics principles *and* also the acquisition of abilities to: design their own experiments, reason from the data, construct explanatory models, solve complex problems, work with other people, and communicate [4-7]. Should we spend time on the development of these latter abilities

CP951, *2007 Physics Education Research Conference*, edited by L. Hsu, C. Henderson, and L. McCullough
© 2007 American Institute of Physics 978-0-7354-0465-6/07/$23.00

or this will harm students' acquisition of physics conceptual learning and ability to solve traditional problems? (2) Many experiments indicate that the ability to transfer what is learned in physics to other unstudied physics areas, to other academic disciplines, and to work after academia is lacking. Can students transfer what they learn in our physics design labs to other unstudied areas of physics and to other academic disciplines—the subject of our other two papers?

THEORETICAL FOUNDATION

Transfer As this study is a part of a larger "transfer" study, we briefly describe the theoretical perspectives that informed the design of the whole project. When designing the learning environment for the experimental group, we carefully followed the recommendations of the literature on how to create a learning environment that promotes transfer.

As mentioned above, the purpose of the whole project was to determine if students in design labs are able to transfer the scientific abilities that they learned during one semester into new physics content and into biology. In other words, to find whether they apply the habits of mind learned in the labs when they face a new problem for which they do not have content knowledge or experimental skills.

Transfer refers to the ability to apply knowledge, skills, and representations to new contexts and problems [8-10]. Research shows that achieving transfer is difficult [11]. However, new work of Lobato shows that transfer occurs often and the problem is in its recognition by researchers, not its existence [12]. There are several theoretical models of transfer [13, 14, 12]. The most relevant to this study are direct applications transfer, recognition of affordances, preparation for future learning transfer, and actor-oriented transfer. For any kind of transfer to occur, the learning environment should have such features as: focusing students' attention on pattern recognition among cases and induction of general schemas from a diversity of problems [15]; engaging students in meta-cognitive reflection on implemented strategies [16]; and presenting students with contrasting cases.

DESCRIPTION OF THE STUDY

The study was conducted in the first (fall) semester of an introductory physics course for science majors (the total enrollment was 193; the number of students attending various activities varied through the semester). There were two 55-min lectures, one 80-min recitation, and a 3-hour lab per week. There were two midterm exams, one final exam and two lab exams. All students learned through the same *ISLE* curriculum [2] in large room meetings and in smaller recitations. The lab sections were split into two groups: design labs (4 sections) and non-design labs (4 sections). Students registered for the sections in March of the previous academic year. In the previous years we found no difference in performance of lab sections on exams, thus we can assume that during the experimental year the student group distribution was random. During the semester, students were not informed about the study. At the end, we disclosed the procedure and students signed a consent form allowing us to use their work for research.

To make sure that the design group and non-design group were equal in learning ability, we administered Lawson's test of hypothetico-deductive reasoning in the first lab session [17]. Coletta and Philips [18] found that student's learning gains are strongly correlated with their scores on this test. Our lab sections were statistically the same. To ensure that the treatment was the same too, we used the same three TAs to teach the labs. Two of the TAs taught one design and one non-design section and the third TA taught two of each. All TAs were members of the PER group, highly skilled in the interactive teaching.

Design labs In these labs students had to design their own experiments. The scaffolding was provided through write-up questions that focused their attention on the elements of the scientific process: representing the situation, deciding on the experiment, analyzing experimental uncertainties, etc. Students used self-assessment rubrics to help them write lab reports [3]. A sample write-up for one lab experiment is provided in the Appendix. The TAs did not help students design experiments and when students had difficulties, they asked questions and provided hints but did not answer their questions directly.

At the end of each experiment students had to reflect on the purpose of the experiment, its relationship to their everyday experience, and its place in an overall scientific process. Lab homework that students did after each lab contained reading passages with reflection questions. Student had to analyze stories about historical developments of several scientific theories and applications such as the nature of AIDS, prophylactics, and pulsars. They had to identify the elements of scientific inquiry that are present when scientists answer new questions or apply knowledge. The purpose of the passages was again to help students reflect on the common elements of a scientific investigation.

Non-design labs In these labs students used the same equipment as in design labs and performed the same number (sometimes even more) experiments. The write-ups guided them through the experimental procedure but not through the mathematics. Students had to draw free-body diagrams, energy bar charts and other representations to solve experimental problems but they did not need to think about theoretical assumptions – these were provided to them in the text. These labs were not cook-book labs; we call them non-design reformed labs. These labs had homework as well—mostly physics problems that prepared students to do the next lab. The TAs taught the labs differently. They provided an overview of the material at the beginning of the lab and then later if students had questions, they answered these questions.

To ensure that students' and TAs' behavior was indeed different in the labs, a trained observer used the method described by Karelina and Etkina [19] to keep track of the time spent by a group of students on different activities. The length of this paper does not allow us to elaborate on the details of the method. The observer "timed" one design group and one non-design group each week, observing 20 3-hour labs.

The part of the study reported in this paper relates to students performance on four exams: the lab practical exam, two midterms, and the final. The practical exam had questions related to the lab experiments and questions that probed student understanding of the physics behind the experiments (see the example in Appendix 2). The regular exams had a multiple choice portion and an open ended portion (3 problems per midterm and 5 on the final). During the comparison, the score on the Lawson's pre-test was held as a covariate. Thus the results are for the students matched by their pre-test score.

FINDINGS

Observations of student activities in the labs: The observer found that students in design labs on average spend more time making sense of the experiment and writing the results (see Table 1). The total time was larger too, although both groups officially had 3-hour labs. Non-design students chose to leave early.

Table 1: The average time in minutes that students spent on different activities in the labs: SM-sense making; Writ.- writing; Proc.-Procedure; Rd. –reading; TA – TA help; OT – off task.

Design group							
	SM	Writ.	Proc.	Rd.	TA	OT	Tot.
Labs 1-10	37	66	24	5	18	8	159
s.d.	10	12	13	1.7	16.0	9.2	25.9
Non-design group							
Labs 1-10	14	41	20	4	17	2	96
s.d.	8.4	15.1	10.7	3.2	12.8	1.4	30.8

Lab practical: The lab exam was held during the 6th week of classes. All students took the whole 3 hours to complete the exam. The average score of the design group was of 11.5/15; the average of the non-design group was 8.5/15. The groups were different at the $p<0.001$ level of significance.

Midterm 1: The scores of the design group were slightly lower than the non-design on both the multiple-choice and the free-response parts of the first exam, but the difference was not statistically significant.

Midterm 2: The design group scored significantly higher than the non-design group on exam 2—multiple choice [$p=0.034$] and overall [$p=0.05$] with the pre-diagnostic test as a covariate.

Final exam: The design group scored significantly higher than the non-design groups in the free response questions [$p=0.043$]. More detailed analysis revealed that students in the design group outperformed non-design students on all problems where they had to identify or analyze assumptions that they used in a solution.

The differences between the groups persisted in the second semester when all students had design labs. On the final exam in the second semester students from the fall semester design group significantly outperformed non-design group (Free Response $p = 0.008$, Overall $p = 0.014$]. The average on the exam for the design group was 167, for the non-design was 156 (out of 240).

DISCUSSION

As we discussed earlier, one purpose of the project was to find whether students who design their own experiments could transfer learned abilities to new content. Although we do not describe the transfer experiment in this paper, we note that there were significant differences between the two groups on the transfer tasks (see other papers in this volume). This paper concerns physics learning – do the design students who struggle designing their own experiments, write long and detailed lab reports, analyze historic passages about scientific discoveries, and perform the labs in a totally constructivist environment miss on learning physics concepts and applying them to problem solving? We found that design students mastered the physics content of the labs better than their counterparts in non-design labs and more importantly learned physics content at least as well as the control group. Furthermore, the difference between the groups became even larger at the end of the second semester. The students possibly had developed a more independent way of thinking that helped them in future learning. One counter argument might be that design students spent more time in the labs and thus learned more. However, non-design students had the same time allocated for the labs in the curriculum, they just did not use this time due to the structure of the labs.

We thank John Bransford, Jose Mestre, and Joe Redish for their advice in the design of the project and Michael Gentile for teaching the labs in the course.

Appendix: Design lab: The energy stored in the Hot Wheels launcher

The Hot Wheels car launcher has a plastic block that can be pulled back to latch at four different positions. Your need to determine the elastic potential energy stored in the launcher in each position.
Available equipment: Hot Wheels car, track, launcher, meter sticks, ruler, tape, timer, scale, spring scale, motion detector. Write the following in your lab report:

a) Make a rough plan for how you will solve the problem. Make sure that you use two methods to determine the energy. Include a sketch in a procedure brief outline.

b) In the outline, identify the quantities you will measure and describe how you will measure each quantity.

c) Construct free body diagrams, and energy and/or momentum bar charts wherever appropriate.

d) Devise the mathematical procedure to solve the problem. Decide what your assumptions are and how they might affect the outcome.

e) Perform the experiment and record the data in an appropriate manner. Determine the energies.

f) Use your knowledge of experimental uncertainties to estimate the range within which you know the value of each energy.

Non-Design lab: Energy stored in the Hot Wheels launcher: The Hot Wheels car launcher has a plastic block that can be pulled back to latch at four different positions. Your first task is to determine the elastic potential energy stored in the launcher in each of these launching positions.
Procedure: Launch the car vertically starting at one of the launching positions. By measuring the maximum height the car reaches, you should be able to decide the original elastic energy stored in the Hot Wheels launcher.

a) Measure the mass of the Hot Wheels car.

b) Hold the Hot Wheels car launcher so that it is oriented almost vertical—so the car does not fall out when placed in the launcher. Experiment a little with shooting the car almost vertically up into the air.

c) Place a meter stick beside the launcher and note the position on the meter stick of the front of the car when the car is ready for launch. Hold the launcher firmly and release. Find the vertical distance the car traveled.

d) Repeat this measurement four times. Take the average of the four vertical distance measurements and calculate the standard deviation of the measurements. Calculate the fractional uncertainty in the vertical distance measurement ($\Delta h/h$).

e) Repeat the measurements for the other three positions.

f) *Analysis:* Construct a work-energy bar chart for the process starting with the car resting on the stretched launcher and ending when the car is at its maximum elevation. Apply the generalized work-energy equation

g) Insert your measurement numbers and determine the initial elastic energy of the launcher. Calculate the fractional uncertainty of the elastic potential energy for each launching position—equal to the fractional uncertainty of the vertical distance traveled times the elastic energy for that launching position.

REFERENCES

1. E. Etkina, A. Van Heuvelen, S. White-Brahmia, D. T. Brookes, M. Gentile, S. Murthy, D. Rosengrant, and A. Warren. *Phys. Rev. Sp. T., Pys. Ed. Res.* 2, 020103 (2006)

2. E. Etkina, & A. Van Heuvelen, in Proceedings of *PERC* (Rochester, NY: PERC publishing 2001), p. 17.

3. E. Etkina, S. Murthy, & X. Zou, *Am. J. Phys.* 74, 979 (2006).

4. R. W. Bybee, B. Fuchs, *JRST*, 43, 349-456 (2006).

5. R. Gott, S. Duggan, & P. Johnson. *Research in Science & Technological Education* 17, 97107 (1999).

6. G. S. Aikenhead, *Science Education*, 89, 242-275 (2005).

7. R. A. McCray, R. L. DeHaan, & J. A. Schuck, Eds.. *Improving undergraduate instruction in science: Report of a workshop.* Washington, DC: The National Academies Press, 2003.

8. S. M. Barnett & S. J. Ceci, *Psyc.l Bulletin*, 128, 612-637 (2002).

9. J. D. Bransford, A.L. Brown, and R.R. Cocking, "How People Learn: Brain, Mind, Experience, and School," Washington, DC: National Academy Press, 1999.

10. J. P. Mestre, *Journal of Applied Developmental Psychology*, 23, 9-50 (2002).

11. M. L. Gick, and K. Holyoak, *Cogn. Psych.*, 15, 1-38 (1983).

12. J. Lobato. *Journal of Learning Sciences*, 15, 431-449 (2006).

13. 13.J. Greeno. *Journal of Learning Sciences*, 15, 537-547 (2006).

14. J. D. Bransford, and D. T. Schwartz, "Rethinking transfer: A simple proposal with multiple implication," in *Review of Research in Education* edited by A. Iran-Nejad and P.D. Pearson, Washington DC: AERA, 24, 1999, pp. 61-100.

15. D. Gentner, J. Loewenstein, and L. Thompson, *Journal of Educational Psychology*, 95, 393-409 (2003).

16. R. Catrambone, and K. Holyoak, *Journal of Exp. Psych: Learning, Memory, and Cognition*, 15, 1145-1156 (1989).

17. A. E. Lawson. *JRST* 15, 11-24 (1978).

18. V. Coletta & J. Phillips. *Am.J. Phys.*, 73, 1172 (2005).

19. A. Karelina, & E. Etkina, in *2006 Physics Education Research Conference*, (Syracuse, NY, July 2006) edited by L. McCullough, L. Hsu, and P. Heron, AIP Conference Proceedings, Vol. 883, 93-96, 2007.

Design And Non-design Labs: Does Transfer Occur?

A. Karelina, Eugenia Etkina, Maria Ruibal-Villasenor, David Rosengrant, Alan Van Heuvelen, and Cindy Hmelo-Silver

Graduate School of Education, Rutgers University, New Brunswick, NJ 08901

Abstract. This paper is the second in the series of three describing a controlled study "Transfer of scientific abilities". The study was conducted in a large-enrollment introductory physics course taught via Investigative Science Learning Environment. Its goal was to find whether designing their own experiments in labs affects students' approaches to experimental problem solving in new areas of physics and in biology, and their learning of physics concepts. This paper reports on the part of the study that assesses student work while solving an experimental problem in a physics content area not studied in class. For a quantitative evaluation of students' abilities, we used scientific abilities rubrics. We studied the students' lab reports and answers to non-traditional exam problems related to the lab. We evaluated their performance and compared it with the performance of a control group that had the same course but enrolled in non-design labs instead of design labs. The project was supported by NSF grant DRL 0241078.

Keywords: Design labs; transfer, sense-making, scientific abilities.

PACS: 01.50.Qb; 01.40.Fk

INTRODUCTION

This manuscript is the second of three papers in these proceedings that describes a study whose goal was to investigate the effects of design labs on student learning of physics and their *acquisition and transfer of scientific abilities*. The motivation for the study, the theoretical foundation, and the set-up are described in detail in the paper "Spending time on design: does it hurt physics learning". The experiment was conducted in an introductory physics course for science majors of about 180 students. The number of students varied slightly during the semester. The students attended the same large room meetings and recitations taught via the Investigative Science Learning Environment (*ISLE*, [1]). In the labs they were randomly split into two equal in size groups. In the experimental lab sections ("design students") students designed their own experiments, wrote elaborate lab reports in which they described and explained their experimental procedure, evaluated experimental uncertainties, justified theoretical assumptions, etc. [2] Students in the control lab sections ("non-design" students) performed the same experiments but were guided by the directions in the lab write-up. The assumptions were provided for them, and instructions for evaluating uncertainties were provided.

LAB TASK

To assess how students transfer scientific abilities to an unfamiliar physics content in the same functional context (classification by Barnett and Ceci, [3]), we developed a lab task where both groups designed an experiment and wrote a lab report. In contrast to regular labs which students performed during semester this particular task was identical for the experimental and the control groups. The task involved drag force in fluid dynamics. This physics content was not covered in the course. To minimize the spreading of information among the students we developed four similar versions (Appendix). Students were provided some necessary and some redundant information in the write-up and had access to textbooks and the internet.

The students performed this task during the lab (3 hours) on week 13 of the semester. Prior to this, they performed 10 regular labs, different for the experimental and control groups and described in details in the first paper of the series. The lab sections were spread from Wednesday to Friday. Four experimental sections had labs on Wednesday and Thursday morning; control sections had the lab on Thursday afternoon and Friday morning. The drag force lab was attended by 89 students in each group.

CP951, *2007 Physics Education Research Conference*, edited by L. Hsu, C. Henderson, and L. McCullough
© 2007 American Institute of Physics 978-0-7354-0465-6/07/$23.00

STUDY METHODS

To assess and compare performance of the experimental and control groups we used different methods. One was direct observation of students' activities during the labs. The second was a general impression of students' lab reports. Based on the impression we decided how to code our observations for quantitative analysis. For this we used some of the scientific abilities rubrics developed earlier [4].

To monitor students' activity, a member of our group observed student behavior during the lab (one group from each lab section—eight groups total). The observer timed and recorded all student activities and conversations and later coded his observations using a coding scheme described in [5]. This coding scheme groups all students' activities into several categories: sense-making, writing, procedure, reading, TA's help, and off-task. The reliability of this method was established prior to the study.

To compare the lab reports of the two groups we chose the following abilities: to communicate, to analyze data, to evaluate assumptions, and to evaluate the result by a second method. We did not evaluate the ability to design a reliable experiment as we observed students exchanging information as the week progressed. Thus we suspected that later sections gained an advantage over the earlier groups.

We used students' written lab reports to evaluate the above abilities using the rubrics devised and validated in multiple studies [4]. A rubric describes four levels of performance for a particular ability (0 to 3) and assigns each level a particular score. "0" means missing; "1" – inadequate; "2" – needs improvement; and "3" – adequate. We checked the inter-rater reliability and test-retest reliability of the scoring with several different raters and by rescoring some part of the lab reports. The ICC (intraclass correlation) coefficient was different for different rubrics but always higher than 0.89, which shows an acceptable raters reliability.

FINDINGS

Observation of student behavior: The observations showed that there was a remarkable difference in the behavior of design and non-design students during the drag force lab. First we noticed that the lab took significantly more time for design students. Although the lab tasks were the same, the design groups spent 40 minutes more time in lab room then non-design students. The difference between the lab duration (162±17 min and 120±25 min) is statistically significant ($p = 0.038$). Figure 1 shows that the main contribution to this difference came from time spent on sense-making discussions. The sense-making lasted about 52±10 minutes in design groups and only 15±5 minutes in non-design groups. This difference is statistically significant with the level of significance $p = 0.0007$. The time students spent on other activities was about the same for both groups. There was a slight difference in the time for writing and TA's help but based on our data we cannot say that this difference is significant.

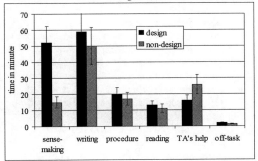

FIGURE 1. The time spent on different activities by students of design and non-design sections. The averaging was over four groups of each type.

General impression: The difference was not only in the time spent on the task. The quality of work was also different. Below we show two examples of the lab reports of two groups of students: one of the best non-design groups and one of the best design groups.

Non-design lab report (Task: version 2)
Determine the velocity of the balloon when air resistance and gravitational force are equal
- place the motion detector on a stand
- place the sensor face downward
- place the helium balloon on the floor
- release the balloon as the motion detector collects data
- on the position-time graph find constant slope segment
- repeat twice more
- find the average velocity
- …determine Reynolds number. You should get a value larger than 10.
- use the equation to solve for drag coefficient …Cd=0.51
- now repeat this procedure for air filled balloon. Make sure to drop the balloon from the level of the motion detector…
- air filled balloon - Cd= 0.61
Drag coefficient for air and helium are indeed different.

Design lab report (Task: version 4)
Part I. We need to know which equation to use based on the Reynolds number…To find the velocity we will have a motion sensor above the helium balloon. The balloon will be released and the motion sensor will measure its upward velocity. *Here is picture of the set-up. The chart is attached*
The velocity was taken 3 times and averaged to allow for random uncertainty….
When the balloon is let go the velocity increases until it reaches terminal velocity, here the net force is zero and acceleration is zero.

When balloon is at rest the net force on it is equal to zero too. *Here are two free body diagrams for balloon at rest and at terminal velocity.* The buoyant force is always the same. Therefore the drag force is equal to the force of the string attaching the balloon to the scale... Cd = 0.43

Assumptions: balloon travels in straight path, balloon is point particle, cross-section is circle, cross-section is level. *Uncertainties are evaluated: diameter, scale, motion detector and random uncertainty of the velocity.*

Part II. Prediction (*of the speed of the air balloon falling to the ground*)

When the air balloon falls it reaches terminal velocity drag force equals the force of the earth. *Here are two free body diagrams for balloon at rest and at terminal velocity ...* We can use the equation ... to get the velocity: V= 0.438±0.021m/s *(the final result incorporates uncertainty)* We will have a motion sensor aimed down and drop a balloon below it. It will record the velocity of the air balloon before it hits the ground. *Picture is here.*

Assumptions: 1. Balloon achieves terminal velocity – otherwise Fe≠Fd; 2. Re>10 – otherwise Fd equation is wrong 3. Cd is the same for air and helium – otherwise calculated velocity will be wrong.

V was measured and averaged over 3 trials (1.476, 1.02, 1.153). V=1.216±0.228m/s

The values do not overlap and therefore are not equal. Some assumptions must have been incorrect.

Scientific abilities rubrics: The reading lab reports reveals the features that make difference in the performance of two groups. The quantitative analysis of the lab reports supported the general impression on students' performance. Figures 2-5 show that there are significant differences in the lab reports of design students and non-design students. Design students demonstrated significantly better scientific abilities than the non-design students.

Evaluating the effect of assumptions: Figure 2 shows that 57 design students (more then 60%) got score 2 or 3 that is they identified relevant and significant assumptions of the theoretical model that they used, whereas only a few non-design students did. Most design students who identified assumptions also evaluated their effect on the result or validated them. Not a single student in non-design section made an attempt to do this.

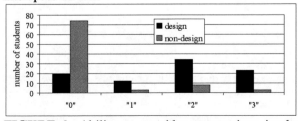

FIGURE 2. Ability to consider assumptions in the theoretical model. The difference is statistically significant with chi-square = 68, *p* < 0.001.

Evaluating effect of uncertainties: During the semester non-design students learned how to identify sources of uncertainties and how to evaluate their effect on the final answer. But only 11 of them (12%) got score 2 or 3 and transferred this skill in the independent experimental investigation (Fig. 3). More then 50% of design students evaluated the effect of experimental uncertainties in this lab.

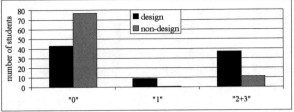

FIGURE 3. Ability to evaluate the effect of uncertainties. Chi-square = 30, *p*<0.001.

Evaluating the result by means of an independent method. A high score on this rubric is possible only when a student discusses the discrepancy between the results of two methods and possible reasons of this discrepancy considering assumptions and uncertainty. As a result design students demonstrated a higher ability to evaluate the result (see Fig 4). We can see that about 64 of design students (72%) got score 2 or 3, i.e. discussed the reasons for the discrepancy while in non-design sections only 38 students (43%) did.

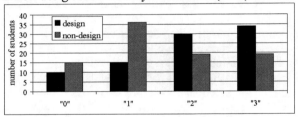

FIGURE 4. Ability to evaluate the result by means of an independent method. Chi-square = 16, *p* < 0.001

Communication: One of the main scientific abilities we want students to develop is an ability to communicate their ideas. This ability includes an ability to draw diagrams and pictures, describe details of the procedure, and to explain the methods. The analysis of lab reports shows that more then 60% of design students drew a picture while only 8% of non-design students did. Figure 5 shows the results of the scoring of the reports using the communication rubric. The difference in their scores is statistically significant (chi-square = 60.6, *p*<0.001).

Understanding of physics: The analysis of the lab reports revealed another interesting feature. Students from different sections demonstrated a different quality of drawing free-body diagrams in spite of the fact that during the semester all students learned to draw FBDs the same way.

In this lab about 22% of non-design students draw incorrect FBDs, (i.e. mislabeled or not labeled force vectors, wrong directions, extra incorrect vectors

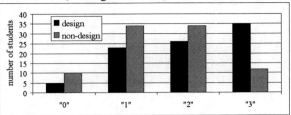

FIGURE 5. Ability to communicate ideas.

present, or vectors missing), while only 2% of design students made a mistake in FBDs. This difference is statistically significant (Chi-square = 18, $p<0.001$).

In addition, we analyzed the consistency of different representations in student work (free body diagram versus mathematics, a picture versus a free body diagram, etc). We found a difference in the number of students who created inconsistent representations: 22% of design students versus 44% of non-design students ($p<0.025$, chi-square = 7.8).

DISCUSSION

Our studies demonstrated a rather high level of transfer of several scientific abilities in the experimental group, supposedly caused by differences in the performance of the two groups in their weekly laboratory work.

Students who were used to designing their own experiments spent much more time on sense making than the students who were used to following clear instructions. That probably resulted in the more profound investigations and more sophisticated lab reports of design students comparing to control group. The observations of the students during the semester showed that on average design students spent 37 min on sense making versus 14 of non-design students.

We found that non-design students' reports resembled their lab write-ups. They gave step-by-step instructions with scarce explanations, rarely showed their reasoning, and did not try to justify the validity of their methods and procedures. Design students tried to satisfy the usual lab requirements: they described the procedure, drew pictures, explained the reasoning, analyzed data, and evaluated results.

The quality of FBD and the level of representation consistency indicate that design students paid more attention to physics understanding and logical reasoning during the lab than non-design students. One explanation is that during the semester, design students had to reconcile different aspects of the phenomenon

and had to make sense of their activity more often then non-design students. That could lead to the higher scores on the rubric evaluating the ability to communicate.

Design students significantly outperformed non-design students in other scientific abilities such as the abilities to analyze data and the ability to identify theoretical assumptions. During the semester design students learned that it was impossible to evaluate the results of their investigations adequately without considering assumptions and uncertainties. The non-design students did not consider evaluating the uncertainties as an important part of the lab, although it was a routine procedure during the semester.

In summary, we found if students consciously plan, monitor, evaluate and reflect on their actions, transfer occurs.

APPENDIX

Complete text of the lab task: Investigation of the behavior of the balloon
Equipment available: *a balloon filled with helium, a balloon filled with air, meter stick, measuring tape, stop watch, motion detector, computer, additional resources.*

Version 1: You hold an air balloon and a helium balloon. Design experiments to determine which physical model best explains their motion if you release them: the model with no air friction, the model with viscous flow or the model with turbulent flow.

Version 2: Design an experiment to determine whether a helium-filled balloon and an air-filled balloon have the same drag coefficients.

Version 3: Use the air balloon to determine its drag coefficient. Then predict the speed of the helium balloon when it reaches the ceiling.

In your report describe the experiment, your analysis and judgment so that a person who did not see you perform the experiment could understand what you did and follow your reasoning. For help we provided some resources on the next two pages of this write-up. You can also use the textbooks available on the instructor's desk or the internet for help.

REFERENCES

1. E. Etkina, & A. Van Heuvelen, Proceedings of *PERC* (Rochester, NY: PERC publishing, 2001, p. 17-21).
2. E. Etkina, S. Murthy, & X. Zou, *AJP*, **74**, 979(2006).
3. S. M. Barnett, and S. J. Ceci, *Psychological Bulletin*, 128 (4), 612-637, (2002).
4. E. Etkina, A. Van Heuvelen, S. White-Brahmia, D. T. Brookes, M. Gentile, S. Murthy, D. Rosengrant, and A. Warren. *Phys Rev. ST Phys, Ed. Res.* **2**, 020103 (2006).
5. A. Karelina, & E. Etkina. In *2006 Physics Education Research Conference*, (Syracuse, NY, July 2006) edited by L. McCullough, L. Hsu, and P. Heron, AIP Conference Proceedings, Vol. 883, 93-96, 2007.

From Physics to Biology: Helping Students Attain All-Terrain Knowledge

Maria Ruibal-Villasenor*, Eugenia Etkina*, Anna Karelina*, David Rosengrant*, Rebecca Jordan† and Alan Van Heuvelen*

*Graduate School of Education, Rutgers University, New Brunswick, NJ 08901
†Ecology, Evolution and Natural Resources Department, Rutgers University, New Brunswick, NJ 08901

Abstract. This paper is the third in a series of three describing a controlled study "Transfer of scientific abilities". The study was conducted in a large-enrollment introductory physics course taught via Investigative Science Learning Environment. Its goal was to find whether designing their own experiments in labs affects students' approaches to experimental problem solving in new areas of physics and in biology and their learning of physics concepts. The part of the project presented in this paper involves students in the experimental and control groups solving a biology-related problem that required designing an experiment and evaluating the findings. We found that students who were in the sections where they had to design their own experiments during the semester were able to transfer the abilities they acquired in physics laboratories to solve a novel biology problem. The project was supported by NSF grant DRL 0241078.

Keywords: physics education research, reformed instruction, transfer, design labs.
PACS: 01.40.d, 01.40.Fk, 01.40.gb, 01.50.Qb

INTRODUCTION

This manuscript is the third of a series of three papers in these proceedings that describes a study whose goal was to investigate the effects of design labs on student learning of physics and their acquisition and transfer of scientific abilities. The motivation for the study, its theoretical foundations, and its methodology are described in the first paper of the series: "Spending time on design: does it hurt physics learning?" The second paper described the "physics" transfer experiment in which students had to solve an experimental problem in an area of physics that they had not studied before. This paper describes a transfer experiment in biology.

The experiments were conducted in a large enrollment algebra–based physics course (with an integrated lab) for science majors. The purpose of these experiments was to test the *hypothesis* that students that learn in a educational environment that resembles scientific inquiry who design their own experiments in a physics lab not only can learn as much physics content as those students who follow the directions in guided write-ups, but also can acquire and transfer scientific abilities better than their counterparts. The *independent variable* in this experiment was the type of learning experiences that the subjects encountered in the lab: they had to complete assignments that were similar in terms of the physics involved but very different in nature. In addition they did not receive the same type of feedback from the TA's. Excluding the laboratories, every one of the participants experienced the same learning environments and worked on the same tasks: the students attended the same large-room meetings and recitations which followed *ISLE* curriculum [1]. The *dependent variable* in this experiment was the extent of student learning and transfer revealed in their performance on the exams and on special experimental tasks. This was measured by exams grades, the coding of lab reports using "scientific abilities" rubrics and the amount of time that students spent on different activities during the labs.

In order to make the comparisons possible, we split the course lab sections into two groups of equal size (about 90 students in each, the number varied slightly during the semester).The students in the treatment group (called design group) designed their own experiments and composed sophisticated lab reports in which they described and explained their experimental procedure, evaluated experimental uncertainties, justified theoretical assumptions, etc. The students in

CP951, *2007 Physics Education Research Conference*, edited by L. Hsu, C. Henderson, and L. McCullough
© 2007 American Institute of Physics 978-0-7354-0465-6/07/$23.00

the control group (non-design students) performed the same experiments but were guided by the directions in lab write-ups. The assumptions instructions for evaluating the uncertainties were provided for them. In contrast, the students in the design group had to struggle and find the answers with thoughtful efforts similar to scientists doing research [2].

The research question in this paper is whether the students who during one semester design their own experiments and are compelled to concentrate on the elements of the scientific investigation, acquire and are able to transfer scientific abilities to a different subject matter better that those students who perform similar lab exercises but do not design their own experiments. This portion of the study focuses on transfer between two different scientific contents (physics and biology) but the same lab environment [3]. Therefore the contexts of the learning tasks and transfer task are very similar but the content differs.

METHODOLOGY

A biology task was given as the final lab exam for the course, for this reason all the students completed the task. Both the treatment and the control groups had to design an experiment to find the transpiration rate of a certain species of plant and subsequently to write a report detailing their experimental procedures, calculations and conclusions. The exact text of the assignment is in the appendix. The researchers selected this lab problem because: a) measuring transpiration is a task simple enough to complete for students with very little plant physiology background; b) students can use multiple measures to determine transpiration rates which gave them some room for inventiveness, evaluation and decision making; and c) students are more willing to accept a biology assignment as a final exam for their physics lab if they perceive that there is a physical basis (evaporation and osmosis) underlying the biological process of transpiration. This last feature of the task as well as the similarity of the contexts may have facilitated the transfer of scientific abilities.

In "Resources for the practical", the handouts provided definitions of transpiration and humidity and also included a table with saturated vapor density or water as a function of temperature (the course did not cover humidity at all). In addition, the students could consult the internet.

During the practical exam students in each lab section worked in the same group of three or four as they did during the semester. During the exam as during the semester, students submitted individual reports for grading. The four treatment sections had the exam earlier in the week than the control sections.

We assessed the extent of transfer that took place between the physics laboratories and the bio task by a) observing students' behavior during the completion of the task and b) analyzing students' lab reports. During the transfer task an observer trained in the method described by Karelina and Etkina [4] measured and recorded the amount time that a group of students from each lab section spent on different behavior patterns. We examined students' reports to find patterns in their length, content and style. In addition we evaluated the quality of the reports using the "scientific abilities rubrics" [5]. All rubrics that we used for scoring can be found at http://paer.rutgersedu/scientificabilities.

FINDINGS

Students' behavioral patterns:

The teams of "design students" spent more time completing the same transfer task than the teams of "non-design students". It took an average of 23.5 minutes more for the design team to finish their reports. However this difference (176±26 min and 153±26 min) is not statistically significant (p=0.1221).

Figure 1 shows the average amount of time that four randomly selected teams from each group spent on six different categories of activities.

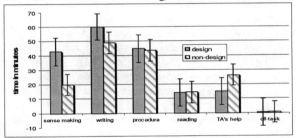

FIGURE 1. The time spent on different activities by teams of students during the final lab exam (biology task).

There is a significant difference (p=0.0026) between the time duration that the subjects spent on sense making. It was 42.75±9.84 minutes for design teams and 19.75±4.50 minutes for non-design teams. In addition, design students spent more time writing their reports and less time receiving help from the TA; however the differences were not significant (p=0.166 and p=0.061 respectively).

Differences in lab reports:

In addition to the differences in the amount of time that the two groups spent on sense making, we found differences in the quality of students' lab reports. Non-design students' reports tended to be shorter on average. They included fewer detailed descriptions of the procedures and fewer pictures and diagrams. Moreover, the reports of non-design students rarely contained any explanations of the advantages and limitations of the methods used, or any justifications for the choice of the approaches and procedures. In order to compare the labs reports from students in the two groups, we used the "scientific abilities rubrics" devised and validated in multiple studies [5] to evaluate students' written lab reports. A rubric describes four levels of performance for a particular ability (0 to 3) and assigns each level a particular score. "0" means missing; "1" – inadequate; "2" – needs improvement; and "3" – adequate. We established the ratings for the sample lab reports with a biology expert and then checked the inter-rater reliability and test-retest reliability of the scoring with two raters and by rescoring some part of the lab reports. The ICC (intraclass correlation) coefficient was different for different rubrics but always higher than 0.85, which shows an acceptable raters reliability.

Identifying assumptions and evaluating their effects: 91% of the non-design students showed no evidence that they had tried to identify the assumptions implicit in their procedure and calculations. Only 6% of the design students were in the same group. We used two rubrics "the ability to identify assumptions" and "the ability to determine specifically the way in which assumptions affect the results" to score two different experiments that students described in the reports. In order to compare the two groups we added the four scores (two per each experiment) and analyzed this aggregate score statistically. [We followed similar procedures when studying the students' abilities to analyze and minimize experimental uncertainties and their abilities to represent and analyze data.] The difference between the two groups was statistically significant (fig. 2). More than half of the design students (53.3%) tried to evaluate the effects of the assumptions that they made on the result or they actually validated their assumptions. Not a single student in non-design group even attempted to do this.

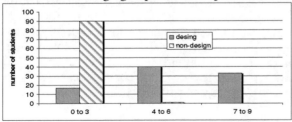

FIGURE 2. The number of students whose reports received aggregate scores for the ability to identify assumptions and the ability to determine the effects of assumptions binned in three categories. The maximum aggregate score that was possible was 12.(Chi-square=119.9, p<0.001)

Identifying, evaluating and minimizing uncertainties: Both groups of students had to evaluate uncertainties

during the semester labs. However, the design group had first to identify the uncertainties, evaluate them, and then to figure out how to minimize them. The instructions in the write-ups in non-design labs included the descriptions of the sources of uncertainty and the minimizing procedures. During the bio practical lab 83.3% of the design students were able to identify correctly most of the uncertainties; 75% of non-design students did not identify any of them.

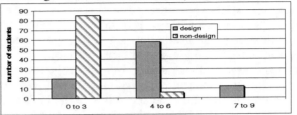

FIGURE 3. The number of students whose reports received aggregate scores for the abilities to identify, evaluate and minimize the effect of uncertainties binned in three categories. The maximum aggregate score was 9. The difference is statistically significant (chi-square=94.49, p<0.001).

Evaluating the result by means of an independent method: When conducting experiments to solve experimental problems during the semester, students in both groups were taught that it was important to perform two independent experiments, to compare the results using experimental uncertainties, and to discuss the possible reasons for the difference. However, only 5.4% of the non-design students evaluated correctly the results including a discussion that referred to both uncertainties and assumptions, while 39% of design students did. Figure 4 presents the results of the scoring of lab reports for this ability.

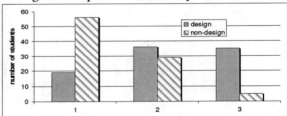

FIGURE 4. The number of students who received scores from 1-3 for the ability to evaluate the result by means of an independent method (nobody received a 0 as the task specifically asked to design two experiments). (Chi-square=42.25, *p* <0.001).

Recording, representing and analyzing data appropriately: These are central abilities for conducting almost any type of research. Most of the lab reports from both groups received scores of 2 and 3 on this ability for the two experiments. However, design students received a perfect or almost perfect score twice as often as non-design students.

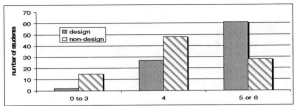

FIGURE 5. The number of students who received scores from 0-6 for the ability to record, represent and analyze data (aggregated for two experiments). The two groups are significantly different (chi-square=28.05, p<0.001).

Communication: This is not a minor ability. Scientists need to be able to communicate their ideas to the other members of the scientific community as well as to the general public. Thus, teaching science involves training students to communicate: to explain their choices, describe completely the procedures, and include good pictures and diagrams. The statistical analysis of student scores on this ability shows that 56% of the non-design students had serious problems describing their experiments while only 17% of the design students did (fig. 6).

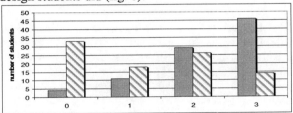

FIGURE 6. The number of students who received scores from 0 to 3 for the ability to communicate scientific ideas. (Chi-square=41.645, p<0.001).

DISCUSSION

The study reported in this paper shows that students who design their own experiments in a physics lab and engage in activities that focus their attention on the elements of scientific investigation, acquire scientific abilities and are able to transfer them to a new content area better than students that follow directions in write ups. Both the learning tasks and the transfer task took place in a very similar context: the same course and the same room but weekly laboratory investigations as opposed to the lab exam.

We found that design students were significantly better than non-design students in the ability to identify assumptions and evaluate their effects; the ability to identify the sources of, evaluate the effects of, and minimize uncertainties; the ability to record and analyze data; and the ability to communicate. These abilities were measured by scoring students' lab reports using the "scientific abilities" rubrics.

Design students spent more time on sense–making (an average of 23min more than non-design students).

That is probably why the reports of design students reflected a more thoughtful take on the task, as they contained more explanations, evaluations, and justifications of the procedures that students selected.

The above results seem to indicate that the design of experiments promotes a more profound and meaningful approach toward laboratory investigations in a particular physics course and possibly in science in general. This new approach promotes in turn the transfer of scientific abilities because students understand their purpose. For instance, uncertainties are not fastidious drill exercises at the end of every experiment but are instead a requisite needed to arrive at well-founded conclusions.

The results of this study have a special relevance since introductory science should *introduce* the practices the scientific community to students. Students need to assimilate the language, methods and quality standards of scientists. The goal of introductory physics courses must be not only to facilitate the learning of physics concepts and their relationships but, equally important, to teach the process and nature of physics through the students' actual practice of the scientific inquiry.

APPENDIX

Application experiment, transpiration rate:
Conduct two experiments to determine transpiration rate using stem cuttings from a single species of plant. *Available equipment:* water, beaker holding plant cuttings, parafilm, tubing, ring stand, graduated pipette, timers, humidity sensor, cup, cup with hole, scissors, and two droppers.

We thank John Bransford, Jose Mestre, and Joe Redish for their advice in the design of the project and Michael Gentile for teaching some labs in the course.

REFERENCES

1. E. Etkina, & A. Van Heuvelen, Proceedings of *Physics Education Research Conference* (Rochester, NY: PERC publishing 2001), p. 17.
2. E. Etkina, S. Murthy, & X. Zou. *American Journal of Physics* 74, 979 (2006).
3. S. M. Barnett, & S. J. Ceci. *Psychological Bulletin*, 128 (4), 612-637, (2002).
4. A. Karelina, & E. Etkina. In *2006 Physics Education Research Conference*, (Syracuse, NY, July 2006) edited by L. McCullough, L. Hsu, and P. Heron, AIP Conference Proceedings, Vol. 883, 93-96, (2007).
5. E. Etkina, A. Van Heuvelen, S. White-Brahmia, D. T. Brookes, M. Gentile, S. Murthy, D. Rosengrant, and A. Warren. *Physical Review. Special Topics, Physics Education Research.* 2, 020103 (2006).

Expert-Novice Differences on a Recognition Memory Test of Physics Diagrams

Adam Feil and Jose Mestre

Departments of Physics and Educational Psychology, University of Illinois at Urbana-Champaign, Loomis Laboratory of Physics, 1110 West Green Street, Urbana, IL 61801

Abstract This study used a recognition memory test and specially constructed pairs of physics diagrams to measure differences between physics experts and novices (who have not taken college physics) in the way simple physics diagrams are encoded in memory. Results show that although physics experts encode aspects and features of physics diagrams that novices do not, in some specific cases novices encode features that experts do not. Physics experts were more likely to encode features of diagrams that were more relevant to the physics depicted. This suggests that the knowledge and experience of physics experts influences the way in which they conceptualize physics diagrams, even in the absence of a question prompt.

Keywords: physics education research, expert novice differences, recognition memory
PACS: 01.40.Fk, 01.40.Ha

INTRODUCTION

Many differences between experts and novices have been identified in a variety of domains, including physics [1]. The current study seeks to identify differences between physics experts and novices in the way simple physics diagrams are conceptualized and encoded in memory. In other words, the study attempts to answer the question, "Do experts and novices focus on different things when looking at the same picture?" In addition, the current study used Signal Detection Theory (SDT), a widely-used method of analysis for certain kinds of tasks, such as the recognition memory test used in this study. A secondary goal of this paper is to provide an example of how SDT can be a useful tool in conducting physics education research.

LITERATURE REVIEW

Expert-novice studies reveal the superiority of expert performance in both verbal and non-verbal tasks. For example, in non-verbal tests of memory experts are able to "chunk" bits of strategically related information into a single unit. This allows experts to outperform novices on many kinds of recall memory tests, such as recalling the layout of chess pieces or the components of complex circuit diagrams [1]. Instead of remembering the location of many individual chess pieces, a chess master can chunk groups of pieces that make up a certain strategic arrangement that he or she is familiar with. Similarly, an electronics technician can chunk a collection of electrical components that make up an amplifier. In his study of chess masters, de Groot noted that much of the thinking done while playing chess is non-verbal, and that chess masters are not used to providing verbal explanations for the moves they make [2], suggesting that intuitive, non-verbal behavior comes with expertise.

It is likely that much of the thought-process of physics experts during problem solving is also non-verbal, despite their ability to explain verbally much better than novices the steps made while solving a problem. The current study does not rely on physics experts and novices verbalizing what they are thinking; rather, a memory test is used as a way to implicitly determine which features of physics diagrams are encoded.

Knowledge organization in memory contributes to the non-verbal prowess of experts. Chi, Feltovich and Glaser [3] studied the contents of schemata among expert and novice physicists and found that experts' schemata were organized around physics principles. In contrast, novices' schemata did include physics principles but not as a central component. The better organization and emphasis on physics principles found in the schemata of physics experts may be due in part to factors other than knowledge structure. Physics experts are consciously aware that physics is built on

CP951, *2007 Physics Education Research Conference*, edited by L. Hsu, C. Henderson, and L. McCullough
© 2007 American Institute of Physics 978-0-7354-0465-6/07/$23.00

relatively few principles and are used to discussing physics concepts in this way.

This study extends explorations of non-verbal skills of experts by implicitly measuring what experts and novices encode when looking at physics diagrams during short exposures. This approach will ensure that any differences found between experts and novices are not a result of factors such as the superior vocabulary or descriptive ability of physics experts.

Memory tests have been used before to study expert-novice differences in various domains [1,4]. These studies have focused on experts' ability to use their extensive long-term memory, which has been formed over years of exposure to various patterns and situations in their domain of expertise, to "chunk" multiple items in a stimulus into a single item. This study uses a different kind of memory test, namely a recognition memory test. In this type of memory test, subjects must decide if the current stimulus is identical to one shown earlier. This type of test will better show differences in the features of diagrams that are encoded by experts and novices.

METHODS

Subjects included 18 novices and 18 experts. Novices were recruited through an educational psychology subject pool. These subjects were both undergraduates and graduate students enrolled in an educational psychology course. The vast majority of these subjects had never taken a college physics course. Experts were graduate students in physics who had recently been TAs in discussion sections of introductory physics courses.

Twenty pairs of images were created. Ten pairs were designed so that the two pictures were slightly different in appearance, but with no meaningful difference in the physics depicted, and ten pairs were designed to include some difference in the physics. Because only simple drawings were shown, the physics differences were fairly subtle. Figures 1-3 are examples of pairs used in the study.

Procedure

A computer program was written to administer the memory test. Subjects were shown a slideshow of 20 images on the screen. Each image was shown for 10 seconds, and a 2-second blank screen was inserted between each image. Subjects were instructed to provide a verbal description of each image as they saw them. These descriptions were audio-recorded and analyzed later. One image was randomly selected from each pair, and those 20 images were shown in a random order.

After the study-portion, subjects were occupied with a worksheet of addition problems for five minutes.

During the test-portion, subjects were shown 20 images and asked to indicate whether each image was "old", meaning it was exactly the same as an image shown before, or "new", meaning it was similar but not identical to an image shown before. Of the 10 pairs from each of the two types, surface-feature changes and physics changes, 4 of the images that were shown at study were also shown again at test, and 6 alternate images were shown. These images were presented in a random order.

Subjects were instructed to say out-loud any differences they remembered for images that they indicated were "new". These statements were also audio-recorded. For example, a subject may say, "Before, it was a cylinder rolling, and now it's a sphere." After making their "old" or "new" judgment, subjects indicated how confident they were with their choice using a 1-5 scale.

FIGURE 1. This diagram pair contains a physically meaningful difference. The angle of the applied force is an important factor in determining the normal and frictional forces on the block, and the acceleration of the block.

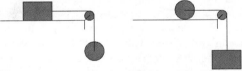

FIGURE 2. This diagram pair contains a physically meaningful difference. The diagram on the right may involve rotational dynamics while the diagram on the left does not.

FIGURE 3. This diagram pair differs only in surface features. Switching the square and circle blocks does not have any effect on the physics depicted.

Signal Detection Theory

Data from this type of recognition memory test is best analyzed using Signal Detection Theory (SDT) [5]. There are four possible outcomes from each trial in the memory test, depending on whether the image shown during the test is the same or different as the one shown during the study-portion and whether the subject responds "old" or "new". Figure 4 demonstrates the four possible outcomes for each trial.

		Picture on test:	
		New	**Old**
Subject's answer choice:	**New**	Correct Rejection (CR)	Miss
	Old	False Alarm (FA)	Hit

FIGURE 4. There are four possible outcomes for each trial because the picture shown can be either old or new and the subject's response can be either old or new. For example, if the picture shown at test is old, meaning the same as shown at study, and the subject responds "old", that trial is scored as a Hit.

In a test situation such as the one in this study, a subject's performance on the test can be influenced by his or her tendency to answer "old" or "new" in situations where the subject is uncertain of the answer. Because there may be differences in bias between experts and novices, it is important to use SDT to analyze the results of the test to help ensure the validity of the findings.

SDT provides a measure of performance that takes into account Hits and False Alarms. This measure, d', indicates how well a subject can distinguish between signal and noise. This measure does not depend on each subject's tendency to answer "old" or "new", which can differ significantly among individuals. In the following sections, d' will be used as the measure of performance on the memory test.

A d' value of 0 represents no ability to distinguish between signal and noise. In this experiment, a subject would have a d' of 0 if he or she had a Hit Rate equal to his or her False Alarm Rate. As other examples, a subject with a Hit Rate of 75% and a False Alarm Rate of 33% would have a d' of 1.1, and a subject with a Hit Rate of 75% and a False Alarm Rate of 17% would have a d' of 1.6.

RESULTS

T-tests were used to determine the significance of the results reported. Results (see Figure 5) show that novices do not perform significantly better on pairs that include physics differences than they do on pairs that differ only on surface-features. Physics experts, however, perform significantly better on pairs that include physics differences than they do on pairs that differ only on surface-features (p = .024). Experts performed marginally better than novices on surface-

feature only pairs (p = .062). Unsurprisingly, experts performed significantly better than novices on pairs that include physics differences (p = .005).

In addition to these main findings, an examination of performance on particular image pairs showed two cases where physics novices outperformed physics experts. Both of these diagram pairs depicted blocks on either side of a double-incline, connected to each other by a string that went through a pulley at the top of the incline. In one diagram, the blocks were at equal heights, and in the other diagram, the blocks were moved so that one hung lower than the other. Figure 6 shows one of these image pairs.

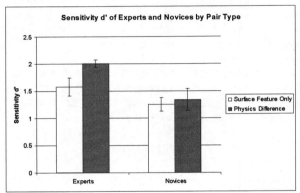

FIGURE 5. Memory test results for physics experts and novices.

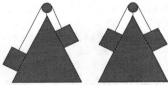

FIGURE 6. Novices were better able to distinguish between these two diagrams than experts, suggesting that the symmetry of the hanging blocks may be a feature considered relevant to novices but not experts in this case.

The descriptions provided by experts and novices during the study-portion of the experiment, in general, were very similar. Because no question prompt was provided, physics experts did not use language describing physics principles or concepts. In most cases, both experts and novices simply gave a description of the surface features of the picture. Some novices did not recognize pictures of circuit diagrams; one subject described a simple resistor circuit diagram as a "heart monitor" because the symbol for a resistor resembles the display of a heartbeat monitor.

Verbal descriptions were coded on a fairly objective criterion: Does the description given for the picture shown during study distinguish between the two pictures of the pair. That is, a description was coded as discriminating if it is only an accurate description of one of the diagrams from a pair. If the description could also be used to correctly describe the other diagram in the pair, it was considered non-discriminating. Figure 7 shows the fraction of discriminating descriptions given by experts and novices. Both groups gave more discriminating descriptions for "physics difference" pairs than for "surface-feature only" pairs ($p < .0001$ for experts, $p = .013$ for novices). Experts gave more discriminating descriptions on "physics difference" pairs than novices ($p = .012$).

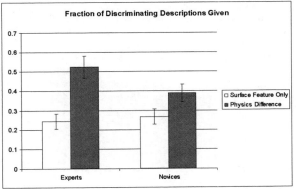

FIGURE 7. The pattern of discriminating descriptions among experts and novices for the two types of diagram pairs is similar to performance on the memory test.

DISCUSSION

The experience and knowledge of individuals causes them to see physics diagrams in a certain way, focusing on different features and eliciting different knowledge depending on expertise. The main difference between experts and novices is that experts have learned which features and relationships shown in a diagram are relevant to the physics and which are not. Because the diagrams used in this study were very simple and no question prompt was given to subjects, it is not sufficient to describe the difference between experts and novices in terms of experts focusing on physics principles and novices focusing on surface features. To a large degree, surface features were all that any of the subjects described while looking at each diagram, so the differences observed between the two groups in this study are due to more subtle mechanisms. The activation of relevant deep structure from viewing surface features in diagrams likely happens automatically for experts, consistent with their hierarchically organized knowledge that bundles together chunks of related contexts, concepts/principles, and problem solving strategies.

Even though both experts and novices encoded surface features when viewing the diagrams, the two groups apparently encoded different relationships among the objects in the diagrams. For example, when describing the diagram on the left shown in Figure 1, one novice subject said, "A force pushing on a block, pushing it up the wedge." One expert subject said, "A block on an inclined plane with a horizontal force." The only difference between these descriptions is that the expert specified the exact direction of the force and novice did not. Our data indicate that when a subject's description for the diagram viewed during the study-portion can distinguish between the two diagrams in the same pair, the subject is almost always able to correctly reject the similar but different diagram shown on the test. This is true of both experts and novices. In the particular example shown in Figure 1, the direction of the applied force appeared to be an important characteristic for an expert to encode since it has implications for the behavior of the system.

The finding that novices outperformed experts on specific instances, such as the diagram pair shown in Figure 6, suggests that some relationships among objects in a specific context are considered irrelevant by experts and hence not encoded in memory, which is perfectly reasonable since attention and memory are limited. The ability to disregard irrelevant information makes any learning or problem-solving task much easier. On the other hand, the symmetry of the situation in Figure 6 was an attractive characteristic for the novice to encode, despite its irrelevance to the behavior of the system.

REFERENCES

1. *How People Learn*, edited by J. Bransford, A. Brown, and R. Cocking (National Academy Press, Washington, D.C., 1999).
2. A. de Groot, *Thought and Choice in Chess* (Mouton, The Hague, the Netherlands, 1965).
3. M.T.H. Chi, P. Feltovich, and R. Glaser, *Cognitive Science* **5**, 121-152 (1981).
4. D.E. Egan and B. J. Schwartz, *Memory and Cognition* **7**, 149-158 (1979).
5. T. D. Wickens, *Elementary Signal Detection Theory* (Oxford University Press, New York, 2002).

Coordination of Mathematics and Physical Resources by Physics Graduate Students

Ayush Gupta, Edward F. Redish, and David Hammer

Department of Physics, University of Maryland, College Park, MD 20742.

Abstract. We investigate the dynamics of how graduate students coordinate their mathematics and physics knowledge within the context of solving a homework problem for a plasma physics survey course. Students were asked to obtain the complex dielectric function for a plasma with a specified distribution function and find the roots of that expression. While all the 16 participating students obtained the dielectric function correctly in one of two equivalent expressions, roughly half of them (7 of 16) failed to compute the roots correctly. All seven took the same initial step that led them to the incorrect answer. We note a perfect correlation between the specific expression of dielectric function obtained and the student's success in solving for the roots. We analyze student responses in terms of a resources framework and suggest routes for future research.

Keywords: complex number resources, manifold framework, graduate education, and physics education research
PACS: 01.40.-d, 01.40.Fk, 01.30.Cc

INTRODUCTION

Development of expertise from a novice state is a core topic of research in learning sciences [1-2]. While there are many studies that characterize novice and expert knowledge [2-5] as the two ends of the spectrum, relatively less work has been done in characterizing the intermediate stages. Expert and novice knowledge have been described within two principal frameworks: the unitary approach, in which entire concepts are considered as individual static cognitive structures [2-4], and the manifold approach, in which one's conception is determined by activation of various resources depending on the context [5-7].

In this paper we explore an intermediate level between experts and novices, from a manifold perspective. We look at how advanced graduate students coordinate very familiar mathematics knowledge in the less familiar domain of a plasma physics problem.

TARGET PROBLEM AND SOLUTION

We analyzed students' homework for a graduate level survey course in plasma physics [8]. Of the 23 students enrolled in the course, 18 students consented to participate in the study. In this paper, we look at their solutions to a homework problem roughly four weeks into the course. The details of the question are provided in Fig. 1.

Given the velocity distribution function of the plasma constituents, the dielectric function $\varepsilon(\omega,k)$ can be derived as a function of the frequency, ω, and the wave-number, k, by a contour integration of a function of the given velocity distribution function in the complex-v plane [see reference 9 for general method]. The derivation treats the frequency ω as complex. The imaginary portion of ω, however, is not just a mathematical trick to select correct boundary conditions (as in non-absorptive quantum mechanical scattering [10]). Rather, it represents absorption in the medium (damping) [9]. Alternatively, one could evaluate the integral for $\varepsilon(\omega,k)$ using the method of calculating the principal value of the integral and including the contribution due to deviation from the real axis, thus allowing for a complex ω. The result can be expressed as either of the following.

$$\varepsilon(\omega,k) = 1 - \frac{\omega_p^2}{(\omega + ikv_{th})^2} \quad \text{(1a)},$$

$$\varepsilon(\omega,k) = \left(1 - \frac{\omega_p^2(\omega^2 - k^2v_{th}^2)}{(\omega^2 + k^2v_{th}^2)^2}\right) + i\frac{2\omega_p^2\omega kv_{th}}{(\omega^2 + k^2v_{th}^2)^2} \quad \text{(1b)}.$$

CP951, *2007 Physics Education Research Conference*, edited by L. Hsu, C. Henderson, and L. McCullough
© 2007 American Institute of Physics 978-0-7354-0465-6/07/$23.00

Both expressions are correct and one can be obtained from the other by the standard procedure for rationalizing the denominator in Eq. 1(a), to bring the explicit '*i*' into the numerator. However, in this case 'rationalization of the denominator' is misleading: Although the expression in Eq. 1(b) looks like an expansion into real and imaginary parts, it is not; ω remains a complex variable. For convenience, we refer to the form of $\varepsilon(\omega,k)$ in Eq. 1(a) as the "compact-form" and that in Eq. 1(b) as "expanded-form". The compact form is directly obtained by a particular choice of the contour during the complex integration, while the principal-value method yields the expanded form. Of course, one could expand or simplify $\varepsilon(\omega,k)$ at any intermediate step. The roots of the equation $\varepsilon(\omega,k) = 0$ are obtained by setting the right side of Eq. 1(a) to zero. They are $\omega = \pm\omega_p - ikv_{th}$.

Obtain the dielectric function $\varepsilon(\omega,k)$ for the following three choices of one-dimensional distribution functions (a) $f_o(v) = \dfrac{n_o v_{th}}{\pi} \dfrac{1}{v_{th}^2 + v^2}$ (Omitting the other two for the purposes of this paper), where v_{th} represents the typical velocity. Make graphs of the real and imaginary parts of dielectric constant for real ω and k positive.
Solve for the roots of $\varepsilon=0$ in each case. Show that all give $\omega=\pm\omega_p$ in the $k=0$ limit, but have different corrections for small k. Explain why some roots are damped and others are not. (Hint: Distribution (a) can be solved using contour integration.)

Figure 1. Problem Statement (courtesy Thomas M. Antonsen, Jr.)

ANALYSIS OF STUDENTS' SOLUTIONS

The pattern of students' success in solving for the roots of $\varepsilon(\omega,k)$ and the expression of $\varepsilon(\omega,k)$ they obtained is summarized in Table 1. Of the 18 consenting students, one did not submit the homework and another student's homework was incomplete and could not be analyzed. Of the 16 homeworks analyzed, 7 obtained $\varepsilon(\omega,k)$ in the expanded form only. These students did not write down the compact form at any intermediate step. The other 9 derived $\varepsilon(\omega,k)$ in the compact form, either directly or by simplifying the expanded form that they got at an intermediate stage. Of these 9 students, 6 drew explicit attention to both the forms (either writing both forms simultaneously or

by expanding $\varepsilon(\omega,k)$ to make the plots of real and imaginary parts of $\varepsilon(\omega,k)$ for real ω and positive k) before solving for the roots.

We observed a 100% correlation in our data between obtaining the dielectric function in the compact form and being able to correctly solve for its roots. All students who used the compact form of dielectric function equated the entire right side of Eq. 1(a) to zero to solve for the roots correctly. It should be noted that these students did not need to treat ω as a complex variable in order to get to the correct solution: the complex roots came out just following algebraic manipulation of the equation. All students who only had access to the expanded-form expression, considered the first and second (without the '*i*' factor) terms of Eq. 1(b) as the real and imaginary parts of $\varepsilon(\omega,k)$ respectively. They solved for the roots by equating the first and second terms individually to zero (Some only set the first term to zero for computing roots).

What led students to make the erroneous assumption that the expanded form separated the real and imaginary parts of $\varepsilon(\omega,k)$? One explanation could be that the form of Eq.1(b) activated a familiar complex-algebra resource of solving an equation of the form

$$z = a + ib = 0 \qquad (2).$$

Typically, in this case a and b are real and the solution step involves equating a and b to zero individually. This in combination with the complex nature of ω not being explicit could have led students to the erroneous step. Another explanation could be that some students were simply not aware that ω could be complex and so were more likely to make the error of identifying the individual terms in Eq. 1(b) as real.

Table 1. Pattern of expression for $\varepsilon(\omega,k)$ obtained and success in solving for roots.

Form of $\varepsilon(\omega,k)$ obtained	Solved correctly for roots	Solved incorrectly for roots
1(a) only	3	0
1(a) and 1(b)	6	0
1(b) only	0	7

We looked at the solutions in detail to analyze students' explicit use of ω as real or complex elsewhere in their solution. The students who drew ω/k as complex in the *v*-plane, or those who explicitly noted that ω can have an imaginary part, calculated it, or discussed damping based on the imaginary part of ω were listed as treating 'ω as complex,' evidence they were aware at some point of ω as complex. Those who drew ω/k on the real v-axis or made no explicit mention of ω as complex were listed as treating 'ω as real'. The results are presented in Table 2: Whether students treated ω as complex elsewhere in their solution did not determine whether they were able to solve for the roots correctly. 71% (5 of 7) of students using the dielectric function in only the expanded form showed some awareness of ω as complex. Of this group, 3 students tried to justify their solution by stating that the imaginary part of ω is very small but 2 other students from the same group went on to explicitly compute the imaginary part of ω in order to determine damping. The 10 students who showed awareness of ω as complex were evenly divided in solving for the roots correctly and incorrectly – the division being determined by whether they had access to the compact form of the dielectric function. Of the 6 students who did not make any explicit reference to ω as complex, the majority (4 out of 6 or 66%) did solve for the roots correctly.

Table 2. Pattern of students treating ω as real or complex and success in solving for roots

	Treated ω as complex	Treat ω as real
Solved correctly for roots	5	4
Solved incorrectly for roots	5	2

These results seem to suggest that the specific form of the dielectric function obtained had a much greater influence on how students interpreted the terms in that expression and solved for the roots than a conception about the real/complex nature of ω.

We suggest a resources-based explanation for this. The homework involves not only non-trivial calculations (algebra, complex integrations) but also coordination of newly learned physics information. The expanded form of epsilon could activate the stable resource of solving familiar equations of the form Eq. 2. It is likely that the newly learned association of the frequency, ω, with a complex variable is relatively less stable and has a lower cuing priority. Also, ω as written in Eq. 1(b) does not naturally cue that it is a complex variable, just as one would be likely to interpret an equation of the form $ax+by=0$ as involving a,b as real constants and x,y as real variables.

Thus we posit that in the context of solving for the roots of $\varepsilon(\omega,k)$, these students fell into treating ω as real, although at other moments of the solution they understood ω to be complex, as was presupposed in the question ("why are some roots damped?") and was a part of explicit discussion in class. That is, the

real/complex nature of ω is not stable in the context of this homework. Some of the students did not hesitate to consider ω as complex or even to compute the imaginary part of ω, in response to the question on damping, right after solving for the roots as if ω was real.

Students who wrote the compact form of $\varepsilon(\omega,k)$ could access other methods for solving the expression $\varepsilon(\omega,k)=0$ – solving the equation $z=0$ where z is not in the $a+ib$ form – leading to a separate set of algebraic steps which led them to the correct solution of the roots. This is apparent by the fact that none of those who had access to *both* forms (6 students) made the erroneous step in interpreting the terms in the dielectric function expression. These students chose to use the compact form of the dielectric function to solve for roots (perhaps simply for the relative algebraic ease of using the compact expression).

DISCUSSION

There are a number of constraints imposed by the nature of the data. We cannot be sure that the chronology of the submitted written material represents the chronology of student thinking, or that it represents all the possibilities that the student might have considered. Analyzing submitted homework also hides from the researchers' eyes the influence of collaborative work among students or that of the professors' comments during office hours. This makes it difficult to attribute a stable conception to the students regarding whether ω is real or complex. There could also be issues of epistemological framing [6] that we have not explored. Do students see the physical significance of the frequency ω as a complex variable when they are solving the integrals or do they consider it a mathematical procedure? Future studies that observe students *in situ* solving homework problems along with follow-up interviews could shed further light on how advanced students coordinate their resources.

We would also like to clarify that we are not calling for instructionally encouraging students to use one or the other form of $\varepsilon(\omega,k)$ for similar problems. The two different expressions for $\varepsilon(\omega,k)$ only lead to slightly different paths that the students follow but did not seem to reflect any demonstrable difference in the way they conceptually treated the problem. Rather, the instructional implication is that even in advanced courses we need to pay attention to how students treat the interplay of physical concepts and mathematics.

CONCLUSIONS

We investigate how advanced graduate students – near experts in simple complex variable manipulations – coordinate their knowledge of complex numbers for solving a problem in plasma physics – a context on which they are not so expert. Our results suggest that a manifold or resources-based framework could be used productively to understand students' solutions and their mistakes. We suggest more detailed future studies that include interviews for better characterization of their knowledge.

There are numerous studies that explore expert and novice knowledge and model them as static and unitary or as dynamic, emergent, and manifold cognitive structures. In either case, modeling the development of expertise is a central topic of research. There are obvious difficulties in monitoring the trajectories of individuals from naïveté to expertise. Indeed one cannot even predict that the starting novices will develop into experts. The alternative is to study a cross-section of the population at various stages of development. This work aims at providing some insight and motivating more detailed future research.

ACKNOWLEDGEMENTS

We thank Thomas M. Antonsen, Jr. for letting us reproduce from the course homework assignment and study participants for providing access to their written work. We also thank the members of Physics Education Research Group at the University of Maryland for discussions. This work was supported in part by NSF grants REC 0440113, DUE 05-24987 and DUE 0524595.

REFERENCES

1. D. P. Maloney, "Research on problem solving," in *Handbook of Research on Science Teaching and Learning*, D. L. Gabel, Ed. (Simon & Schuster MacMillan, 1994) 327-354; L. Hsu, E. Brewe, T. Foster, & K. Harper, "Resource Letter RPS-1: Research in Problem Solving," *Am. J. Phys.* **72**, 1147-1156 (2004).
2. Carey, "Science education as conceptual change", *J. Appl. Dev. Psych.* **21(1)**, 13-19 (2000)
3. Michael McCloskey, "Naïve theories of motion," in Dedre Gentner and Albert L. Stevens, Eds., *Mental Models* (Erlbaum, Hillsdale NJ, 1983), pp. 299-324.
4. M.T.H. Chi, P.J. Feltovich and R. Glaser, "Categorization and representation of physics problems by experts and novices," *Cognitive Science* **5**, 121-152 (1981)
5. A. A. diSessa, "Toward an Epistemology of Physics," *Cognition and Instruction*, **10** (1993) 105-225.
6. David Hammer, "Student resources for learning introductory physics," *Am. J. Phys., PER Suppl.*, **68**:7, S52-S59 (2000); D. Hammer, A. Elby, R. E. Scherr, & E. F. Redish, "Resources, framing, and transfer," in *Transfer of Learning: Research and Perspectives*, J. Mestre, ed. (Information Age Publishing, 2004).
7. M. Wittmann, "Using resource graphs to represent conceptual change," *Phys. Rev. ST-PER* **2** 020105, 1-15 (2006).
8. Phys761 Plasma Physics I: Survey, *Dept. of Physics, University of Maryland, College Park* (Fall 2006). Instructor: Thomas M. Antonsen, Jr.
9. R. Fitzpatrick, "Landau Damping" in *Introduction to Plasma Physics: A Graduate Level Course* (Richard Fitzpatrick 2006) pp 213-221. (available online at http://farside.ph.utexas.edu/teaching/plasma/lectures/node80.html)
10. G. B. Arfken and H. Weber, "Quantum Mechanical Scattering" in *Mathematical Methods for Physicists* (Elsevier Academic Press, Sixth Edition, 2005) pp 469-471.

How Elementary Teachers Use What We Teach: The Impact Of PER At The K-5 Level

Danielle Boyd Harlow

Gevirtz Graduate School of Education, University of California, Santa Barbara, CA 93106-9490
dharlow@education.ucsb.edu

Abstract. This paper presents an investigation of how a professional development course based on the Physics and Everyday Thinking (PET) curriculum affected the teaching practices of five case study elementary teachers. The findings of this study show that each teacher transferred different content and pedagogical aspects of the course into their science teaching. The range of transfer is explained by considering how each teacher interacted with the learning context (the PET curriculum) and their initial ideas about teaching science.

Keywords: Teacher education, science education, physics education research, professional development, inquiry, elementary science, elementary teacher education.
PACS: 1.40E-, 1.40eg, 1.40Fk, 1.40G-, 1.40J-, 1.40Jh.

INTRODUCTION

It is not surprising that elementary school teachers face challenges in their efforts to teach physical science in ways that are consistent with authentic scientific inquiry. Attaining the sort of conceptual understanding necessary to teach through methods of inquiry-based instruction is difficult even for secondary school teachers who ideally major in a science content area [1]. Elementary teachers take only a few science and science methods courses, making these few courses vital to shaping their understanding of science content and the nature of scientific inquiry.

Research on teacher learning indicates that courses in which teachers have the opportunity to learn in the way they are expected to teach may be useful for teachers trying to implement new teaching methods [2]. As a result, several inquiry-based physics curricula have been designed by the PER community with the needs of future and practicing elementary teachers in mind [3]. While there is evidence that these courses result in greater conceptual understanding as measured by written tests, there is less evidence that teachers *use* what they learn in these courses when they teach science to their elementary school students. In fact, we have little understanding of what teachers *do* with the skills and knowledge gained in such courses once they return to their elementary classrooms. Only by understanding what teachers do when they are teaching, can we really understand how our courses can help elementary teachers.

Research indicates that learners may transfer unexpected or even unintended aspects from a learning situation into a target context and many contemporary scholars of transfer argue that understanding *what* learners transfer is more important that understanding *whether* learners transfer specific knowledge and skills [4][5][6][7]. Using such an understanding of transfer, I aimed to identify what teachers (who are learners in our physics courses) transferred from a physics course for teachers into their teaching practices.

THE LEARNING CONTEXT

The teachers who participated in this study were enrolled in a 6-week (2.5 hours/week) physics course offered as a professional development course for practicing K-5 teachers. The course was entitled Magnetism and Electricity for Elementary Teachers (MEET) and was adapted from the magnetism and electricity units of the Physics and Everyday Thinking (PET) curriculum [8]. PET differs from other curricula in its content as well as in its focus on elementary students' ideas, the nature of science, and learning about how one learns physics. Through hands-on and computer-based activities and small group and whole class discussions, the teachers develop their understanding of physics.

CP951, *2007 Physics Education Research Conference*, edited by L. Hsu, C. Henderson, and L. McCullough
© 2007 American Institute of Physics 978-0-7354-0465-6/07/$23.00

During the magnetism unit of MEET, the elementary teachers proposed and developed models of magnetism that explained why an iron nail that had been rubbed by a magnet behaved differently than an iron nail that had not been rubbed by a magnet (a nail that has been rubbed by a magnet will itself act like a magnet while a nail that has not been rubbed by a magnet will act like an ordinary piece of non-magnetized ferromagnetic metal). Through the series of activities in this unit, the elementary teachers enrolled in the class not only learned physics content associated with magnetism but also about the evidence-based nature of scientific models. In the subsequent unit, the elementary teachers developed an understanding of current electricity through exploration of simple series and parallel circuits.

METHODS

This study was designed to answer the question, "What do teachers transfer from MEET to their science teaching and what factors are involved in the transfer?" I studied five teachers in detail through qualitative research methods including video-taped observations and interviews. Data were collected in three stages: 1) Prior to the first day of class, I video taped each teacher teaching two science lessons of his or her choice; 2) During the MEET course, participating teachers' interactions in small group and whole class discussions and activities were video taped; and 3) After MEET, the teachers were asked to teach two science lessons that aligned with the topics of the course and these lessons were also video taped. Interviews about the teachers' experiences with science and about the decisions they made while teaching were also conducted pre- and post-MEET.

The observations and interviews resulted in a total of 60 hours of video data, 40 of which were relevant to the questions under investigation. These 40 hours were transcribed and these transcripts of observations were

the main source of data because they represented what the teacher actually *did* with his or her students and therefore represented science to his or her elementary students. In particular, to identify what teachers transferred from MEET to their teaching practices, I looked for evidence that the teachers used ideas and content learned in MEET to shape how they taught.

Teachers draw on their understanding of science content and of the nature of scientific inquiry when transforming an activity to be appropriate for a particular class of students and when responding to unexpected science ideas expressed by their students during science lessons. Such improvisational uses allows one to see how teachers *used* ideas rather than just that they recalled ideas. Thus, only teachers' use of ideas learned in MEET to solve new problems in the context of teaching science to elementary school students counted as transfer in this study. From observed instances of improvisation in the context of teaching science, I identified the requisite knowledge of the teacher to make the decisions and then examined pre-measures to identify whether evidence existed that this knowledge was part of his or her pre-MEET teaching repertoire. Only knowledge used in post observations that was not present in pre-measures was considered transferred.

THE TEACHERS

Five of the ten teachers enrolled in MEET participated in this study. The five case study teachers ranged in teaching experience from a first year teacher to a teacher who had been teaching for 29 years. Their frequency of teaching science also varied greatly. One teacher rarely had the opportunity to teach science while another teacher taught science twice a day.

As shown in table 1, based on the criteria used to identify transfer, three teachers exhibited evidence that they transferred aspects of MEET into their teaching practices and two showed little evidence of transfer.

Table 1: Transfer claims and evidence by teacher

Teacher	Transfer Claim	Evidence
Ms. Carter 12th yr teacher	Physics content, model based reasoning, pedagogical strategies	Recognized testable student models and proposed ways of testing, asked students to predict based on models and support with evidence, self report/observations of changes in teaching strategies.
Ms. Bailey 3rd yr teacher	Used content to teach consistent with prior idea of inquiry	Observations of post science lessons followed format described in pre-interviews. Used MEET activity to fit her stated inquiry goals.
Ms. Allen 1st yr teacher	Transferred physics content and idea that students can create science ideas	Self-reports of how science content helped implement science lessons, recognition that an activity done in MEET would address problem students came up with in post-MEET lesson.
Mr. Dugan 13th yr teacher	Little or no evidence of transfer	Taught topics he was familiar with prior to MEET. Followed provided lesson plans closely.
Ms. Evans 29th yr teacher	Little or no evidence of transfer	Taught topics she was familiar with prior to MEET. Used activities as she had in the past.

Next, I briefly describe the three teachers for whom I observed transfer and what he or she transferred from MEET into her post-MEET teaching.

Ms. Carter: Prior to MEET, Ms. Carter (a 12th year teacher) was apprehensive about teaching and learning physical science. Her teaching, while hands-on, often represented science as vocabulary and fact-based. After MEET, Ms. Carter's teaching appeared very different. During her observed teaching about magnetism after MEET, Ms. Carter's students proposed models to explain why iron nails which had been rubbed by a magnet act like a magnet, while ordinary (unrubbed) nails do not. They then tested and revised these models. While both the models presented by the students and the experiments done to test and revise the models of magnetism differed from those proposed by the elementary teachers in MEET, the tests were appropriate for the models presented and thus Ms. Carter's teaching showed that she used her new knowledge about magnetism and about the nature of scientific models as well as pedagogical practices modeled in MEET to guide her students in developing knowledge of magnetism [9].

Ms. Bailey: In interviews prior to MEET, Ms. Bailey indicated that she believed that science should be about children asking questions and investigating these questions to find out the answer. After MEET, Ms. Bailey adapted a MEET activity to meet these goals. Like Ms. Carter, Ms. Bailey asked her students to draw models which would explain why a magnet is different from a non-magnet. However, rather than have her students test and revise these models, she used it as an exploration activity to inspire students to ask questions. Her students then shared their questions and listed steps they might take to answer them.

Ms. Allen: As a first year teacher, Ms. Allen was looking for strong guidance in how to teach physical science. She used many of the accompanying K-5 lessons that accompany PET/MEET and learned from implementing these activities that her students were capable of creating science knowledge. When her students came across a puzzling situation (two of three light bulbs not lighting in a large series circuit), she identified a MEET activity which she believed would help her students figure out their problem. In order to connect her students' question to the activity, she had to draw on her new understanding of circuits.

The other two teachers, Mr. Dugan and Ms. Evans did not show evidence of transfer.

Mr. Dugan: Mr. Dugan was a strong science teacher prior to MEET. He routinely tested new science kits developed by his school district and had much science experience to draw on. After MEET, he taught two science lessons about topics in magnetism that he understood prior to MEET and followed lesson plans with little deviation. Therefore, the data collected provide little evidence for transfer.

Ms. Evans: Like Mr. Dugan, Ms. Evans was a science teaching expert and taught her lessons after MEET on topics she was previously comfortable with and used lessons plans she had in the past. Thus there is little evidence of transfer.

These claims and the evidence to support these claims are summarized in table 1.

FINDINGS

As described in the previous section, the five participating teachers can be divided into two groups: those for which there is evidence of transfer and those for whom there is not. Considering first the teachers who show evidence of transfer, clearly there are significant differences in the ways that the three teachers used what they learned in MEET to teach elementary school science. However, it is also important to note that there are also some interesting commonalities. In all three cases, the teacher expressed dissatisfaction with some aspect of her science teaching prior to beginning MEET and was seeking something from the course. Ms. Carter was dissatisfied with her ability to teach physical science in general and expressed conflicting views about how to best teach science. Ms. Bailey was confident in her ability to teach science and had strong ideas about how to teach through inquiry, but felt that she lacked the content knowledge to implement such teaching in physical science topics. Ms. Allen, a new teacher, expressed concern about her science teaching because of a lack of experience and lack of guidance in the science curriculum used at her school. Each of these teachers began MEET with ideas about what they hoped to gain from the course. Furthermore, each of these teachers subsequently identified aspects of MEET that were useful in addressing their perceived deficiencies and transferred these aspects. Thus, each of these teachers learned physics (and inquiry) *mindful of their future use of this knowledge* and transferred *only what they sought.*

Mr. Dugan and Ms. Evans, the two teachers for whom there is little evidence of transfer, differed from the other three teachers in important ways. Both of these teachers were familiar and comfortable teaching inquiry-based science and both were considered expert science teachers by other teachers in their districts prior to enrolling in MEET. Furthermore, neither Mr. Dugan nor Ms. Evans had expressed dissatisfaction in their science teaching. They enrolled in MEET for other reasons (e.g., at the request of the district science coordinator and to earn professional development credits). Both Mr. Dugan and Ms. Evans expressed how much they enjoyed the course and how much they learned about the more advanced topics in the course, but neither attempted to *use* anything new from the course in their teaching of these topics to their elementary students.

The comparison of the similarities and differences across and between the two groups leads to three major findings: 1) All teachers who transferred were dissatisfied with their own science teaching in some way and the teachers who were not dissatisfied did not show evidence of transfer, 2) teachers who transferred, transferred different things from MEET into their elementary teaching practices, and 3) teachers transferred what they came in desiring to know.

DISCUSSION AND IMPLICATIONS

The finding that the teachers who transferred were dissatisfied with their teaching is interesting because it would follow that, to facilitate the use of the science ideas that we teach and methods we use to teach science, it may be useful to help practicing teachers recognize dissatisfaction in their own teaching prior to courses like MEET and to provide tools to help them evaluate their own teaching. When teaching physics, we recognize the importance of eliciting and using students' ideas about science. For example, in MEET, the elementary teachers proposed initial models for magnetism and revised their models because evidence which challenged their model was collected. Perhaps challenging teachers' assumptions about how well their model for teaching children works would create an opportunity for them to consider changes that would enhance their teaching.

Even more provocative is the finding that the teachers knew precisely what they thought they needed and took exactly the part of MEET that fit their predetermined need. One implication of this is that *target-minded learning* (knowing what one seeks from a learning context) may increase the likelihood of transfer. Target-minded learning may facilitate transfer because teacher education courses afford the opportunity to learn so many different things. The teacher who is focused on what she needs is more likely to find it.

A further implication of this study is that it may be useful to explicitly connect physics instruction to teachers' current classroom contexts. This aligns with research in the professional development literature that suggests situating professional development in teachers' classrooms [10]. Requiring teachers to try out new ideas or to practice what they have learned about content or inquiry may help them identify how salient aspects of the course can be directly used in their teaching practices. This sort of direct application of physics and understanding of authentic scientific inquiry is challenging as the physics instructor must not only have a strong understanding of physics and the nature of science but also of the pedagogical challenges that face elementary teachers. This suggests a need for greater collaboration among education and content specialists to support the teaching of content courses for teachers.

Contemporary theories of transfer claim that much more than we expect may be transferred from a learning situation and that transfer is context-dependent [4]. Understanding what teachers actually do in the context of teaching science in their elementary classroom is vital to understanding what teachers take away from our courses and how they use it in their teaching.

Current reform initiatives [11][12] challenge teachers to create learning environments in which students are active participants in science. Such teaching methods place greater intellectual demands on teachers. New and experienced teachers will need help adopting inquiry-based instructional methods. The findings of this study suggest that by teaching teachers through inquiry, we can help them to use these strategies in their own instruction. However, in the case of such a complex learning environment, we need to help teachers identify the pedagogical and content aspects of the physics instruction that are relevant to their particular teaching situation so that they can identify how what they are learning will help them reform their teaching practices.

REFERENCES

1. S. Wilson, L.S. Shulman, A. & A. Reichart, *Exploring Teachers' Thinking*, 194-124, (1987).
2. Grossman, Smagorinsky, & Valencia, *Amer. Journal of Education*, **108**, 1-25 (1999).
3. some examples include Goldberg, F., Robinson, S., Otero, V., *Physics for Elementary Teachers*, 2007; L. McDermott, *Physics by Inquiry*, 1996; and AAPT, *Powerful Ideas in Physical Science*, 1996.
4. J. Mestre, *Transfer of Learning: Issues and Agenda*, NSF Report, 2003.
5. J. Lobato, *Educational Researcher*, **31 (1)**, 17-20 (2003).
6. K. Beach, *Rev. of Research in Educ.*, **24**, 101-140 (1999).
7. see also many of the articles found in J. Mestre, *Transfer of Learning from a Multidisciplinary Perspective*, 2005.
8. Goldberg, F., Robinson, S., Otero, V., *Physics and Everyday Thinking*, 2007. Note that the early version of PET was called *Physics for Elementary Teachers*.
9. D. Harlow & V. Otero, *PERC Proceedings* (2006). Note that in this earlier work, her pseudonym was Ms. Doty.
10. See, for example, R. Putnam & H. Borko, *Educational Researcher*, **29**, 4-15, 2000.
11. AAAS, *Benchmarks for Scientific Literacy*, 1990.
12. NRC, *National Science Education Standards*, 1996.

Student Categorization of Problems – An Extension

Kathleen A. Harper[*], Zachary D. Hite[○], Richard J. Freuler[○], and John T. Demel[○]

[*]Department of Physics, The Ohio State University, 191 W Woodruff Ave, Columbus, OH 43210
[○]First-Year Engineering Program, The Ohio State University, 244 Hitchcock Hall,
2070 Neil Ave, Columbus, OH 43210

Abstract. As part of gathering baseline data for a study on problem categorization, first-year engineering honors students who had recently completed a two-quarter sequence in physics were interviewed. The primary task in this interview was much like the problem categorization study described by Chi et al. There were, however, at least two distinct modifications: 1) in addition to the problem statements, solutions were included on the cards to be sorted 2) the problems were written such that they could also be grouped according to the nature of information presented in the problem statements and/or the number of possible solutions. The students in this baseline study, although similar in background to the novices described by Chi et al., in many ways performed more like experts. Several possibilities for this behavior are discussed.

Keywords: interdisciplinary problem solving, STEM, problem categorization, teaching problem solving
PACS: 01.40.Fk, 01.40.gb, 89.20.Kk

INTRODUCTION

Previous work proposed a problem categorization scheme that is based upon the nature of the information provided in the problem statement (insufficient, exactly sufficient, or excess) and the nature of the problem's answer (no answer, one answer, or multiple answers). [1] This scheme is independent of the problem's content and is connected to problem-solving skills that students will use in the "real world," such as approximating, estimating, assuming, filtering, or optimizing.

As part of collecting baseline data to assess the effectiveness of implementing this scheme in a course for honors engineering students at The Ohio State University, an interview task was developed based upon the groundbreaking work of Chi, Feltovich, and Glaser. [2] In their study, subjects were asked to sort cards containing problem statements. The expert physicists in the study (PhD physics students) sorted the cards according to the principals needed to solve the problem, or the problem's deep structure. The novices (students who had just completed a semester of introductory physics) sorted the cards by the surface features of the problems.

This paper describes some unanticipated results from this baseline study, along with a small probe to gather more information on what led to these preliminary results.

CONTEXT OF THE STUDY

The subjects in the current interview sample were 3rd-quarter students enrolled in Ohio State's Fundamentals of Engineering for Honors (FEH) program. [3] The FEH program consists of coordinated courses in engineering, math, and physics throughout the students' first year of college. Ohio State is on the quarter system. The instructors meet on a weekly basis to discuss opportunities to link the courses. The bulk of students participate in one of two math sequences, depending upon their level of calculus readiness. Almost all students take two nontraditionally taught physics courses, which will be described in more detail below; some take a third quarter, and even students with AP credit are strongly encouraged to take at least one of the physics courses. All students in the program take engineering courses in graphics and computer programming, and the program culminates with a quarter-long design-and-build project in the spring quarter.

The physics courses include nontraditional elements such as interactive lectures in the style of Etkina and Van Heuvelen's ISLE program, conceptual questions with personal response systems, cooperative group problem-solving exercises, group midterms, and lab exercises utilizing VPython. [4,5]

CP951, 2007 Physics Education Research Conference, edited by L. Hsu, C. Henderson, and L. McCullough
© 2007 American Institute of Physics 978-0-7354-0465-6/07/$23.00

THE INTERVIEW TASK

A set of 18 problems was developed for subjects to sort. These problems were written such that subjects would be able to group them by deep structure, surface features, the nature of the information provided, or the nature of the answer. To make sure that all of these sorting options were equally represented on the cards, each card included the problem's solution. A sample card is shown in Figure 1.

Problem:

A 1-kg cannonball is launched so that it slides up the deck of a pirate ship angled 30 degrees to the horizontal. The ball's initial velocity is 5 m/s. How fast is the cannonball going after it has traveled 3 meters up the deck? Assume the deck is frictionless.

Solution:

Apply conservation of energy.

Initially: $K_i = \frac{1}{2}mv_i^2 = \frac{1}{2}(1kg)(5m/s)^2 = 12.5$ J

At 3 meters up the ramp:

$U_{g,3m} = mgh_{3m} = mg(3m)\sin 30°$

$= (1kg)(9.8m/s^2)(3m)\sin 30° = 14.7$ J

$K_i + U_{g,i} = K_{3m} + U_{g,3m}$

$\Rightarrow 12.5J + 0 = K_{3m} + 14.7J \Rightarrow K_{3m} = -2.2$ J

Negative kinetic energy isn't possible. The cannonball can't get 3 m up the ramp!

FIGURE 1. Sample Card For Card Sorting Task

Notice that it involves the surface features of a cannonball, pirate ship, and inclined plane, and contains the deep structure of energy conservation. The nature of the provided information is that there is excess information (since the mass is not needed to solve the problem). The nature of the answer is that there is no solution to the question asked. All problems were written with a fair amount of context, since the students were accustomed to seeing problems of this sort in their physics courses. The problems contained a wide distribution within each of these features, as summarized in Figures 2 through 5.

Concept	# of Problems Containing
Energy Conservation	3
Newtons Laws and Kinematics *	3
Projectile Kinematics	5
Momentum Conservation	3
Relative Velocity	4

* These problems required application of both concepts in the solution.

FIGURE 2. Card Distribution According to Deep Structure

Surface Feature	# of Problems Containing*
Car	5
Pirates	4
Ball	5
Block	3
Inclined Plane	6

*a few problems have multiple features

FIGURE 3. Card Distribution According to Surface Features

Nature of Provided Information	# of Problems Containing
Insufficient	4
Exactly Sufficient	9
Excess	5

FIGURE 4. Card Distribution According to Provided Information

Number of Answers	# of Problems Containing
None	5
One	8
More than One	5

FIGURE 5. Card Distribution According to Nature of Answer

INITIAL RESULTS

The initial interviews were of ten honors engineering students at Ohio State. These students had completed rigorous courses in mechanics and electricity and magnetism, and a few were enrolled in a course on waves, quantum mechanics, and optics at the time of the interviews. The subjects were selected from a pool of volunteers, such that they would represent a range of physics achievement, as measured by their final course grades.

As part of the interview, the students were asked to sort the cards into categories while thinking aloud. [6] If a subject indicated that he or she could think of multiple ways to sort the cards, they were asked to sort the cards first in the way they liked the best, then to repeat the sort according to the other criteria they had considered. Subjects were also asked to name the categories into which they had sorted the cards.

All but one of the students sorted the problems according to the deep structure, largely titling the groups with similar language to that used in Figure 2. Some grouped all kinematics problems together, and a few talked about how many dimensions were involved. One grouped all conservation problems together. The tenth subject sorted based on his perceived difficulty of the problem, which involved whether trigonometry was needed to solve the problem, the length of the text, and whether the problem contained an impossible situation.

One striking aspect of the results was that the amount of time these students had spent learning physics was much closer to that of the novices in Chi's study than the experts, but that the performance of the vast majority on the task was more like what one might expect from her expert subjects. Based upon prior work, it had been expected that there would be more variation in the sample than this. The expectation was that there would be some students appearing much like novices, others like experts, and several somewhere in transition. [7]

SUBSEQUENT RESULTS

One possibility for this behavior was that the initial statement in the solution ("Apply x...") was viewed as such a compelling feature of the card that subjects relied heavily on it. Given that not all of the students named their groupings strictly according to the concepts listed in the solutions, it seemed somewhat unlikely that this was the only cause, but it was something that should be probed to make sure. To test this theory, the experiment was repeated with a comparable group of students from the same population, but this time this initial statement identifying the concept was removed. Other than this modification, the cards remained identical.

The results from the second group were largely the same. In this case, all subjects sorted based on the problem's deep structure. The most noticeable difference apparent in the initial analysis of the data is that the subjects tended to group the problems slightly differently. In this second set of interviews, the subjects were more likely to group all of the projectile kinematics and relative velocity problems together in one group designated "kinematics" or sometimes "position and velocity." Additionally, students were less likely to mention the kinematics in the problems that also involved Newton's laws. Some took friction into account when sorting the Newton's law problems. The majority easily recognized the energy and momentum conservation problems. The most unique approach in this set was by a student who initially sorted based on methods (energy and work or kinematics), then sorted again based on the surface features. When asked which method he would choose if he had to pick one, he said that he would use the more conceptual-based approach as his primary sorting.

INTERPRETATION

The results of the two studies together clearly show that the students in the interview pool sorted the problems in a way that is consistent with the experts in Chi's study. There are several possibilities for what drove this behavior that need to be considered in more detail before broader conclusions can be drawn.

One is that something in the way the physics courses were taught to these students made them behave more expertly. Given that the population is honors students and that the physics sequence has a many-year history of achieving significant student gains on conceptual evaluations like the FCI and CSEM, this is not an incredibly farfetched possibility. Could it be that the exposure to alternative problem types is having an impact on this aspect of their problem solving? Could some other factors in the course be fostering this maturity in viewing problems?

Of course, another possibility is that because these students are honors students, they already possessed some expert-like problem-solving skills when they arrived at the university, and that the courses had nothing to do with it. As part of the larger study, there are already plans to repeat this interview with incoming students in the fall, and the answer to this question will be known shortly.

A third possibility is that the differences between these cards and those in the Chi study were significant enough to cause this difference in behavior. The major differences are the more involved contexts of these problems and the provided solutions. Perhaps one or both of these changes caused students to ignore the surface features. Again, more should be known about these possibilities once the fall interviews are conducted. If the students in the fall behave similarly, the team may elect to repeat the interviews without the solutions on the cards and see if the results are different.

Closely related to this, the possibility exists that the modifications made to the task make it different enough from the Chi study that experts may behave differently than expected, as well. At some point in the future, the task may be repeated with physics faculty.

Until these areas are fully probed, it would be unwise to make any sweeping statements or generalizations about the findings to this point.

FUTURE DIRECTIONS

Regardless of the answers found by probing the possibilities above, the data collected thus far give a solid baseline for the larger problem categorization project. This study will continue in the next academic year, as the new categorization strategy is introduced to the students. The card sorting task will be repeated several times during this period. As mentioned above, incoming students with similar profiles to those already interviewed (in terms of gender, class rank, ACT scores, AP credit, etc.) will be interviewed within the first few days of their college experience.

Comparable groups will be selected to perform the card sort at the conclusion of the mechanics and electricity and magnetism courses. At that point, it will be seen if students exposed to the new categorization scheme look at the nature of information provided in the problem or the number of possible answers when sorting problems into categories.

The interview protocol also includes a series of questions about problem solving skills and strategies, and analysis of these questions will shed additional light on the student behavior observed during the card sorting task.

ACKNOWLEDGMENTS

This work is sponsored by NSF grant DUE-063367. Thanks to Fritz Meyers and Stuart Brand for assistance in arranging the interviews. We also appreciate the constructive comments of the reviewers.

REFERENCES

1. K. A. Harper et al., *Proceedings of the 2006 Physics Education Research Conference*, McCullough, Hsu, & Heron, eds., 141-143 (AIP, 2007).
2. M. T. H. Chi et al *Cognitive Science* **5**, 121-152 (1981).
3. J. T. Demel et al., *Proceedings of the 2004 ASEE Conference* (ASEE, 2004).
4. For more information on ISLE, see http://www.rci.rutgers.edu/~etkina/ISLE.htm.
5. P. Heller et al., *Am J. Phys.* **60**, 627-636 (1992); P. Heller et al., *Am J. Phys.* **60**, 637-644 (1992).
6. K. A. Ericsson & H. A. Simon, *Protocol Analysis* (MIT Press, 1996).
7. K. A. Harper, *Proceedings of the 2003 Physics Education Research Conference*, 129-132 (AIP, 2004).

Students' Ideas of a Blender and Perceptions of Scaffolding Activities

Activities

Jacquelyn J. Haynicz and N. Sanjay Rebello

Department of Physics, 116 Cardwell Hall, Kansas State University, Manhattan, KS 66506-2601

Abstract. Research has shown that students can be motivated to learn science by demonstrating its connection to everyday life. We investigated students' understanding of an everyday blender. We have previously reported on students' progression through a series of hands-on activities designed to facilitate learning about how the blender works [1]. Here, we report on the ideas about the blender expressed by students after completing the sequence of activities and the students' perceptions of the activities themselves.

Keywords: electricity, motors, everyday context, students' conceptions, physics education research
PACS: 01. 40.Fk

INTRODUCTION

Studies have shown that students are motivated to learn when they see a connection between learning and everyday life and that student learning is enhanced by real-life contexts [2]. However, there typically is a gap between educators' learning goals and students' perceptions. We investigated students' ideas of a blender after completing a sequence of scaffolding activities and their perceptions of the activities. Our research questions are:

- What are the ideas about the blender expressed by students after completing a sequence of hands-on activities about electromagnetic motors in an interview setting?
- How do students perceive the value of these demonstrations in relation to the blender?
- Are there significant differences in the ideas or perceptions expressed by students enrolled in various levels of introductory physics?

THEORETICAL FRAMEWORK

In this pilot study, we were interested in examining what students transfer from either their previous knowledge or knowledge constructed while interacting with hands-on activities during the interview to the context of the blender. We adapted Lobato's [3] actor-oriented perspective, examining everything students transfer, including spontaneous intuitive knowledge [4] and attunement to affordances [5]. This perspective on transfer of learning was consistent with the idea that learners construct their own knowledge. The teaching interview was based on Vygotsky's social constructivism by which learning occurs within a Zone of Proximal Development, in which the learner can learn with the assistance of a more knowledgeable individual. [6].

METHODOLOGY

We conducted semi-structured, individual teaching interviews with 12 students. The participants in this study came from three different introductory physics courses: conceptual-based, algebra-based and calculus-based. Four students, two male and two female, were randomly selected from a pool of volunteers from each course. The teaching interview is a mock instructional setting during which students interact with a sequence of activities to facilitate learning [7]. The phenomenological approach was used to analyze the students' ideas about the blender after they had completed the activities [8].

First students saw a household blender with the back carved out and motor visible. Next, they saw individual pieces of the blender motor to help activate their prior knowledge. In the first activity, a rail gun (RG) was used to help students realize that motion results from a magnetic field and current. Next, students interacted with a permanent magnet board motor (PMB), shown in Fig. 1.

The PMB consisted of a rotor, two moveable permanent magnets and copper strips that served as brushes. Students connected a battery to the brushes

CP951, *2007 Physics Education Research Conference*, edited by L. Hsu, C. Henderson, and L. McCullough
© 2007 American Institute of Physics 978-0-7354-0465-6/07/\$23.00

FIGURE 1. Permanent Magnet Board Motor (PMB)

and examined the effect of various orientations of the magnets. The PMB was designed to help students see the necessity of magnets in the motor since the permanent magnets were easily identifiable and could be entirely removed. Next, students were given an electromagnet board motor (EMB), which was similar to the PMB, except the permanent magnets were replaced with coils to which students needed to attach batteries in order to produce electromagnets. The EMB was included to help students identify the electromagnets in the blender. Last, students explored the effect of a permanent magnet and an electromagnet on a compass in the electromagnet coil activity (EC). The EC was included to assist students unable to identify the electromagnets in the EMB. All four activities were used both as motors and generators; however, the focus was on motors alone.

RESULTS & DISCUSSION

At the end of the teaching interview, the students were asked a series of wrap-up questions designed to elicit their ideas about the blender and their perception of the activities. The results are presented below.

Ideas About the Blender

At the conclusion of the teaching interview, most students (11 out of 12) were asked to explain how they thought the blender worked after having completed the preceding activities. Since the activities focused on what caused the motor to spin and what created the magnets, we analyzed their responses in terms of their descriptions of the mechanisms for producing (a) magnets or magnetic field and (b) spinning in the blender. In addition, we identified the key words that students used in describing how the blender worked.

Causes of Magnets or Magnetic Field

Students' ideas about the source of the magnetic field in the blender varied. The most common response, given by four out of 11 students, was that the magnets were a result of "current." For example, one student stated, "You've got your current running through these coils setting up a... magnetic field."

This response was given by three students enrolled in the calculus-based physics course, but only one other student overall.

The other responses included "charge," "electricity," the blender being "powered up" and having "oppositely charged bundles of wire." Three students did not discuss the mechanism for creating electromagnets in their ideas about the blender.

Causes of Spinning of Blender

Students also offered a variety of ideas as to what caused the blender to spin. Six students attributed spinning to something to do with magnetism, such as the magnetic field, magnetic force or oppositely poled electromagnets. For instance, one student explained, "that causes a magnetic force within and because this [rotor] is inside here [electromagnets] that magnetic force gets this [rotor] spinning which in turn gets the entire blender spinning." All four calculus-based physics students attributed spinning to magnetism, as did one student in each of the other courses.

Two of the above students also included a role for current in their mechanism for spinning. One student pointed to the rotor and stated that the current was always in the same direction through it and the other attributed spinning to currents trying to align.

Other reasons given for the blender's spinning included "electricity," "metals reacting upon each other," "charge," "electrons moving," the switch connecting the blender's parts with the outlet cord and the "power source." Two students, both from the calculus-based class, specifically mentioned that the blender worked in the same way as the EMB activity.

Keywords Used by Students

In describing their ideas about the blender, students used the following keywords or phrases. The keywords were identified by searching the transcripts of students' responses for commonly occurring words. Table 1 indicates how many students in each course used each keyword or phrase. Here, the keyword "magnetism" includes all references to "the magnets," "magnetic force" and "magnetic field."

TABLE 1: Keywords Used by Students

Keyword	Concept-based (N=4)	Algebra-based (N=3)	Calculus-based (N=4)
Charge	2	1	0
Electricity	2	1	0
Magnetism	3	3	4
Electromagnetism	0	0	1
Source/Supply	0	1	2
Current	0	1	4

Several trends were noticeable from the table. The term "magnetism" appeared in nearly all the students' responses. Only one student (who was enrolled in the conceptual-based physics course) did not use this term. The term "current" was used by all four students in the calculus-based course, but by only one other student. Additionally, the term "electromagnet" was only used in one student's description of how the blender works.

Students in the algebra-based course used the largest variety of keywords, with at least one student mentioning five of the six identified keywords. The terms "source" or "supply" and "current" were not mentioned by any of the conceptual-based students, while the terms "electricity" and "charge" did not appear among the calculus-based students' responses.

Perceptions of the Activities

At the conclusion of the teaching interview, 11 of the 12 students were asked a series of questions designed to assess their perceptions of the activities that had been used. To determine which activities the students felt were similar to the blender, they were asked to identify the activities most similar to and most different from the blender. To determine which activities the students found to be most useful in thinking about how the blender worked, they were asked to identify which was the most useful and which activity, if any, they would leave out of the sequence of activities. Finally, students were asked if they would change anything about the order of the activities.

Most Similar or Most Different

The EMB was chosen by nearly all students (10 of 12) to be most similar to the blender. The remaining two students could not decide between the EMB and PMB, or the EMB, PMB and RG. The students were more divided on which activity was the most different from the blender. Four students each selected EC or RG, and three students cited both as the most different.

Table 2 displays the reasons given for choosing an activity as most similar or most different. The category "spinning" includes references to both "spinning" and "motion" and the category "magnets" includes any reference to magnets or magnetism, but not "electromagnets," which was a different category. The most common reason given was the presence of spinning or motion. For instance, one student said, "I would choose this one [EC] as the uh, the most different simply because there's no motion involved." The second most common reason had to do with magnets, such as the student who stated, "I think this one [PMB] helped me a lot because um I didn't really

see where the magnets were and so once I saw that these were magnets and it was causing it to spin I could see the magnets were being produced here [coils in blender]." We labeled the categories as to whether they referred to structural or functional connections and found nearly equal references to both.

TABLE 2: Reasons for Similar/Different (S= Structure; F= Function)

Reason	S or F	Concept-based (N=4)	Algebra-based (N=3)	Calculus-based (N=4)
Same pieces	S	2	3	0
Coils	S	1	2	1
"Motor"	S	0	1	4
Spinning	F	4	2	1
Magnets	F	2	1	3
Electromagnet	F	0	0	1
Works same	F	1	1	0

The only obvious difference between groups was that all of the calculus-based physics students cited the presence of a "motor" (by which they meant rotor) as a reason for choosing an activity as the most or least similar to the blender.

Most Useful or Least Useful

The students were again divided on the topic of which activity was the most useful. Four students stated that the PMB was most useful, while three chose the EMB and another four could not decide between the two.

Many of the students (5 out of 12) stated that they would remove the EC from the sequence of activities. Three more students said they would leave out both the EC and the RG. One student said he would choose not to show the EC used as a generator and three stated that they would not remove any of the activities.

The most common reason given for eliminating the EC was that the knowledge it helped them gain was repeated or redundant. For example, one student stated, "It just seemed like it was added information that I already learned from all this [previous activities]." Interestingly, this reason was given even by some students who needed the EC to recognize the electromagnets in the EMB or the blender.

Another common reason, given by four students, for choosing an activity as most or least useful again had to do with magnets. For example, one student said, "This one [PMB] was useful just in knowing that, okay, they're blatant magnets here." There were no obvious differences between the groups in choosing which activities were most or least useful.

All but three students chose to alter the order of the activities in some way. Table 3 below shows the various orders suggested by students, as well as the frequency with which they were suggested. The most common reason for changing the sequencing was that the EC would assist the transition from the PMB to the EMB, as suggested by one student who said, "I kind of see how the transition from the magnet to the coil wires would work."

TABLE 3: Preferred Sequence of Activities

Sequence	Frequency
RG, PMB, EC, EMB	4
RG, EC, PMB, EMB	1
PMB, EMB	3
RG, PMB, EMB	2
Do not change	2

CONCLUSIONS

After completing the sequence of scaffolding activities in a teaching interview setting, the most commonly given mechanism for creation of the magnets or magnetic field in the blender was current. The most commonly given mechanism for the creation of spinning was the magnetic field. Both of these responses were given most frequently by students enrolled in the calculus-based introductory physics course. There was no evidence that students knew the meaning of these terms or their function in a blender.

The largest variety of keywords was used by students in the algebra-based course. Students in the calculus-based course were more likely to use the terms "current" and "source" or "supply," probably because they had covered the material related to these terms recently in their course. On the other hand, students in the conceptual-based course had not yet covered this material and were more likely to use common terms such as "electricity" and "charge." The term "magnetism" was used by all but one student.

In deciding which activity was either most similar to the blender or most helpful in understanding the blender, students tended to focus on the presence of spinning and magnets in the activity. If we look again at Table 2, we see that students in the conceptual and algebra-based courses were more likely to focus on structural or low-level functional (i.e. spinning) similarities and differences. Finally, most students chose to alter the sequence of activities, moving EC earlier in the sequence or removing it entirely. Students commonly reasoned that this change was necessary because the EC provided redundant information as the final activity.

LIMITATIONS & FUTURE WORK

Due to the small size of this sample, we should be cautious in drawing generalizations from the data presented. In the future, we plan to interview more students to see if the observed trends still hold.

In this paper we have focused only on analyzing students' responses after they have completed the entire sequence of activities. Our next step will be to analyze students' ideas as they progress through each activity and to map out their learning trajectories. Additionally, we will consider students' suggestions gathered from our data and implement some of them, such as the preferred sequence of activities, and examine the effect on students' learning trajectories. Finally, we will also connect the learning activities with material already covered in introductory physics courses to determine where these learning experiences should optimally be placed to help students learn.

ACKNOWLEGEMENT

This work is supported in part by NSF grant REC-0133621. Any opinions, findings, and conclusions or recommendations expressed in this material are those of the authors and do not necessarily reflect the views of the National Science Foundation.

REFERENCES

1. J.J. Haynicz, P.R. Fletcher and N.S. Rebello. "College Students' Ideas About Some Everyday Electrical Devices" in *National Association for Research in Science Teaching Annual Meeting Proceedings*, San Francisco, CA, 2006..
2. P.R. Pintrich and D. Schunk in. *Motivation in Education: Theory, Research and Application* by Merrill Prentice-Hall, Columbus, OH, 1996..
3. J.E. Lobato, *Educational Researcher* **32**(1) 17-20, 2003.
4. D. Hammer and A. Elby, "On the form of a personal epistemology," in *Personal Epistemology: The Psychology of Beliefs about Knowledge and Knowing*, edited by P.R. Pintrich and B.K. Hofer, Lawrence Erlbaum: Mahwah, N.J. 2002, pp. 169-190.
5. J.G. Greeno, J.L. Moore and D.R. Smith, "Transfer of situated learning," in *Transfer on Trial: Intelligence, cognition and instruction*, edited by D.K. Detterman and R.J. Sternberg,Ablex: Norwood, NJ, 1993, p. 99-167.
6. L.S. Vygotsky, *Mind in Society: The Development of Higher Psychological Processes* by Harvard University Press, Cambridge, MA, 1978.
7. P.V. Engelhardt, et al., "The Teaching Experiment - What it is and what it isn't," in *Physics Education Research Conference Proceedings*, Madison, WI, 2003
8. F. Marton, *Journal of Thought*. **21**, 29-39 (1986).

Promoting Instructional Change in New Faculty: An Evaluation of the Physics and Astronomy New Faculty Workshop

Charles Henderson

Department of Physics, Western Michigan University, Kalamazoo, MI, 49008, USA

Abstract. An important finding of Physics Education Research (PER) is that traditional, transmission-based instructional approaches are generally not effective in promoting meaningful student learning. Instead, PER advocates that physics be taught using more interactive instructional methods. Although the research base and corresponding pedagogies and strategies are well-documented and widely available to physics faculty, widespread change in physics teaching at the college level has yet to occur. Since 1996, the Workshop for New Physics and Astronomy Faculty has been working to address this problem. This workshop, jointly administered by the American Association of Physics Teachers, the American Astronomical Society, and the American Physical Society with funding from the National Science Foundation, has attracted approximately 25% of all new physics and astronomy faculty each year to an intensive 4-day workshop designed to introduce new faculty to PER-based instructional ideas and materials. This paper describes the impact of the New Faculty Workshop as measured by web-based surveys of 527 workshop participants and 206 physics and astronomy department chairs. Results indicate that the NFW is quite successful in meeting its goals and that it may be significantly contributing to the spread and acceptance of PER and PER-based instructional ideas and materials.

Keywords: Dissemination, Educational Change, Higher Education
PACS: 01.40.Fk, 01.40.Jp

INTRODUCTION

There are approximately 9000 full-time equivalent physics faculty at 797 degree-granting physics and astronomy departments in the United States [1]. Roughly 300 new physics faculty are hired into tenure-track positions at the assistant professor level in a typical year [1-4]. These new faculty often have little preparation for their roles as teachers and frequently struggle with their teaching responsibilities. Thus, an appropriately developed program for new physics faculty has the potential to help support new faculty in their teaching while at the same time promoting the spread of instructional strategies and materials based on Physics Education Research (PER).

The Physics and Astronomy New Faculty Workshop (NFW) brings together physics and astronomy faculty in their first few years of a tenure-track faculty appointment at a four-year college or university to a 4-day workshop at the American Center for Physics in College Park, MD. During the workshop participants are introduced to some of the basic findings of PER as well as specific instructional strategies and materials based on these findings.

Presentations are made by leading curriculum developers and PER researchers.

The primary goals of the NFW are to:

1. reach a large fraction of the physics and astronomy faculty in tenure-track appointments prior to their receiving tenure;

2. help participants develop knowledge about recent developments in physics pedagogy and the assessment of changes in pedagogy; and

3. have participants integrate workshop ideas and materials into their classrooms in a way that has a positive impact on their students and their departments.

DESCRIPTION OF THE NFW

Attendees at the NFW are nominated by their department chair and the only cost to the department is transportation to College Park, MD. The workshop runs from approximately 4:00 pm on a Thursday to noon the following Sunday and contains roughly 12 hours of programming each full day. Presentations and discussions include a mix of large group sessions and small group sessions. It is important to note that the workshop presenters are leading and well-

CP951, *2007 Physics Education Research Conference*, edited by L. Hsu, C. Henderson, and L. McCullough
© 2007 American Institute of Physics 978-0-7354-0465-6/07/$23.00

respected curriculum developers since research suggests that the reputation of the reformer and/or their institution can have an important impact on how a reform message is received [5].

NFW Attendees

During its first 11 years in operation, 759 faculty have participated in the NFW. This represents roughly 25% of the assistant professor faculty hires during these years [1-4].

NFW participants represent 344 distinct colleges and universities. This is 43% of the 797 degree-granting physics and astronomy departments in the US [1]. In addition, 170 departments (21% of all departments in the US) have had more than one faculty member attend the NFW.

Cost

The NFW was funded with two grants from the National Science Foundation (NSF #0121384 and NSF #9554738). A total amount of $742,000 was spent during the first 11 years reported on in this paper [6]. This works out to a cost per participant of $978 (not including transportation cost to College Park, MD).

DATA COLLECTED

During spring 2007, a web survey was administered to the 690 NFW participants who could be located and who were still in academia. Of these, 527 (76%) completed the web survey before analysis began. Also, during spring 2007, a web survey was administered to all 794 US physics and astronomy department chairs using an email list provided by the American Institute of Physics. The survey was completed by 206 department chairs (26%). Approximately 53% of survey respondents reported having a faculty member from their department attend the NFW.

RESULTS

The NFW improves participants' knowledge of PER-based teaching techniques and interests participants in trying these techniques. Table 1 shows that most NFW participants indicate familiarity with the specific PER-based approaches discussed at the NFW after the workshop. In addition, nearly all (93.7%) of the NFW participants report being interested in incorporating some of the workshop ideas into their teaching right after the workshop. This suggests that they had formed a positive opinion of the instructional techniques presented at the workshop.

TABLE 1. Summary of participant responses to web survey question #17.

17. Please rate the following:

	I currently use it	I have used it in the past	I am familiar with it but have never used it	Little or no Knowledge
Astronomy Tutorials	8.7%	5.0%	30.2%	**56.1%**
Collaborative Learning	**39.2**	17.2	23.0	20.6
Cooperative Group Problem Solving	**47.2**	21.9	22.9	8.0
Interactive Lecture Demonstrations	**46.1**	24.2	23.4	6.3
Just-In-Time Teaching	22.9	18.0	**50.9**	8.2
Peer Instruction	**54.1**	21.4	22.4	2.1
Realtime Physics	5.2	7.5	**46.6**	40.7
Personal Response Systems	32.6	15.0	**43.7**	8.7
Physlets	19.7	21.4	**41.3**	17.5
Tutorials in Introductory Physics	13.1	20.9	**45.8**	20.3

Participants report instructional changes to more alternative modes of instruction. Department chairs agree. Figure 1 shows participant responses to survey questions 12 (rating of teaching style prior to NFW) and 13 (rating of current teaching style). This shows that there was a large shift to more alternative teaching styles after the NFW. Only 1% of participants rate their post-NFW teaching style as highly traditional.

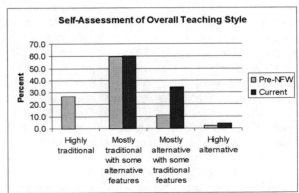

FIGURE 1. Participant self-assessment of their overall teaching style.

In addition, 70.7% of participants rate their teaching style as more alternative than other faculty in their department, and only 4.0% rate their teaching style as more traditional than other faculty in their department. The department chair survey corroborates the participant self-report data -- 58.6% of department chairs rate NFW participants as having a more alternative teaching style than other faculty in the department and only 2.9% of department chairs rate NFW participants as having a more traditional teaching style than other faculty in the department.

Finally, 32.1% of participants report a "considerable" or "full" change in their teaching style since participating in the NFW. An additional 64.4% reported some change in their teaching style. Only 3.5% reported no change in their teaching style. Department chairs again corroborate the participant self-report data. Most department chairs (72.4%) report that NFW participants have made changes to their teaching as a result of the workshop. It is reasonable that this percentage is smaller than the percentage of faculty who reported making a change in their teaching since a department chair is unlikely to be aware of all changes made in the teaching practices of faculty in their department.

Participants report improved student learning. Department chairs agree. Most participants (64.7%) believe that the NFW has had a considerable or larger positive impact on their students and only 1.3% say that the NFW has not had a positive impact on their students. Similarly, most department chairs (72.6%) who have sent faculty to the NFW indicate that the workshop has led to improved student learning in classes taught by workshop participants.

Faculty report discussing NFW ideas with their colleagues and that some colleagues have made changes as a result of these discussions. Department chairs agree. If one of the ultimate goals of the NFW is to change the culture of physics teaching in the US, it is not enough for NFW participants to just learn about PER-based instructional materials and strategies and make changes to their own teaching. They must also bring these ideas into interactions with colleagues in their home departments. There is evidence that this has taken place. Most participants (86.8%) say that they have discussed NFW ideas with their colleagues. Based on their written descriptions, these discussions often occur as a result of pre-tenure teaching observations made by colleagues or during formal presentations at a department colloquium or faculty meeting.

Many NFW participants (39.8%) report that their colleagues have made changes in their teaching as a result of these discussions. Many department chairs (51.0%) also believe that NFW attendees have influenced other faculty in the department.

THE DANGER OF SELF-REPORTED DATA

Table 1 shows that participants report high levels of use of one or more PER-based instructional strategy. One danger of the use of such self-report data is that a respondent may say that they are using a particular instructional strategy when an outside observer would think otherwise. One way to estimate the degree of over-reported use is to compare the general statements of instructional style made on Table 1 with more detailed descriptions of instructional activities. In one part of the web survey, NFW participants were asked to select a particular class that they had taught frequently since the NFW and to identify the frequency with which they engaged in particular instructional activities. Table 2 shows the responses for the 192 participants (36.4%) who said that they used Peer Instruction on question #17 (Table 1) and also opted to report on the details of an introductory level class that they had taught. These instructors report instructional patterns quite different from traditional instruction, which would rarely, if ever, involve students working on quantitative or qualitative problems nor would students ever engage in pair or small group discussions. But, to what extent are the reported instructional styles indicative of Peer Instruction?

TABLE 2. Summary of self-described instructional activities of NFW participants who are teaching an introductory course and indicated that they currently use Peer Instruction.

22. During the most recent time you taught the course, over the semester or quarter, how frequently did/do you use the following teaching strategies during the lecture portion of your course?

	Never	Once or twice per semester	Several times per semester	Weekly	Nearly every class	Multiple times every class
Instructor solves/discusses quantitative problem	3%	5%	13%	33%	**36%**	10%
Instructor solves/discusses qualitative problem	4	2	10	25	37	24
Students solve/discuss quantitative problem	9	9	16	**33**	23	10
Students solve/discuss qualitative problem	3	1	12	25	**33**	27
Pair or small group discussion	4	2	15	24	25	**30**
Instructor questions answered simultaneously by entire class	8	2	8	15	26	**40**

Three of the instructional activities are particularly relevant for the non-traditional aspects of Peer Instruction: "Students solve/discuss qualitative problem", "Pair or small group discussion", and "Instructor questions answered simultaneously by entire class". In Peer Instruction, each 1-hour class session involves three or four lecture-ConcepTest segments [7]. During each segment the instructor first lectures on a particular topic. Students then work individually on a multiple-choice ConceptTest (a qualitative problem) followed by a pair or small group discussion with nearby classmates. Finally, the students simultaneously report on their answers using a show of hands, flash cards, or a classroom response system (i.e., clickers). Thus, we would expect faculty engaging in the "pure" form of Peer Instruction to report each of the three relevant activities occurring multiple times each class. This was true for only 37 (19%) of the subsample of 192 participants. If the criteria is loosened a bit to include faculty who report each of these three activities "nearly every class", the number increases to 73 (38%).

Thus, only 19%-38% of faculty who say they are using Peer Instruction report instructional activities that could be consistent with Peer Instruction. A possible explanation for this difference, consistent with other available evidence, is that many of these faculty have not used Peer Instruction "as is", but rather have made significant changes based on some of the ideas of Peer Instruction [8]. Henderson and Dancy call this mode of operation reinvention and, in a small qualitative study, found that this was the most common way that faculty made use of developed curricula [9]. There is not enough evidence available in the survey data to judge whether the reinventions of Peer Instruction engaged in by NFW participants are likely to be productive or not.

CONCLUSIONS

Evidence presented suggests that the NFW has been effective in meeting its goals of introducing new faculty to PER-based ideas and materials and motivating faculty to try these ideas and materials. There is some evidence that NFW participants have also had an influence on other faculty in their departments. Thus, the NFW appears to be contributing significantly to the spread of PER ideas.

The apparent success of the NFW appears to be its ability to give faculty an introduction to PER-based instructional strategies and materials and to motivate many of them to continue to work on instructional improvement after the NFW. Thus, this study suggests that, at least under certain conditions, a relatively short, one-time, transmission-based professional development program can successfully spread research-based ideas and curriculum into higher education. Keys to the success of the program may be that: 1) it is sponsored and run by three major disciplinary organizations; 2) it introduces participants to a wide variety of PER-based instructional strategies and materials; and 3) presentations are made by the leading curriculum developers in PER. Other disciplines may find it useful to implement similar disciplinary-based programs for new faculty. Such disciplinary-based models may also be appropriate for experienced faculty.

ACKNOWLEDGEMENTS

The author would like to acknowledge the NFW Organizers and Advisory Committee who have made the NFW possible: Susana E. Deustua, Warren Hein, Robert Hilborn, Theodore Hodapp, Bernard Khoury, Kenneth Krane, Tim McKay, Laurie McNeil, and Steven Turley.

REFERENCES

1. R. Ivie, S. Guo, and A. Carr, *2004 physics & astronomy academic workforce,* American Institute of Physics, College Park, MD, 2005.
2. R. Ivie and K. Stowe, *1997-98 academic workforce report,* American Institute of Physics, College Park, MD, 1999.
3. R. Ivie, K. Stowe, and R. Czujko, *2000 physics academic workforce report,* American Institute of Physics, College Park, MD, 2001.
4. R. Ivie, K. Stowe, and K. Nies, *2002 physics academic workforce report,* American Institute of Physics, College Park, MD, 2003.
5. J. Foertsch, S. B. Millar, L. Squire, and R. Gunter, *Persuading professors: A study of the dissemination of educational reform in research institutions,* University of Wisconsin-Madison, LEAD Center, Madison, 1997.
6. Personal communication with Warren Hein, June 11, 2007.
7. E. Mazur, *Peer instruction: A user's manual,* Prentice Hall, Upper Saddle River, New Jersey, 1997.
8. C. Henderson and M. Dancy, When one instructor's interactive classroom activity is another's lecture: Communication difficulties between faculty and educational researchers, Paper presented at the American Association of Physics Teachers Winter Meeting, Albuquerque, NM, 2005.
9. C. Henderson and M. Dancy, Physics faculty and educational researchers: Divergent expectations as barriers to the diffusion of innovations, submitted.

Explanatory Framework for Popular Physics Lectures

Shulamit Kapon, Uri Ganiel, and Bat Sheva Eylon

Department of Science Teaching, Weizmann Institute of Science, Rehovot 76100, Israel

Abstract. Popular physics lectures provide a 'translation' that bridges the gap between the specialized knowledge that formal scientific content is based on, and the audience's informal prior knowledge. This paper presents an overview of a grounded theory explanatory framework for Translated Scientific Explanations (TSE) in such lectures, focusing on one of its aspects, the conceptual blending cluster. The framework is derived from a comparative study of three exemplary popular physics lectures from two perspectives: the explanations in the lecture (as artifacts), and the design of the explanation from the lecturer's point of view. The framework consists of four clusters of categories: 1. Conceptual blending (e.g. metaphor). 2. Story (e.g. narrative). 3. Content (e.g. selection of level). 4. Knowledge organization (e.g. structure). The framework shows how the lecturers customized the content of the presentation to the audience's knowledge. Lecture profiles based upon this framework can serve as guides for utilizing popular physics lectures when teaching contemporary physics to learners lacking the necessary science background. These features are demonstrated through the conceptual blending cluster.

Keywords: explanations, popular science, lectures, physics education, contemporary physics, informal science education, conceptual blending.
PACS: 01.40.Fk

INTRODUCTION

The following analysis is part of a study which explores the features, design, and learning outcomes of explanations in popular physics lectures. These scientific explanations are interesting since they provide a 'translation' that bridges the gap between the specialized knowledge that formal scientific content is based on, and the audience's informal prior knowledge. We term these explanations TSEs, Translated Scientific Explanations, as they translate the scientific findings which emerge from the context of specialized knowledge and vocabulary into lay language without corrupting their meaning. The noun 'translation' has been used explicitly when referring to popular scientific writing [1].

The study involved a comparative examination of three exemplary popular physics lectures that were given by practicing physicists known to be excellent popular lecturers. The lectures are explored from three perspectives: the explanations in the lecture (as artifacts), the design of the explanation from the lecturer's point of view, and the learning that takes place by the audience. An explanatory framework for popular physics lectures, based upon the first two perspectives was developed. The paper presents the framework briefly and focuses on one of its main

features: a cluster of categories that is referred to as the conceptual blending cluster.

The process of conceptual blending, or blending, is a fundamental cognitive process, characterizing (as a prerequisite) many cognitive phenomena, including categorization, generation of analogies and metaphors. Conceptual blending processes yield a new conceptual structure through composition, completion, and elaboration of inputs and existing schemes. Such a new conceptual structure is original as it is not present in any of the inputs [2].

Good popular physics lectures can be used as an instructional source of contemporary scientific ideas for audiences that lack a formal scientific background (e.g. high school students). The TSE explanatory framework may be used to create a lecture profile that can highlight the need for mediation tools and help better incorporate contemporary science through these lectures into high school instruction and learning [3]. The analysis of exemplary lectures can also enrich our understanding of how to craft explanations, especially abstract and complex ideas, where the students cannot have the entire necessary background.

Previous studies of scientific lectures in general aimed to provide a list of principles for the good lecturer (e.g. [4]). However, the explanatory view, especially in a popular scientific lecture, still needs to be explored. Physics provides a major challenge for

CP951, *2007 Physics Education Research Conference*, edited by L. Hsu, C. Henderson, and L. McCullough
© 2007 American Institute of Physics 978-0-7354-0465-6/07/$23.00

popularization, as it is the most hierarchical of the natural sciences. The fact that physics progress and achievements during the last 100 years are hardly ever included in most high school curricula may be attributed to this feature.

METHODOLOGY

The data sources included three videotaped exemplary popular physics from different domains of physics, and three stimulated-recall interviews [5] with the lecturers. The lectures were selected using criteria that considered content, explanations, and lecturer.

The analysis was carried out using a grounded theory approach resulting in a hierarchical theoretical categorization induced from the data [5] [6], in the following manner: Lectures were segmented thematically (content) and structurally (e.g. analogy/organizational means, etc). Analysis unit varied from 20 seconds to 3 minutes. Interviews were studied through a mapping analysis [5]. An iterative comparison of the analysis of the lectures and the interviews, and the relevant ideas from the literature (e.g. research on analogies) yielded a grounded theory explanatory framework of TSE's in popular physics lectures.

RESULTS AND DISCUSSION

A grounded theory explanatory framework emerged from the analysis. This framework demonstrates that TSEs are derived from four clusters of categories. The categories connect the emerging theory (category definition) to the data (all the instances for this category in each lecture). The following four clusters of categories were common to all three lectures:

1. Conceptual blending - Explanatory elements that explain the novel in terms of the known. A characteristic category is the metaphor.
2. Story - Elements that deliver scientific ideas through means that are common in literature (fiction). A characteristic category is the narrative.
3. Content - Elements that reflect a judicious choice of content: what to include, what to omit, and means to achieve this goal. A characteristic category is the selection of topics.
4. Knowledge organization - Elements that manage knowledge. A characteristic category is structure.

Table 1 defines each category of the conceptual blending cluster, and presents a list of the number of instances of each category in every lecture. Examples are presented below.

TABLE 1. Definition of Categories Forming the Conceptual Blending Cluster, and the Number of Various Instances of Each Category in Every Lecture: Q= Quantum Physics, P= Particle Physics, A= Astrophysics.

Category	Definition	Q	P	A
Positive analogy	A systematic mapping between two situations: target (the novel situation) and source (the familiar one). The mapping is governed by causal, mathematical, and functional interrelations [7] that the two situations share [8].	2	1	5
Negative analogy	The analogy consists in properties or relations that the source and target do not share [8].	2	-	1
Bridging Analogy	Intermediate analogy between source and target [9].	1	-	-
Visual Analogy	Graphical resemblance between analogical elements, where a transfer of the relations occurs in similar visualizations (source-target) [10].	1	2	3
Metaphor	Structuring one concept in terms of another. Unlike analogies, metaphors do not necessarily map source-target directly; similarities can be associative [11].	2	-	2
Category Extension	Categorization is a classification and labeling of things into groups on the basis of their properties [12]. Extending a category means enrichment of its properties while retaining its original characteristics [2].	1	3	-

To illustrate some of the categories forming the conceptual blending cluster, we present examples from one lecture - "Quantum mechanics in a nutshell" [13]. The lecturer tried to explain that a quantum particle evolves through many parallel histories (can be in many places simultaneously), and its behavior at a given time is the sum of these histories. This peculiar behavior takes place under very restrictive conditions: the quantum particle cannot "leave a mark" (a proof of its passing) in any of the parallel paths. The lecturer

stressed that this peculiar behavior is documented experimentally (the Aharonov Bohm effect was presented).

The lecturer said that although electrons move in a conductor like cars move on a road, a car can never be on two roads at the same time while an electron can, under the right conditions (small dimensions, low temperature, and "no mark left" – as in the "Aharonov Bohm" device). This is a negative analogy (Table 1). There is a direct mapping between source and target but the correspondence is between opposite relations (the ability to be in two places in the same time vs. not being able to do that).

This mapping was also cued graphically as depicted in Figure 1, which shows some of the illustrations that were used in this context. Figure 1 is an example of a visual analogy (Table 1). The graphic resemblance leads to a transfer of relations.

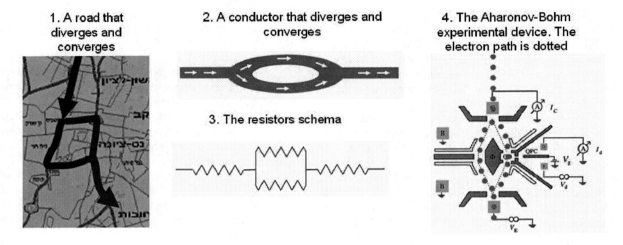

FIGURE 1. Visual Analogy – Excerpts from the Quantum Mechanics Lecturer's PowerPoint Presentation [13].

What about the phrase "leave a mark"? A child with dirty shoes leaves marks on the floor... This is a metaphor (Table 1) because one concept was structured in terms of another.

The lecturer presented sound waves as a known example of a wave that propagates through many paths simultaneously, stressing that the behavior of sound at a given point (e.g. our ear...) is the sum of all these "parallel histories". Immediately afterwards he made a category extension (Table 1): He considered particles as waves, while stressing the similarities under the necessary conditions.

The lecturers regarded all the conceptual blending elements as 'analogies'. All of them stressed that these 'analogies' are the most important explanatory mechanism in popular scientific explanations and the most challenging and time consuming element to craft: *"It can take days, but I know I must have an analogy there, something that will connect them to what I'm saying. Sometimes I drop a whole subject if I do not find the right analogy...people won't understand me, and it will be just empty words"* (interview).

Table 1 shows that conceptual blending elements were present in all three lectures. However, the degree of usage differed, and a lecture profile could be created. The content of the quantum mechanics lecture was the most distant from the audience's background and the most abstract one. This lecture was the richest in conceptual blending different category instances. The particle physics [14] lecture presented a historical and phenomenological survey and its content was the least abstract of all the three lectures. This lecture used only a small number of conceptual blending elements. We think that when the degree of abstraction of the subject is high the use of blending categories becomes more frequent. However, this claim requires further investigation. One could argue that this difference is due to the lecturer's style. However, the content of popular lectures that are delivered by practicing physicists is usually connected in some way to the lecturer's expertise. Clearly the lecturer's style plays its part, but one cannot exclude the influence of the subject that is presented.

The literature [2], [7], [11] suggests that conceptual blending categories provide a way to understand an unfamiliar or novel idea through the use of terms and inner connections of familiar situations or processes. This is achieved through an appeal to the following: a. visual resemblance, as in the visual analogy category; b. proximity of meaning, as in the category extension category; c. proximity of process, as in some of the positive analogy, negative analogy, and metaphor categories, especially in the instances where the source and the target are distant. This cluster's categories therefore expand available prior knowledge resources.

However, can listeners indeed decipher the information that lecturers provide through conceptual blending? Our findings (to be published elsewhere) show that often they do not. This is in accordance with the literature. For example, work on instructional use of analogies (e.g. [15]) suggests that to augment learning from popular lectures, mediation is needed.

Distant source-target instances can create misconceptions due to inappropriate mapping and deduction. This may suggest that explicit mapping and a definition of constraints may be necessary when the lecture profile shows high use of conceptual blending categories.

Although the conceptual blending elements are important, they do not stand alone. These explanatory elements are merged and enhanced by the usage of explanatory elements from the other clusters. Their impact also derives from this merger. One example is the effect of merging the narrative category of the story cluster with the positive analogy category as found in the astrophysics lecture. We classified an instance as a narrative if we could identify a narrator, who tells us about a protagonist involved in an event that happened in the past, and serves as main explanatory tool [16].

An important idea that was delivered in the astrophysics lecture was that star evolution is the underlying source of the diversity of atoms. The lecturer explained the evolution of stars through a positive analogy between the evolution of stars and the evolution of humans. This beautiful analogy was presented through a narrative about an alien that comes to earth, looks at the people, and tries to figure out how to describe them to his master (quoting from the lecture [17]): *"'I have come to earth and I have seen 125 people all different'. He [the alien] is smart enough to understand that the younger guy is the same as the older guy. In time, the little guy will get older and become like the older guy. It might not be the same object, the same person, but they evolve. This is an evolution. The same is true for stars: When we look at the sky we see different types of stars. We can recognize that some of them are young, and some of them are older, even if they are not the same objects. Just as you [audience] are not the same human beings".* We characterize this narrative as having an explanatory agent attribute because it 'carried' the analogy stars ~ humans. It created a 'time for telling' [18] which enhanced the impact of the analogy.

CONCLUSIONS AND IMPLICATIONS

Conceptual blending is one of the four clusters of categories from which TSEs, Translated Scientific Explanation, in popular physics lectures are derived. The other three are Story, Content, and Knowledge organization. These clusters were found in all three lectures. The explanatory framework may suggest *how* the lecturer can fit the content of the presentation to his/her audience's prior knowledge when formal prior knowledge is lacking. This process was demonstrated in this paper through the use of the conceptual blending cluster, but it should be noted that all the other clusters take part in this process. A full description of the 'Translated Scientific Explanation' (TSE) explanatory framework is in preparation.

Differences among lectures emerge when the categories that make up each cluster are examined. The explanatory framework can be used to create a lecture profile. Such profiles could be used as a methodological design tool to guide the utilization of popular physics lectures to teach contemporary physics to learners who lack a sufficient academic science background (high school physics students for example). The lectures can be used as an instructional source, where their profiles suggest possible mediation tools.

REFERENCES

1. J. Turney, *Public Underst. Sci.* **13**, 331-341 (2004).
2. G. Fauconnier, and M. Turner, *The Way We Think: Conceptual Blending and the Mind's Hidden Complexities*, New York: Basic Books, 2002.
3. R. Millar and J. Osborne, *Beyond 2000 - Science Education for the Future*, London: King's College London, School of education, 1998.
4. N. Movshovitz-Hadar and O. Hazzan, *Int. J. Math. Educ. Sci. Technol.* **35**, 793-812 (2004).
5. A. Shkedi, *Multiple Case Narrative: a Qualitative Approach to Studying Multiple Populations*, Amsterdam: John Benjamins, 2005.
6. S. B. Merriam, *Qualitative research in practice: examples for discussion and analysis,* San Francisco: Jossey-Bass, 2002.
7. D. Gentner and K. J. Holyoak, *Am. Psychol.* **52**, 32-34 (1997).
8. M. Hesse, *Models and analogies in science*, New York: University of Notre Dame Press, 1966.
9. J. Clement, *J. Res. Sci. Teach.* **30**, 1241-1257 (1993).
10. D. L. Schwartz, *J. Res. Sci. Teach.* **30**, 1309-1325 (1993).
11. G. Lakoff and M. Johnson, *Metaphors We Live By*, Chicago: University of Chicago Press, 1980.
12. G. Lakoff, *Women, fire, and dangerous things: What categories reveal about the mind*, Chicago: University of Chicago Press, 1990.
13. A. Stern (A lecture given in the Amos de-Shalit popular science lectures series at the Weizmann institute of science, 2002).
14. H. Harari (A lecture given on the occasion of the 50th anniversary of the physics department of the Weizmann institute of science, 2004).
15. Z. R. Dagher, *Sci. Educ.* **79**, 295-312 (1995).
16. S. P. Norris, S. M. Guilbert, M. L. Smith, S. Hakimelahi, and L. M. Phillips, *Sci. Educ.* **89**, 535-563 (2005).
17. D. Zajfman, (A lecture given on the occasion of the 50th anniversary of the physics department of the Weizmann institute of science, 2004).
18. D. L. Schwartz and J. D. Bransford, *Cognition Instruct.* **16**, 475-522 (1998).

Research-based Practices For Effective Clicker Use

C. Keller[*,†], N. Finkelstein[*], K. Perkins[*], S. Pollock[*], C. Turpen[*], and M. Dubson[*]

[*]University of Colorado at Boulder
Department of Physics, Campus Box 390 Boulder, CO 80309
[†]i>clicker (Bedford, Freeman, and Worth Publishing)
33 Irving Place, 10th Floor
New York, NY 10003

Abstract. Adoption of clickers by faculty has spread campus-wide at the University of Colorado at Boulder from one introductory physics course in 2001 to 19 departments, 80 courses, and over 10,000 students. We study common pedagogical practices among faculty and attitudes and beliefs among student clicker-users across campus. We report data from online surveys given to both faculty and students in the Spring 2007 semester. Additionally, we report on correlations between student perceptions of clicker use and the ways in which this educational tool is used by faculty. These data suggest practices for effective clicker use that can serve as a guide for faculty who integrate this educational tool into their courses.

Keywords: Clickers, Personal Response Systems, Classroom Response Systems
PACS: 01.40.gb, 01.50.-i, 01.50.H-, 01.40.Fk

INTRODUCTION

Since being introduced six years ago in one introductory physics course at the University of Colorado at Boulder, clicker[1] use has spread extensively, with nearly half the undergraduate population using clickers in one semester. Although their use is becoming more prevalent, it is not known how this tool is used by faculty at the campus level, nor do we understand student perceptions and attitudes towards this tool.

Research on clickers remains a popular topic within PER and other science and education communities. (For example, see [1, 2, 3]; for an extensive literature review, see [4].) The purposes of the present study are to identify common faculty pedagogical practices regarding clicker use across the variety of disciplines at one institution, study student perceptions towards this tool, and look for correlations between faculty practices and student perceptions. We seek to identify effective clicker uses across these varieties of disciplines and environments in which they are employed. An ultimate goal will be to correlate

faculty practices with student learning—others have already demonstrated correlations between student attitudes and beliefs and content learning gains in other contexts (e.g., see [5] and [6]). However, studying student learning gains is beyond the scope of the present work. In this piece, we present limited results from an extensive study of dozens of faculty and thousands of students.

When correlating faculty practices with student attitudes, we find that the students' perception of the utility of clickers improves as faculty encourage peer-discussion and succeed in getting students to discuss with each other in lecture. Additionally, students find conceptual questions slightly more useful than factual recall or calculation-oriented questions.

INSTITUTIONAL USE OF CLICKERS

During the Spring 2007 semester at the University of Colorado at Boulder, clickers were used by 70 faculty in 94 lecture sections, with an average enrollment of 144 students in each lecture section. Although this breadth of use represents a tiny fraction of all faculty on campus (3%), it represents a significant fraction of the student body due to the high average enrollment of courses using clickers. In this semester, clickers were used by 10,011 unique

---

[1] We opt for the term "clicker," whereas others use "personal response system," "voting machine," and a myriad of other terms (see Reference [4]).

students, which include 9,941 undergraduates and 70 graduate students. Students using clickers made up 44% of all undergraduate students and 1.6% of graduate students. Despite the widespread use among the undergraduate student body, there is still opportunity for clicker use to expand. Only 28% of departments on campus are using clickers and only 24% of large lecture sections (where the enrollment is greater than 100) are using clickers. We see some departments that currently use clickers in all large lecture courses that they offer, such as Physics, Astrophysics, and Chemical Engineering. Other large departments, such as Psychology and Sociology, use clickers in 1 and 2 large courses (out of 16 and 7, respectively).

We find the majority of courses using clickers to be in STEM fields. The total number of courses that used clickers in STEM fields was 63, while there was 10 in Business, 6 in Social Sciences, and 1 in Humanities.

Of the 94 lecture sections[2] using clickers, 79 of these are using i>clicker[3] and the remaining 15 are using H-ITT[4]. Of all unique students who used clickers, 70% used clickers in one course only, while 29% used clickers in 2 courses, and fewer than 1% used clickers in either 3 or 4 courses.

FACULTY PRACTICES

Faculty using clickers were given two different online surveys. The first survey was given at the start of term approximately 3 weeks after the beginning of classes, and 54 faculty responded to 16 multiple-choice and long answer questions. The second online survey was given at the end of term, and 69 responses were collected to 15 multiple-choice and long answer questions. To access both surveys, see [4]. Questions from both surveys probed how faculty used clickers in their own courses and on their experience and beliefs surrounding clickers. The results presented in this section were collected from both faculty surveys. Observational data of classroom practices demonstrate some similar trends as seen in faculty self-reported data presented here for a subset of physics courses using clickers. [7]

The majority of faculty using clickers has little or no experience using this tool—59% of the respondents are using clickers for the first time or have only one prior semester of clicker experience.

Some slight variation does exist in how frequently faculty use clickers, in terms of the average number of

questions given per day and the overall percentage of class days when clickers are used throughout the semester (see Table 1 & 2). However, the majority of faculty asks 3 to 4 questions per day and use clickers 90 – 100% of class meetings over the semester.

TABLE 1. Average number of questions given per class meeting, reported by faculty

Number of Clicker Questions	% of Courses [standard error][†] N=69
1 to 2	19 [5]
3 to 4	65 [6]
5 to 6	13 [4]
7 or more	3 [2]

[†]Bracketed numbers in tables are estimated standard error of the mean.

TABLE 2. Percent of all classes days when clickers are used, reported by faculty

% of class days	% of Courses, N=69
< 50%	7 [3]
50 - 75%	4 [3]
75 - 90%	20 [5]
90 - 100%	68 [6]

In addition to reports of frequency of use, we examine how clickers are used. Figure 1 reports the frequency of use of several broad categories of clicker questions.

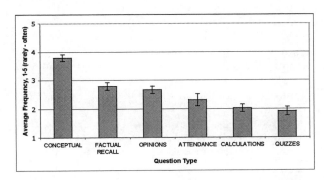

FIGURE 1. Types of clicker questions used, reported by faculty.

We see some variation in the extent to which faculty encourage discussion among their students, and the extent to which students do discuss with their peers in lecture (according to faculty). The majority of faculty claim to encourage discussion and claim to succeed at getting a large fraction of students to discuss in lecture (see Table 3).

---

[2] Note that many courses have multiple lecture sections; hence the difference between 80 courses and 94 lecture sections.
[3] http://www.iclicker.com
[4] http://www.h-itt.com

TABLE 3. Extent of peer-discussion, reported by faculty

Type of Discussion	% of Faculty
Do not allow discussion	3 [2]
Do not encourage discussion, & *small* fraction of students discuss	6 [3]
Do not encourage discussion, & *large* fraction of students discuss	6 [3]
Encourage discussion, & *small* fraction of students discuss	22 [5]
Encourage discussion, & *large* fraction of students discuss	63 [6]

STUDENT PERCEPTIONS & PRACTICES

Approximately one month prior to the end of term, an online survey was distributed to students in courses where clickers were currently being used. To access the student survey, see [4]. Of the 10,011 students using clickers, 3,697 responses were collected. The 11 multiple-choice questions on the survey probed students' attitudes and beliefs about clickers and asked them to respond to how clickers are currently being used in their courses. Of the 80 courses using clickers, data from students were collected in 51 courses. The data presented in this section are a summary of results from the student survey.

Overall, we find students' experience with clickers to be positive. 56.4% of students responding had a favorable attitude towards the *utility* of clickers in their respective courses (compared to 22.9% that were neutral and 20.7% that had unfavorable attitudes) and 55.3% of students had a favorable attitude towards the *enjoyment* of clickers (compared to 24.5% that were neutral and 20.2% that had unfavorable attitudes). These favorable results for clickers are noted elsewhere (for example, see [8]).

TABLE 4. Student discussion practices during clicker questions, reported by students

Student Practice	% of Students, N=3,697
Usually not allowed to talk with other students	2 [0.03]
Rarely use a clicker in this course	1 [0.01]
Guess the answer and do not check with other students	2 [0.03]
Actively think about the question independently and arrive at an answer without speaking or listening to other students	19 [0.3]
Listen to other students' answers and/or reasoning	18 [0.2]
Actively participate in discussions with other students around me	59 [0.4]

When asked what they normally do during the delivery of a clicker question, most students claim to be actively participating in discussions with their peers (see Table 4).

The utility of different types of clicker questions was rated. Students found conceptual questions the most useful (3.92 ± 0.02, on a scale of 1–5, negative to positive), followed by factual recall (3.51 ± 0.02), and numerical calculations (3.32 ± 0.02).

CORRELATIONS BETWEEN FACULTY PRACTICES & STUDENT ATTITUDES

The final goal of the present work is to study what impact faculty practices regarding clickers have on student attitudes. We begin to determine these relations by correlating faculty use of this tool with student perception of clickers.

It is commonly argued that encouraging discussion among students is of greater benefit than passively using clickers in lecture [2]. We find there to be a strong relationship between the extent of peer-discussion in lecture and students' attitude towards the utility of clickers. Of course, students do not uniformly agree within a single course when asked to what extent their instructor encourages discussion. Taking the mode of student responses to be an accurate representation of how instructors encourage student discussion, the average fraction of students with a favorable attitude towards the utility of clickers is 66% in courses where instructors encourage discussion and get a large fraction of students to do so (see Table 5).

TABLE 5. Average percent of students who have favorable attitudes towards clicker use, listed by classes where plurality of students reported use of clickers in stated fashion.

Type of Discussion	% Favorable
Does not allow discussion	38 [5]
Does not encourage discussion, & *small* fraction of students discuss	37 [8]
Does not encourage discussion, & *large* fraction of students discuss	36 [22]
Encourages discussion, & *small* fraction of students discuss	55 [8]
Encourages discussion, & *large* fraction of students discuss	66 [3]

Similarly, we examine how the role of students during a question correlates with their perceived utility of the clickers. We see a trend toward students finding clickers more useful as they become more active in lecture, with 64% of the students who claim to be actively participating in discussion having a favorable attitude towards the utility of clickers (see Table 6).

We also see a strong correlation between the instructor's experience with clickers (i.e., the number of prior semesters where an instructors has taught with clickers) and student perception of utility (r=0.52), suggesting that faculty become better over time at effectively using this tool and that novice faculty may need more assistance.

In addition to students finding conceptual questions more useful than other types of questions, the fraction of students in a course who claim to be actively participating is correlated with the average student rating of utility of conceptual clicker questions within a course (r=0.43), suggesting that conceptual questions are most useful when students discuss the questions with their peers.

TABLE 6. Percent of students from each student role who have favorable attitudes toward the utility of clickers.

Student Role	% Favorable
Does not apply--usually not allowed to talk with other students	28 [5]
Rarely use a clicker in this course	16 [6]
Guess the answer and do not check with other students	17 [4]
Actively think about the question independently and arrive at an answer without speaking or listening to other students	51 [2]
Listen to other students' answers and/or reasoning	45 [2]
Actively participate in discussions with other students around me	64 [1]

CONCLUSION

This study presents self-reported faculty pedagogical practices and student perceptions of clicker use at a large research university across many disciplines. In addition to shedding new light on how this tool is being used by faculty and its corresponding perception by students, we wish to use these data to study how faculty practices impact student behaviors and views.

Student attitude is strongly impacted by the extent to which faculty encourage and succeed in generating peer-discussion during the administration of clicker questions. Students have a much more positive attitude towards the utility of clickers if faculty encourage discussion and are able to get a large fraction of students discussing. Likewise, student attitude is also improved when students are actively participating in discussions with their peers, as opposed to being passive or working independently.

We wish to move beyond providing discipline-specific and/or anecdotal suggestions with regard to the use of clickers. Rather, we seek to provide research-based evidence of their effective use. At this point, we can make two suggestions to faculty that are supported by this work: 1) encourage students to discuss with their peers during clicker questions and create environments that get students to discuss; and 2) ask conceptual questions appropriate for most students' level of knowledge.

Other research-based suggestions exist, but this is a topic for future work. For example, how do faculty practices differ between the highest and lowest rated courses, as measured by student perception of the utility of clickers? How useful to students are clickers compared to other course resources? How do varying methods of awarding credit for clicker use affect students' attitudes? Future work will seek to answer these questions to provide guidelines on the effective use of clickers to the expanding population of novice clicker-using instructors.

ACKNOWLEDGMENTS

This work has been supported by the NSF (DUE-CCLI 0410744 and REC CAREER# 0448176), AAPT/AIP/APS (Colorado PhysTEC program), University of Colorado, and i>clicker (Bedford, Freeman, and Worth Publishing). The authors express their gratitude to the faculty and students who participated in this study, in addition to Shaun Piazza and the rest of the Physics Education Research Group at the University of Colorado at Boulder.

REFERENCES

1. Banks, David A. (ed.) *Audience response systems in higher education: applications and cases.* Information Science Publishing, 2006.
2. E. Mazur, *Peer Instruction.* Prentice Hall, New Jersey, 1997.
3. Caldwell, J. CBE—Life Sciences Education, Vol. 6, Spring 2007.
4. http://www.colorado.edu/physics/EducationIssues/keller/clicker.html
5. N.D. Finkelstein and S.J. Pollock, Phys. Rev. ST Phys. Educ. Res. 1, 010101 (2005).
6. K.K. Perkins, W.K. Adams, N.D. Finkelstein, S.J. Pollock, and C.E. Wieman. "Correlating student attitudes with student learning using the Colorado Learning Attitudes about Science Survey." PERC Proceedings, 2004.
7. Turpen, C., Finkelstein, N., and Keller, C., "Understanding Faculty Use of Peer Instruction," PERC Proceedings 2007 (this volume).
8. d'Inverno, R., Davis, H., and White, S., "Using a personal response system for promoting student interaction." Teach. Math. Appl. 22(4), 163-169.

Expert and Novice Use of Multiple Representations During Physics Problem Solving

Patrick B. Kohl and Noah D. Finkelstein

Department of Physics, University of Colorado, Campus Box 390, Boulder, CO 80309

Abstract. It is generally believed that students should use multiple representations in solving certain physics problems. In this study, we interview expert and novice physicists as they solve two types of multiple representations problems: those in which multiple representations are provided for them, and those in which the students must construct their own representations. We analyze in detail the types of representations subjects use and the order and manner in which they are used. Somewhat surprisingly, both experts and novices make significant use of multiple representations. Some differences emerge: Expert use of multiple representations is more dense in time, and novices tend to move between the available representations more often. In addition, we find that an examination of multiple representation use alone is inadequate to fully characterize a problem-solving episode; one must also consider the purpose behind the use of the available representations.

INTRODUCTION

PER has a long history of research into problem solving, especially as it pertains to physics novices (see Maloney[1] or Hsu[2] for reviews). Much of this research has investigated novice (and sometimes expert) use of representations during problem-solving episodes. "Representations" is used broadly, and can refer to whether the problem solver is representing surface features of a problem or the deeper physics.[3,4] The term can also refer to the use of multiple representations together, such as pictures, free-body diagrams, and equations. Note that in this paper, we will use the term "representations" to refer to representations on paper, as opposed to mental representations. Much of the work on multiple representation use during problem solving has focused on how to promote this use during instruction.[5,6,7,8] Some prior work exists in PER on multiple-representations problem solving at the scale of single problems,[9,10] For instance, Larkin and Simon[11] find that expert representations tend to focus on "physical" features such as energy, while novices are more likely to attend to surface features (such as shape). Outside of PER, Kozma[12] has compared chemistry experts in the workplace to novices engaged in academic tasks.

A few significant questions remain regarding the differences between expert and novice use of representations during physics problem solving. Do the major differences lie in the kinds and frequencies of representation use, or do they lie in the patterns and sequences of representation use? We might also expect there be differences in *why* experts and novices use representations. Without a clear picture of how experts and novices differ in their approaches to multiple-representation problem solving, it will be difficult to properly bridge the expert/novice gap with instruction.

In our study, we begin to address these issues in detail. We interview ten introductory physics students and five graduate students, providing novice and expert samples. In these interviews, the subjects solve a variety of problems using multiple representations in the areas of kinematics and electrostatics. We then perform a fine-grained analysis of these interviews, with the goal of characterizing the major differences and similarities between the two samples in terms of success, frequency and sequence of representation use, and purpose towards which these representations are applied.

We present two main results in this paper. First, as one would expect, the experts solve their problems in less time, often using the same set of representations as novices but in a shorter time. Surprisingly, however, the novices interviewed do not show the reluctance to use a variety of representations that one might expect. Indeed, their solutions were often more complex in terms of number of translations between representations. Second, experts differ quite noticeably from novices regarding their applications of representations, with more careful analysis and self-checking, and less weakly-directed, unplanned work. These results together allow us to begin to paint a picture of what distinguishes a novice from an expert in terms of representation use, a picture likely to have consequences for instruction.

CP951, *2007 Physics Education Research Conference*, edited by L. Hsu, C. Henderson, and L. McCullough
© 2007 American Institute of Physics 978-0-7354-0465-6/07/$23.00

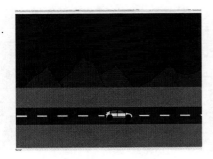
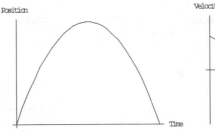
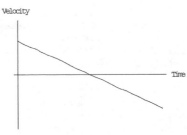

FIGURE 1. Four different representations of the motion of a car. The car problems studied required students to group these representations together as appropriate.

A car starts in motion and undergoes constant acceleration in the –x direction.

METHODS

Five of our interviewed novices were from CU's first-semester large-lecture introductory algebra-based physics course from the fall of 2005, which we refer to as Physics 201. The other five novices were drawn from the second semester of this course in the spring of 2006, Physics 202 (a course we have described previously[8]). Our five expert problem-solvers were physics graduate students, usually in the first year or two of their program. Interviews typically lasted about an hour.

Our problem-solving interviews used two types of multiple representations problems: those in which students were provided with a variety of representations, and those in which students had to generate additional representations on their own. Students from Physics 201 solved problems of the first type. In these, which we refer to as the car problems, students were given sets of representations of the motion of a car. This included a set of graphs of position versus time, a set of graphs of velocity versus time, a set of Flash animations depicting a moving car, and a set of written descriptions of a moving car. The students were instructed to make as many groups as possible of members from the various sets; that is, they were told to select position graphs, velocity graphs, animations, and written descriptions that all corresponded to one another. They were also told that not all members of each set would be used in all groups, and that it was possible to find groups of fewer than four elements. Figure 1 shows a sample grouping of two graphs, a written description of motion, and a screenshot from a flash movie.

The 202 novices solved the five electrostatics problems seen in Ref. 8. All of these problems involved numerically calculating either a force or a charge, usually with Coulomb's law. One explicitly required the production of a free-body diagram as part of the answer, while all five (especially the fifth, or challenge problem) were made easier by drawing a picture and/or an FBD. The challenge problem involved calculating the tension in strings holding two charged, attracting balls.

The experts solved all of the problems given to the 201 and 202 novices, as well as one problem designed to be challenging. The expert problem is the pulley problem used by Larkin in Ref. 3. We will reserve analysis of this pulley problem for a long-format paper. All of the problems are located on the web.[13]

All interviews were coded in two main ways. First, we coded representation use as a function of time. We divided the interview episode into ten-second blocks and noted which of the available representations (pictures, FBDs, written language, math, movies, position graphs, velocity graphs) students made use of or reference to in each block. It was possible for more than one representation to be present in each block of time.

Second, the parts of the interviews corresponding to the solutions of the five electrostatics problems were coded in an additional way. Here, we coded for the kinds of activities students engaged in, including reading, translation, analysis, exploration, planning, implementation, and verification. We adapted the activity categories and rubric from Schoenfeld,[14] with input from the authors and a member of the CU PER group unrelated to the project. We performed interrater reliability checks for all of the codings in this paper.

We also defined a complexity parameter describing each of the coded solutions. Briefly, this parameter reflects the number of translations between representations made by a subject during a problem-solving episode. For instance, a solution that begins with a written description, moves on to a picture, then a free-body diagram, and finally mathematics, would be

TABLE 1. (right) Performance of expert and novice students on the car problems, electrostatics (E1-4), and the challenge problem. Car problem figures show the average number of correct/incorrect representation groups created. Electrostatics figures show the number of students solving each problem correctly/incorrectly.

	Car	E. 1	E. 2	E. 3	E. 4	Challenge
Expert	7.6 / 0	4 / 1	4 / 1	5 / 0	5 / 0	5 / 0
Novice	5.0 / 1.0	3 / 2	3 / 2	1 / 4	0 / 5	1 / 4

TABLE 2. (right) Measures of the representational complexity (Comp) of student solutions and of the time density of their translations between representations (Dens, translations per minute), averaged across experts (E) and novices (N). Higher Comp numbers represent more translations between representations during a solution.

	Comp, E	Comp, N	Dens, E	Dens, N
Car Probs.	5.2	7.4	4.6	3.4
Elect. 1-3	7.2	6.4	2.1	1.3
Elect. 4	6.0	10.3	2.7	2.3
Challenge	11.8	15.8	2.4	1.5

assigned a complexity of 3 representing the three shifts in representation. Should the student return to a previous representation, that shift would increase the coded complexity. A small number indicates a simple, linear sequence of representations used, while a larger number represents a more complex and iterative approach.

DATA

There are many options available for characterizing the solutions studied here. In this paper, we focus on a small subset of these options for brevity.

In Table 1, we see expert and novice performance on the problems studied. As is expected, the experts outperformed the novices on all problems, though the novices are at least partly successful on all but electrostatics problem 4.

In Table 2, we characterize the patterns of representation use present in student solutions with two parameters. The "Comp" columns display the complexity parameter, averaged over all experts and all novices for the car problems, electrostatics problems 1-3, electrostatics problem 4, and the electrostatics challenge problem. The novices show a tendency (not absolute) towards more complex solutions, in the sense that they move back and forth between representations more often (if not as quickly) over the course of their solutions.

In the "Dens" columns of Table 2, we see the average number of translations between representations per minute made by experts and novices over the course of their solutions. Experts appear to be consistently moving more quickly between the representations available. Data on the number of representations used in total (not shown) suggests that both expert and novice problems solvers were using a wide variety of representations, both in the car problems in which this was required and in the electrostatics problems where it was not. Most novices and experts used at least a picture and equations in all electrostatic problem solutions, and some used free-body diagrams as well.

In Figure 2, we consider the purposes towards which the individuals applied the representations they used. Here, we focus on the electrostatics problems in which students had to generate their own problem representations. The graphs show the fraction of the codeable time intervals occupied by each of the activities listed. The biggest differences between experts and novices appear in the Reading, Analysis, and Exploration categories. Experts spend a greater fraction of their time reading the problem, though we believe this to be an artifact of their generally shorter solution times. More significant is the time spent in Exploration by the novices. Analysis represents a directed, systematic attempt to more fully understand the problem. Analysis generally is explicitly goal-oriented, trying to figure out some specific aspect of the problem. Reasoning does not have to be correct to be coded as analysis. Exploration represents less focused behavior, where the student is searching for options or trying things out with little direction or expectation of moving forward. It is generally not clear what they expect to find through their efforts. On average, the experts spent 43% of their time on Analysis and 1% of their time on Exploration, while the novices spent 25% of their time on Analysis, and 16% of their time on Exploration. Note that the difference between Analysis and Exploration was the major focus of interrater reliability checks.

DISCUSSION AND CONCLUSION

In this paper, we presented example data describing the differences between expert and novice problem solvers when handling multiple representations. Some of our results were expected: Experts were more successful in solving problems that required the use of multiple representations, finished faster, and moved more quickly between the representations available. However, we were surprised to find that these novice problem-solvers were just as likely to use multiple representations extensively in their solutions. Novice

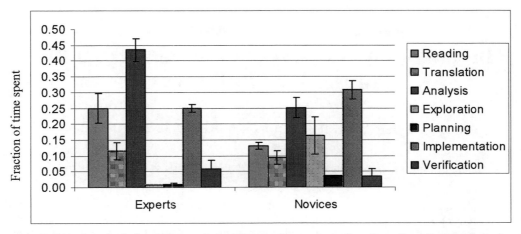

FIGURE 2. Average fraction of solution time spent in various ways by experts and novices. Error bars represent the standard error of the mean.

solutions also included on average more translations between the different representations at hand, matching our qualitative observation that students 'flailed about' when stuck, whereas the experts proceeded more systematically.

Over the course of this study, it appeared to us that considering only the number, kind, and correctness of the representations used would not fully characterize the differences between the experts and novices. In response, we have coded and analyzed the behaviors that our subjects engaged in while using these representations. The time spent in different activities was quite similar except in the categories of Analysis and Exploration. Experts spent much more of their time pursuing specific goals or sub-goals even when they did not know exactly how to proceed, as with the more challenging pulley problem not presented here. Novices still spent some time in this manner (which is encouraging), but were much more likely to engage in exploration with no clear purpose, perhaps hoping to strike a more correct approach.

The analysis presented here is preliminary, but suggests some key differences between expert and novice problem solvers. A number of studies (including refs. 5-8) have demonstrated success in getting novice physicists to use multiple representations, but usage alone is insufficient. We believe that careful characterization of how and why students use multiple representations will provide clearer paths for instruction, and suspect that attending to meta-level skill sets (like knowing what different representations are useful for) may result in more expert-like use of problem-solving time and greater success.

ACKNOWLEDGEMENTS

This work was supported in part by an NSF Graduate Fellowship, and an NSF CAREER grant. Special thanks to the PER group at CU-Boulder and to N. Podolefsky for aid in coding.

REFERENCES

1. Maloney, D.P., in *Handbook of Research on Science Teaching and Learning*, D. Gabel, Editor. MacMillan Publishing New York, NY. 1994.
2. Hsu, L., et al., *Am. J. Phys.*, **72**, 1147, 2004.
3. Larkin, J.H., in *Mental Models*, D. Gentner and A. Stevens, Editors. 1983, Lawrence Erlbaum: Mahwah, New Jersey.
4. Chi, M., P. Feltovich, and R. Glaser, Cognitive Science, **5**, 121. 1980.
5. Dufresne, R.J., R.J. Gerace, and W.J. Leonard, *Phys. Teach.*, **35**, 270. 1997.
6. Heller, P., R. Keith, and S. Anderson, *Am. J. Phys.*, 1992. **60**, 627. 1992.
7. Heuvelen, A.V. and X. Zou, *Am. J. Phys*, **69**, 184. 2001.
8. Kohl, P.B., D. Rosengrant, and N.D. Finkelstein, *Phys. Rev. ST Phys. Educ. Res.* **3**(1), 010108. 2007.
9. Rosengrant, D., E. Etkina, and A.V. Heuvelen, in *Proceedings of the 2005 PERC*. AIP Conference Proceedings. 2005.
10. Singh, C., *Am. J. Phys.*, **70**(11), 1103. 2002.
11. Larkin, J. et. al., *Science*, 208(20), 1335. 1980.
12. Kozma, R.B. and J. Russell, *J. Res. Sci. Teach.* **34**, 949. 1997.
13. http://per.colorado.edu/kohl/problems.doc
14. Schoenfeld, A.H. in *Cognitive science and mathematics education*, A.H. Schoenfeld, Editor. 189. Lawrence Erlbaum Associates: Hillsdale, NJ. 1987.

Investigating the Source of the Gender Gap in Introductory Physics

Lauren E. Kost, Steven J. Pollock, and Noah D. Finkelstein

Department of Physics, University of Colorado at Boulder, Boulder, Colorado 80309, USA

Abstract. Our previous research showed that despite the use of interactive engagement (IE) techniques at our institution, the difference in performance between men and women on a conceptual learning survey persisted from pre to posttest. This paper reports on a three-part follow-up study that investigates what factors contribute to the gender gap. First, we analyze student grades in different components of the course and find that men and women's course grades are not significantly different ($p > 0.1$), but men outscore women on exams and women outscore men on homework and participation. Second, we compare average posttest scores of men and women who score similarly on the pretest and find that there are no significant differences between men and women's average posttest scores. Finally, we analyze other factors in addition to the pretest score that could influence the posttest score and find that gender does not account for a meaningful portion of the variation in posttest scores when a measure of mathematics performance is included. These findings indicate that the gender gap exists in interactive physics classes, but may be due in large part to differences in preparation, background, and math skills as assessed by traditional survey instruments.

Keywords: gender, conceptual learning, introductory physics, physics education research
PACS: 01.40.Fk, 01.40.G-, 01.40.gb, 01.75.+m

INTRODUCTION

Recently, there has been great interest in the gender gap, the performance difference between men and women, in science. Several studies have suggested that both men and women learn more in interactive and engaging educational environments, but these techniques may disproportionately benefit women. [1,2] Researchers at Harvard were able to show that a pre-instruction gender gap was eliminated when fully interactive engagement techniques were employed. [3]

In our previous work [4], attempting to replicate the Harvard study, we found that engaging students in interactive educational environments did not always reduce the gender gap – IE techniques are not sufficient to reduce the gender gap. Our results suggest that a variety of factors are likely to contribute to men and women's differential performance. Not only which practices are used but how they are enacted appears to be critical. Furthermore, we hypothesize the content background (physics and math performance as assessed by standard measures, such as conceptual surveys) of the students plays a significant role [5], and this is the subject of our current investigation. Our preliminary follow-up studies involve three parts, and we report on each here.

We first question whether males and females perform differently on each component of our own introductory courses. By analyzing exam, homework, participation, and total course grades by gender, we find that there are differences between men and women on exams, homework, and participation, but these differences offset one another resulting in no difference between men and women's course grades.

In the second component of the study we ask: is the difference that we see in the posttest scores between men and women a result of gender, or is the variation due to some other factor? We compare posttest scores of men and women who have similar pretest scores. We see no substantial differences between men and women who score similarly on the pretest, indicating that pretest score is more relevant than gender.

The third part of the study asks what other factors influence the outcome of the posttest. Preliminarily, we are interested in math skill and gender. Using multiple regression analysis we determine that gender does not account for a meaningful portion of the variation in posttest scores. These results persist when we explicitly include a measure of mathematics performance. Our findings are in line with Meltzer, who found that mathematical skill correlated with student conceptual learning gains. [5]

CP951, *2007 Physics Education Research Conference*, edited by L. Hsu, C. Henderson, and L. McCullough
© 2007 American Institute of Physics 978-0-7354-0465-6/07/$23.00

TABLE 1. Analysis of students' course grades. Each column contains the difference between the average scores for men and women ($<S>_M - <S>_F$). Standard error of the mean is shown in parenthesis. Those courses for which there is no participation listed included it in the homework grade, and participation alone could not be extracted. The * indicates statistically significant (via two-tailed t-test, $p<0.05$).

	Participation	Homework	Exams	Course Grade
Spring 04	-6.0 (1.4) *	-5.6 (1.6) *	5.2 (1.5) *	1.1 (1.2)
Fall 04		-7.0 (1.9) *	4.0 (1.5) *	0.4 (1.4)
Spring 05		-6.0 (1.6) *	3.9 (1.4) *	0.5 (1.2)
Fall 05		-5.6 (1.8) *	3.9 (1.4) *	0.6 (1.3)
Spring 06	-2.0 (2.0)	-3.3 (1.8)	4.6 (1.5) *	1.7 (1.3)
Fall 06	-7.5 (1.6) *	-2.2 (1.7)	4.4 (1.4) *	1.8 (1.3)
Spring 07	-2.7 (1.6)	-2.0 (2.0)	3.5 (1.3) *	1.5 (1.3)
Average	-4.6 (0.8) *	-4.5 (0.7) *	4.2 (0.5) *	1.1 (0.5) *

METHODS

The data in all studies were collected from seven offerings (spring 2004 to spring 2007) of the first semester, calculus-based mechanics course at the University of Colorado (CU). All seven classes used interactive engagement techniques, some to a higher degree than others. Each of the seven classes employed student discussions around ConcepTests[6], online homework systems[7], and voluntary help-room sessions on problem-solving homework. Four of the seven classes used *Tutorials in Introductory Physics* [8] during a one-hour per week recitation, while the remaining three classes held more traditional recitation sections.

Several measures were used to assess student performance in the course and preparation or background. We analyzed students' homework, exam, participation, and course grades as measures of student performance in the physics class. Conceptual performance was assessed using the Force and Motion Concept Evaluation (FMCE) [9]. Only students with matched pre and posttest data are included, N ~ 2100 students. An applied math test was used to assess students' math skills upon entering the physics course. This test has been administered for many years to incoming engineering students at CU to identify at-risk students going into the Calculus I courses. We have these data for the subset of students who took a calculus course in the Applied Math department (N = 965). We note that these assessments only measure student performance on these instruments – however we use them as a proxy measurement of student understanding of physics concepts upon entry and exit. (Although we recognize these instruments may be measuring more, such as test taking ability, and may differ by gender.)

RESULTS

Course Grades Analysis

Our initial investigation into the gender gap examines course grades to determine if men and women perform differently on any components of the course. For each of the seven semesters of the mechanics course men and women's scores are averaged on homework, participation, exams, and total course grade. In all of our introductory courses exams make up 60% - 65% of the course grade, homework counts for 25% - 35%, and participation makes up the remainder. The difference between the average men and women's scores in each component ($<S>_M - <S>_F$) is calculated for each class. These differences, along with the average differences, are shown in Table 1. For several courses the participation grade was included in the homework grade and could not be extracted.

In the past seven semesters there has been no significant (via two-tailed t-test, $p<0.05$) gender difference in total course grade. Men outscore women by about 4 points on exams and women outscore men by about 5 points each on homework and participation. These differences offset one another and result in course grades that are not significantly different.

Matched Pretest Analysis

The second part of the study compares students who have similar pretest scores. Students are binned by FMCE pretest score (each bin contains about equal numbers of students, N ~ 420), and then the average FMCE posttest score is calculated for men and women in each bin. The results are plotted in Fig. 1. The same trends that are described below exist for a range of reasonable bin sizes.

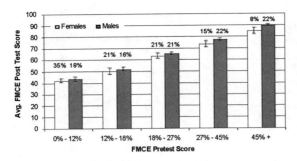

Figure 1. Average FMCE posttest scores for women and men with matched FMCE pretest scores. The percentages above each bar represent the percentage of the women (or men) from the total in each bin. The error bars represent the standard error on the mean. There are no significant differences between men and women in any individual bin.

We observe that students who have similar pretest scores have similar posttest scores, regardless of gender. There were no statistically significant differences (as measured by two-tailed t-test, p>0.1) in any individual bin, i.e. between men and women who scored similarly on the pretest. Though the differences in each bin are not significant, males consistently score higher than females in all bins. We also see a correlation (r = 0.562) between FMCE pre and posttest score.

The error bars on the plot are only a lower limit on the actual error. These error bars do not account for sources of error other than statistical error, such as systematic error or sampling bias. They also do not account for differences in class practices from semester to semester and across different instructors. Regardless, we find that the differences between men and women are not significant. These same trends exist for each individual class.

We do find that a higher percentage of the women fall into the low pretest bins. The percentages above each bar in the plot (Fig. 1) represent the percent of the women, or men, from the total that fall into that bin. 56% of women versus 34% of men fall into the lowest two pretest bins, while 23% of women versus 44% of men fall into the highest two pretest bins.

The same trend exists for *normalized* learning gain; students with similar pretest scores have similar normalized gains, regardless of gender. We also see a correlation, albeit weaker (r = 0.281), between FMCE pretest score and normalized gain.

Multiple Regression Analysis

Our results above suggest that the FMCE pretest could be a significant factor in predicting students' posttest score. Using multiple regression analysis, we can determine what other pre-factors are significant in

predicting the posttest score. Preliminarily, we are interested in math skill and gender.

To analyze the impact of math skill and gender on posttest score, we perform a stepwise multiple regression analysis where variables are included in the regression one at a time. We are interested in whether the addition of variables, applied math test score and gender, accounts for the variation in posttest. [10] The stepwise multiple regression analysis includes only those students for whom we have FMCE pre and posttest data and applied math test data, N = 965.

We start by including only the FMCE pretest as an independent variable. The results are shown in Table 2. The FMCE pretest alone accounts for only 34% of the variation in posttest scores. As a measure of students' math skills, we include an applied math test as an independent variable in the regression analysis. This regression has a multiple R value of 0.651. The additional variation accounted for by the math test is significant (via F-test, p<0.01). The FMCE pretest and the applied math test together account for 42% of the variance in posttest scores.

TABLE 2. Multiple Regression. The FMCE posttest is the dependent variable and the FMCE pretest, the applied math test, and gender are the independent variables.

	Multiple R	R^2
FMCE pretest	0.583	0.340
FMCE pretest and applied math test	0.651	0.424
FMCE pretest, applied math test, and gender	0.655	0.429

We then include gender along with the FMCE pretest and applied math test in the regression of the FMCE posttest. The multiple R value for this regression is R = 0.655, accounting for 43% of the variance in posttest scores. The additional variation accounted for by gender is very small (although still statistically significant via F-test, p<0.05). Gender contributes to less than 1% of the variance in the observed posttest scores. We find that the explicit inclusion of gender as a variable is not a principal factor in accounting for variation in students' posttest scores when other measures of students' background and preparation are included.

The multiple regression analysis is repeated looking instead at the normalized learning gain. Although the multiple R values were lower (as we expect since posttest score is more highly correlated with pretest score than normalized gain is), the same trends exist. The correlation of normalized gain with pretest alone is R=0.351. When adding math scores, the multiple R value becomes 0.487. Upon adding gender, it stays nearly the same, R=0.493. Thus, the multiple regression analysis shows that explicitly including gender does not additionally account for much of the variation in the learning measures.

DISCUSSION AND CONCLUSIONS

The analysis of course grades suggests that which course practices we engage in and how they are enacted do not differentially affect men and women in overall course grade. When we look at men and women with similar pretest scores, we see only small differences in average posttest scores. This suggests that the gender gap we observe is due in large part to a gap in measured background and preparation between men and women. Women come in with lower pretest scores, and because of the significant correlation between pre and posttest scores, women have lower posttest scores.

Finally, when we conduct a multiple regression analysis, the applied math test and FMCE pretest score predict the posttest score just about as well as when we explicitly include gender as an additional variable. These results again suggest that the gender gap is due in large part to differences between men and women in measures of preparation and background.

In one sense, it may be interpreted that gender does not play a role in measures of student achievement – the variation in FMCE posttest score may be attributed to other variables, notably pretest score and math achievement [11]. Furthermore, overall course grades are similar for male and female students. Such a stance would suggest that there is no explicit gender bias in the classes observed. Both men and women show learning gains from pre to posttest. Nonetheless, in these classes we observe a gap in performance by gender and observe instances where over the course of instruction, this gap is increased [4].

Another frame of analysis is that of implicit bias – that is, those components of a class that are most heavily weighted and essential for success disproportionately favor male students. While course grades are neutral overall, male students are more likely to score higher on exams (which are weighted more heavily in a typical class). While the classes studied are introductory courses with no expectation of prior knowledge of physics, those students who arrive to the class with greater background knowledge (higher pretest scores) are more likely to achieve high posttest scores and greater learning gains. The class favors those students with stronger physics and math backgrounds – in this case, male students.

Such an arrangement of a class (or any social environment) plays to certain student backgrounds and when those backgrounds are correlated with particular demographic groups, it demonstrates bias. That is not to say this is an explicit or purposeful bias, but rather one that is the codified structure of systemic cultural bias. [12] Tatum refers to this as a 'smog of bias' [13] and others to the privileged preparation of some group (at the expense of others) as an "accumulated disadvantage" [14].

Recognizing that student preparation in physics or mathematics is a means by which this bias is propagated allows us as researchers and educators to proactively address the challenges of the gender gap in physics. Simply enacting research based reforms, or supporting current practices (the *status quo*) may improve aggregate student learning gains, but may also be promulgating the disparity of performance and lack of equity in our educational system.

ACKNOWLEDGMENTS

Thanks to PhysTEC (APS/AIP/AAPT), NSF CCLI (DUE0410744) and NSF CAREER (0448176). Thanks to Prof. Crouch, the CU Physics Department, and the PER at Colorado group for ongoing support.

REFERENCES AND NOTES

1. P. Laws et al. *Am. J. Phys.* **67**, S32 (1999).
2. M. Schneider *Phys. Teach.* **39**, 280 (2001).
3. M. Lorenzo et al. *Am. J. Phys.* **74**, 118 (2006).
4. S.J. Pollock et al. *Phys. Rev. ST PER* **3**, 010107 (2007).
5. D.E. Meltzer *Am. J. Phys.* **70**, 1259 (2002).
6. E. Mazur, *Peer Instruction: A User's Manual*, Upper Saddle River, NJ: Prentice-Hall, 1997.
7. CAPA, http://www.lon-capa.org/; Mastering Physics, http://www.masteringphysics.com/.
8. L.C. McDermott and P.S. Schaffer, *Tutorials in Introductory Physics*, Upper Saddle River, NJ: Prentice-Hall, 2002.
9. R.K. Thornton and D.R. Sokoloff, *Am. J. Phys.* **66**, 228 (1998).
10. Lomax, R.G., *An Introduction to Statistical Concepts for Education and Behavioral Sciences*, Mahwah, NJ: Lawrence Erlbaum Associates, 2001.
11. We also note that the identified variables only account for 42% of the variance in the data. Examining additional sources is the subject of current work.
12. Dancy, M., "The Myth of Gender Neutrality" in *2003 Physics Education Research Conference*, AIP Conference Proceedings 720, New York: American Institute of Physics, 2004, pp. 31-36.
13. B.D. Tatum, *Why are All the Black Kids Sitting Together in the Cafeteria? And Other Conversations About Race*, New York: Basic Books, 1997.
14. V. Valian, *Why So Slow: The Advancement of Women*, Cambridge: The MIT Press, 1999.

Exploring the Intersections of Personal Epistemology, Public Epistemology, and Affect

Laura J. Lising

Department of Physics, Astronomy, and Geosciences, Towson University, Towson, MD 21252

Abstract. This paper discusses an approach to exploring the divide between students' stances toward their own learning and their perceptions of what is productive for the scientific community (sometimes called "personal epistemology" and "public epistemology"). The possible relationship between this divide and students' science- and course-related affect (e.g. preferences, motivation, emotions) will also be discussed. Previous research and theory indicate certain methodological considerations in study design and analysis, particularly attention to survey context, both with respect to attempting to tease apart epistemology from course expectations, and in considering differences between stated and enacted epistemologies and, similarly, beliefs vs. resources. A survey instrument designed to explore personal/public epistemology splits and affective variables will be described and preliminary results will be presented.

Keywords: Personal Epistemology, Nature of Science, Affect, Motivation
PACS: 01.40.Fk, 01.40.Ha

INTRODUCTION

In recent years, science education researchers have paid increasing attention to understanding the nature, influences, and impact of students' approaches to learning science and their ideas about the nature of science (NOS),[1-2] with more and more evidence pointing to the complexity of these ideas and approaches.[2-9] One important complexity is the sometimes vast difference between what students think is productive for their own learning (their "personal epistemologies") and what they feel they need to do in a particular course (their "expectations").[3-4] Another divide is between students' personal epistemologies and their "public epistemologies" (perceptions of NOS, or stances on knowledge that a student perceives as appropriate for the scientific community.)[5-6] Students' epistemologies also vary widely across different scientific domains.[7] Clearly these are cognitive constructs that are highly sensitive to context.

The richest and clearest data on these complexities come from observational studies where students' epistemological stances can be inferred from their actual learning behavior. This methodology allows researchers to determine epistemology and context simultaneously, and also look for other influencing factors. Survey methods can be useful but must be very carefully devised and analyzed.

PREVIOUS RESEARCH

One of the earliest relevant qualitative studies is Hammer's[3] case study from introductory physics in which a student describes having to curtail her learning to survive the course. This is an example of the personal epistemology/expectations divide.

Other qualitative studies bear on the personal/public epistemology divide and the possible link to affect. Hogan[5] analyzed interviews and classroom discourse to study children's "personal frameworks" for science learning, finding that students' "learning-referenced" frameworks (similar to personal epistemology) had a much larger impact on their learning behavior than their "discipline-referenced" frameworks (similar to public epistemology). The affective variables she measured ("self-referenced" frameworks) such as interest in science, success attribution, etc., correlated less directly with observed learning behaviors, but Hogan suggests that future research look at the inter-relations among the three constructs. In Lising and Elby's[6] observational case study of an undergraduate physics student, the subject displayed a sophisticated understanding of how scientists learn despite her often unproductive practices in her own learning. There were indications that for this student, the divide had something to do with affect. She did not identify with physics, stating that she was not "a physics person." She expressed low confidence in her ability to use

CP951, *2007 Physics Education Research Conference*, edited by L. Hsu, C. Henderson, and L. McCullough
© 2007 American Institute of Physics 978-0-7354-0465-6/07/$23.00

common sense and everyday experiences in reasoning about physics problems, claiming that it seemed there was always "something that I've overlooked."

Other studies have also shown relations between affective and similar variables and students' learning approaches.[8-9] Of particular interest is the belief in "natural talent," which can be considered to lie in both the affective and epistemological domain and can have a huge negative impact on learning behaviors.[10]. Although motivation and other emotional factors have been studied extensively in the science education literature, to date no targeted work has looked at the intersections of these factors and the divide between a students' personal and public epistemological stances. The present study attempts to make a start at this.

Survey methods have proved to be an interesting way to collect a broader set of data. Building on Hammer's[3] qualitative study, Elby[4] surveyed students about their actions and attitudes towards their physics course, and then asked what advice they would give a hypothetical student who only wanted to learn the material well and not worry about grades. He found that students' advice indicated a much more sophisticated understanding of productive learning behaviors than their own actions would seem to indicate (personal epistemology vs. expectations). The Colorado Learning about Science Survey (CLASS)[10] was used by Gray et al.[11] by having students mark both their responses and the responses a physicist would give.[9] They found that students often chose the more productive response in the "physicist" case, and that splits were larger for women than men. In preliminary follow-up interviews, asking subjects about the reasons for their disparate responses, these authors found some indications of epistemological, expectational, and affective factors. For example, students explained that physicists like physics more and may have different learning styles. These data are useful for understanding how students interpret the CLASS survey and provide more insight into how interwoven are these various factors.

METHODOLOGICAL CONSIDERATIONS

Elby's result makes it clear that, in looking at the personal/public epistemology divide, effort must be made to avoid measuring expectations in lieu of personal epistemology. Although with a survey it is not possible to manipulate context precisely in order to absolutely isolate different epistemological constructs, careful attention to teasing apart these different issues could make the data easier to interpret and use.

For instance, some of the personal/public epistemology splits suggested by the Gray et al. survey data are difficult to discern due to the fact that most of the individual CLASS items as well as the survey as a whole do not specify a context, so responses are likely to reflect a mix of personal epistemology and expectations from students' varied classroom experiences. Gray et al. measured the largest splits on an affective item ("I enjoy solving physics problems") and two items that were statements about what the students actually do, which tends to probe expectations more strongly than epistemology. It can be expected that students split heavily on expectations vs. public epistemology since physics courses tend to stand in such contrast to scientific research environments.

It is also important to have a theoretical viewpoint with which to approach the analysis, even with primarily quantitative survey data. A resource view of epistemology[12] predicts the observed context- and domain-sensitivity of personal epistemologies, describing this as a natural outcome of a cognitive structure whereby, rather than stable, unitary constructs (stable "beliefs"), epistemologies are a collection of resources for knowing and learning that are activated by internal and external factors. This implies that surveys and interviews can best characterize students' epistemologies when they vary the context so the analysis can look for patterns of both consistencies and inconsistencies in students' responses and behavior.

A resource-type cognitive structure also accounts for differences between students' stated epistemological stances and those that they display in action, even when the statements are made in a seemingly identical context (since statement and action are inherently different contexts.) Thus a survey only can measure students' stated stances, and raise questions about their stances in practice.

PRELIMINARY SURVEY

The survey developed for this study contains three sections: a section with 10 items set in a personal epistemology context; a second section with 10 parallel items set in a public epistemology context; and a third section with an array of affective items. Most of the epistemological items were adapted from the widely-used surveys.[11,13,14] A sample of instructions and items from the survey illustrates the different contexts and items.

Question Set 1. Suppose you were taking a physics class and your goal was to learn the material well and didn't have to worry about grades. That may or may not seem far-fetched to you, but please do your best to answer the following questions as if you were in this situation.

Item 2. Remembering lots of facts and formulas would be the most important thing for you in understanding physics.

Question Set 2. Now think about what scientists do. Please answer the following questions with scientists in mind. (You'll notice they are almost the same set of questions as above, but about what scientists do.)

Item 12. Remembering lots of facts and formulas is the most important thing for scientists in understanding physics.

Question Set 3. Now forget about the hypothetical situation in Question Set 1 and the scientists in Question Set 2. Please answer the following questions about yourself.

Item 22. If there is such a thing as natural talent in physics, I am not one of those people.

Because the first context is rather contrived, the survey also contained an item to provide some indication as to whether or not students were able to successfully consider that scenario. There was also an item to probe domain-sensitivity and several items to collect demographic information.

The survey was administered in an algebra-based introductory physics course. Forty-six surveys were returned completed. Some preliminary results from this small sample are described below. Interviews, refinement of the survey, collection of a larger sample, and further interviews and/or observations are planned for the future.

During the analysis the raw data were collapsed from a 5-point to a 3-point agree/neutral/disagree scale (i.e. "agree" includes "agree" and "strongly agree" responses.)

Differences Between Stated Personal and Public Epistemologies

Splits in one direction. There were two item pairs where the means for the class showed statistically significant differences between the students' stated public and personal epistemologies. Item Pair 5/15 (see Table 1) showed the largest split (p=0.01 from a χ^2 analysis). For this item, one would expect some difference due to the fact that scientists have more knowledge and resources with which to evaluate controversies. However, it is not productive for students to think that they have "*no way* to evaluate." For this item, a full 12 of the 46 students (26%) made the full split, choosing disagree for the scientists (the productive response), while choosing agree for themselves.

The other large split was with Item Pair 6/16 (p=0.03), concerning alternate ideas (see Table 1.). Almost all of the students (45 of 46) chose agree for the "scientist" item, suggesting a sophisticated understanding of this aspect of scientific knowledge, yet 14 of 46 students chose neutral or disagree for their own learning (although it is still impressive that 32 students chose agree for themselves .)

Bimodal Splits. Several item pair responses showed high degrees of splitting with bimodal splitting patterns. On Item Pairs 1/11, 2/12, and 4/14, over 40% of the students split (see Table 1.) Students who split were almost equally likely to split in either direction and were more likely than not to split drastically (from agree to disagree or vice-versa rather than to or from the neutral position.) Two of these item pairs involve real-life experiences and the third involves facts and formulas, which might be counter-posed in some students' minds with common-sense conceptual resources like their real-life experiences. These data suggest that some students might perceive that, in understanding physics, real-life experiences are only necessary for students, who perhaps don't think like real scientists. Conversely, other students might perceive that real-world experiences are only useful for scientists, whereas students can only hope to learn the simplified or abstracted material. This possible dichotomy merits further exploration.

TABLE 1. Notable splits between personal and public epistemology (statistically significant mean differences or splitting percentages over 40%)

Item Pair	Item in Personal Epistemology Form	Percentage Splitting	Splitting Pattern	p-value for χ^2 analysis
1/11	1. Physics is related to the real world, but you could understand physics without thinking about that connection.	48%	bidirectional	
2/12	2. Remembering lots of facts and formulas would be the most important thing for you in understanding physics.	43%	bidirectional	
4/14	4. Suppose you read something in your physics textbook that seemed to disagree with your own experiences. But to understand physics well, you shouldn't think about your own experiences; you should just focus on what the book says.	48%	bidirectional	
5/15	5. When it comes to controversial topics in physics, there would be no way for you to evaluate which scientific studies are the best.	59%	unidirectional	0.01
6/16	6. To understand physics, it would be important for you to be able to understand and explain not only what I think is the right answer, but also what other answers my peers might have.	30%	unidirectional	0.03

Affect Results

On all but three of the ten affect items, over 50% of the students responded with negative affect. The two items with the strongest negative response (both over 80%) concerned the existence of natural talent and if the student felt lacking in that type of talent. On only one item was there an average positive affective response.

To study correlations between the personal/public splits and affect, affective responses were compared for negative splitters (who expect scientists' learning to be more sophisticated than theirs), non-splitters, and positive splitters. These data suffer most from the small statistics and the lack of range in affective response, and almost no differences were noticeable with p less than 0.1. Further data collection should allow the analysis to capture subtleties that are only accessible with larger statistics and qualitative data.

The only statistically significant affect correlation measured was with Item Pair 8/18 (splitting at 33%):

Item 8: If you were working in a group and got the right answer to a problem, convincing your peers should be easy. The correct answer would sort of "speak for itself."

This item was correlated with affect Item 22 (student's identification of self as one with natural talent.) Perhaps feeling lacking in "natural talent" could impact students' confidence in convincing their peers compared to scientists' ability to do the same.

Other Data

Item 31 was designed to probe the difference between personal epistemology and course expectations, although it also is an indirect indicator of the success of the contextualization of Question Set 1.

Item 31: If I were to go through and redo Question Set 1 but instead of thinking about how I might go about learning physics just for the sake of learning, I thought about how I actually go about learning physics these days, my answers would probably be significantly different.

Twelve of the 46 students agreed with this item, indicating that the contextualization was at least partially successful.

CONCLUSIONS

Including NOS instruction in courses has been shown to impact stated public epistemologies, but the assumption that this will translate into more sophisticated personal epistemologies is problematic.

To understand this better we must look more closely at the divide between personal and public epistemology and try to tease it apart from the more externally-driven splits between expectations and epistemology, and also explore what other factors might influence this complex landscape. If we discover there are affective issues influencing some of these splits, there may be targeted interventions that are beneficial. In addition to overall course reform, we might, fwor example, we might make effective use of free writes, learning reflections, stories about students and scientists, and readings about the malleability of intelligence. These might help students bridge the gap between what they think can be productive for others into what can be productive for their own learning in science.

ACKNOWLEDGMENTS

I would like to thank the students who participated in this research, the course instructor, and David May and Pamela Lottero-Perdue for helpful comments.

REFERENCES

1. N. Lederman, "Nature of Science: Past, Present, and Future," in *Handbook of Research on Science Teaching*, edited by S. K. Abell and N. Lederman, Mahwah, NJ: Lawrence Erlbaum, 2007, pp 831-880.
2. B. K. Hofer and P. R. Pintrich (eds.), *Personal Epistemology: the Psychology of Beliefs about Knowledge and Knowing*, Mahwah, NJ: Lawrence Erlbaum, 2002.
3. D. Hammer, *The Physics Teacher* **27**, 664-671 (1980).
4. A. Elby, *American Journal of Physics* **67(7)**, s52-s57 (1999).
5. K. Hogan, *Science Education* **83(1)**, 1-31 (1999).
6. L. Lising and A. Elby, *American Journal of Physics* **73**, 372-382 (2005).
7. B. K. Hofer, *International Journal of Educational Research* **45(1-2)**, 85-95 (2006).
8. C. S. Dweck, *Self-Theories: Their Role in Personality, Motivation, and Development*, Philadelphia, PA: Psychology Press (1999).
9. A. H. Schoenfeld, *Cognitive Science* **7**, 329-363 (1983).
10. W. K. Adams, K. K. Perkins, M. Dubson, N. D. Finkelstein, and C. E. Weiman, *Physical Review Special Topics – PER* **2**, 010101 (2006).
11. K. E. Gray, K. K. Perkins, W. K. Adams, and C. E. Weiman, submitted to *Physical Review Special Topics*.
12. L. Louca, A. Elby, D. Hammer, and T. Kagey, *Educational Psychologist* **39(1)**, 57-69 (2004).
13. E. F. Redish, R. N. Steinberg, and J. M. Saul, *American Journal of Physics* **66**, 212-224 (1998).
14. B. White, A Elby, J. Frederickson, and C. Schwartz: http://www2.physics.umd.edu/~elby/EBAPS/home.htm

Students' Perceptions of Case-Reuse Based Problem Solving in Algebra-Based Physics

Fran Mateycik[1], Zdeslav Hrepic[2], David Jonassen[3] and N. Sanjay Rebello[1]

[1]*Department of Physics, 116 Cardwell Hall, Kansas State University, Manhattan, KS 66506-2601*
[2]*Department of Physics, 600 Park Street, Fort Hays State University, Hays, KS 67601*
[3]*Department of Educational Psychology, 221C Townsend Hall, University of Missouri, Columbia, MO 65211*

Abstract. Problem solving is an important goal in almost all physics classes. In this study we explore students' perceptions and understanding of the purpose of two different problem solving approaches. In Phase I of the study, introductory algebra-based physics students were given an online extra credit problem-solving assignment. They were randomly assigned one of three problem-solving strategies: questioning, structure mapping and traditional problem solving. In Phase II of the study, eight student volunteers were individually assigned to work problems using one of the strategies in two sessions of semi-structured interviews. The first session investigated students' general problem solving approaches a few weeks after they had completed the online extra credit assignment. The second session investigated students' perceptions of problem solving strategies and how they relate to the extra credit assignments. In this article, we describe students' perceptions of the purpose of the activities and their underlying problem solving techniques.

Keywords: problem solving, algebra-based physics, conservation of energy, students' perceptions, physics education research
PACS: 01.40.Fk

INTRODUCTION

Problem solving is regarded as an important cognitive function used in all manner of contexts, including Physics [1, 2]. Research has shown that programs combining several interrelated instructional strategies are more effective than single-strategy programs [3, 4].

In this pilot study, we combine the use of several problem solving strategies for short-term treatment in an introductory algebra-based physics course. Our objective was to gauge how students might perceive certain problem-solving strategies -- their purpose, ease of use and overall value in problem solving and how these compare with traditional strategies that they may already use. In this report, we address the following research questions:

- To what extent do students find these strategies to be useful in problem solving?
- How well are these strategies aligned with students' existing problem solving technique(s)?
- To what extent do students understand the purpose of these strategies?
- To what extent do students find these strategies difficult to implement?

We acknowledge that students can acquire procedural automation of a strategy over long term exploration, assimilating it into their problem-solving repertoire [5]. However, long term implementation of any strategy for a large enrollment class is a logistically difficult task, so it was important to conduct this pilot study to determine how students might respond to the treatment design before implementing it for the long term.

LITERATURE REVIEW

In this study we examine the use of Case-Based Reasoning (CBR) in problem solving. CBR is the process of using analogies to solve real-world problems [6]. Case reuse is a strategy which helps promote CBR by employing problem pairs that share similarities in deep structure [7]. Here we examine how case reuse can be implemented in conjunction with two other strategies: 'Questioning' and 'Structure Mapping.'

The 'Questioning' strategy refers to Graesser's psychological model of question asking [8]. Graesser generated generic questions based upon the level of knowledge the question was looking to answer. A few of these questions were: *What does X mean?*

CP951, *2007 Physics Education Research Conference*, edited by L. Hsu, C. Henderson, and L. McCullough
© 2007 American Institute of Physics 978-0-7354-0465-6/07/$23.00

(Taxonomic), *What does X look like?* (Sensory) and *What causes X?* (Causal) We adapted this questioning strategy for implementation with case reuse such that the questions were asked based upon a pair of analogous problems. Declarative questions such as, *What quantities are given,?* were followed by causal questioning such as, *Which of the following quantities change,?* for each problem in each problem pair. We also added questions that asked students to modify and extend the given problems such that the resulting problems were most similar to one another, thus compelling students to compare and contrast the two problems.

The 'Structure Map' is best described as a visual representation expressing functional interdependency between concepts and quantities [9, 10]. The structure map for this project was produced by three experts knowledgeable in physics education and educational psychology. Students use the structure map by marking quantities that are given in a problem, the quantity that is to be found and quantities that must be calculated to proceed from what is given to what is to be found. Figure 1 shows an example. This strategy too was adapted to include problem pairs.

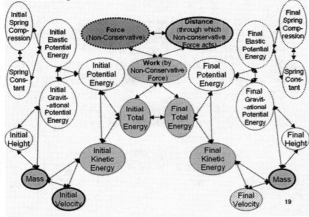

FIGURE 1. A structure map used for the following problem. What average force is exerted by the brakes to stop a 1250 kg car traveling at 30 m/s over a distance of 20 m?

There were two treatment groups: Questioning and Structure Mapping. The control group was asked to solve unpaired problems without being given extraneous help through questioning or structure maps.

METHODOLOGY

Participants for Phase I included 150 students enrolled in algebra-based physics at Kansas State University in spring 2007. All enrolled students were given the opportunity to participate in this study for extra credit and all were evenly assigned into one of three groups at random: Questioning, structure mapping and control. Students were asked to access their assignment online. Each group worked with three types of problem pairs: work-energy theorem problems, potential energy problems and conservation of energy problems. The structure mapping group was also given a training document prior to being asked to complete the three problem pairs. The training document built the structure map using the work-energy domain and worked through an example of using the map with a simple work-energy physics problem. (See Fig. 1 for exact example) Once students assigned to either group involving questioning or structure mapping completed the three problem pairs, they were given three different problems to solve and hand in, one from each type. The control group was asked to solve and hand in six problems, two from each type. All students in the class were also given a transfer problem on their course examination, attempting to assess the influence from previous extra credit exercises.

Phase II involved two 50-minute sessions of semi-structured interviews for each of eight volunteers. The students were selected based upon the extra credit assignment they completed. Two students each were selected from the questioning strategy group and structure mapping group, the remaining four were from the control group. The first interview session investigated students' general problem solving approaches. Students were asked to work through a work-energy problem (Fig. 2). They were allowed to use their course textbook and a calculator. After each student completed their attempt to solve the problem, we asked them questions about their work.

Students returned for a second interview one or two weeks later. The second interview focused on acquiring information about students' perceptions of the strategies. For students in the structure mapping or questioning strategies group, we asked them to recall their extra credit assignment as best they could. If they were capable of recalling any part of the extra credit assignment, they were asked to apply what they remembered to the problem they were provided during the first interview. If they were unsuccessful, they were given a copy of the extra credit to reexamine. Once participants were given time to reacquaint themselves with the assignment, we concentrated our interview questions on students' views about the intended purpose behind the map or question strategy. If time allowed, students were asked to use their strategy while attempting for a second time to solve the problem given in the first interview.

The control group was asked similar questions during their interview sessions, but they spent no time reviewing their own extra credit, since their extra credit assignments did not include either of the strategies. Instead, in the second interview they were first explained one of the two strategies and then asked to apply it to the interview problem from the first interview.

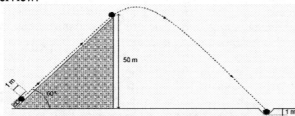

A ball of mass 2kg is held against a spring compressed by 1 m with a spring constant of k = 3 x 10³N/m, and sits at the bottom of a ramp 50 m high. The ramp is inclined at an angle of 60°. When the ball is released, assuming a frictionless ramp and no wind resistance, the ball will hit a pit of sand. The depth the ball sinks into the sand is 1m.

(a)At what speed will the ball leave the ramp?
(b)What is the average force on the ball by the sand?
(c) Is there any information provided in the problem that you did not need to in (a) and (b)? If so, what information?
(d) If 'k' is reduced, what happens to the depth the ball sinks into the sand?

FIGURE 2. Interview 1 Work-Energy Problem.

RESULTS

Phase I showed no statistically significant difference between the three treatment groups. This result is consistent with previous studies which suggest that students given a short term treatment of any problem solving strategy do not show marked improvement [10].

The first session interviews of Phase II were not initially relevant to the treatment groups, but were important to gaining insight into the development of students' strategies for solving problems during the second interview. Students consistently used the *working-backward* approach during the first interview session and so it became apparent when students reverted back to that approach while working the interview problem for the second time [11]. In the second interview session, several themes emerged regarding the students' perceptions of these strategies which are discussed below.

Questioning Strategy

The questioning strategy was worked out by four of the eight interviewees. All four students determined the strategy was purposeful and similar to their own problem solving techniques. When asked to explain the possible purpose of the questioning strategies, all interviewees replied similarly,

QS1: "the purpose? Mm... to help us um visualize the problem, um to help us think of what we should take into account, and get us thinking of what shouldn't be taken into account because there's answers on there [questioning strategy] that don't apply to the problem, so it helps us to decide what we

can apply to the problem to help figure out what we need to know about the problem"

Interviewees were also asked to explain how the questioning strategy varied from their own strategies. Responses showed students believed the strategy was designed to mimic good question asking procedure and said that they already ask themselves the same or similar questions when they solve problems.

QS2: "Well, I always ask myself this, which is what I am given, or what's implied in the problem. So like this, it's talking about how far the arrow went, I'd have to take in gravity, but it's not given in the problem, but I know what it is..."

Responses also showed question miscommunication when asked to identify from a list, concepts, laws and/or theories applicable to the problem. Three out of four responses reflected interviewees' use of equations in the process of identification of concepts.

QS3: "Umm, on this one (question 2), I usually try to find the equation I'm using from what I'm given or I try to find an equation with a lot of what I'm given in it and try to see if there is something missing that I need...."

Overall, students responded affirmatively when asked if they felt the questioning strategy was helpful for solving physics problems and if they felt comfortable using the strategy. Furthermore, all students who used the questioning strategy recognized that the problems were paired in the extra credit assignment. Three of the four students articulated reasoning for the paired problems:

QS4: "Umm... they're both dealing with the same uh... work and energy but they're showing it in different ways, like this one is using a spring compression in order to move the arrow and this one is just using a human just throwing it and it tells you the initial speed, but it's still using the same type of equations."

Structure Mapping

The structure map was highly regarded, yielding praise from all four of the interviewees who used the strategy. When asked to compare this strategy with the one that they used, all students found that the structure mapping strategy was quite different from their own problem solving technique, nevertheless they found structure mapping helpful in understanding *"what you need for a problem."* None of the students expressed any difficulties about using the structure map. Three students liked how the map represented all of the problem information. Two students liked the way all quantity relationships were apparent.

SS1: "..half the time its hard for me to figure out what equation to use, but like when you figure out like what it gives you and then how to figure out what equation to use from the arrows, helps, like it doesn't give you the equation but it tells you what you need in order to figure out how to get the answer."

Overall, students felt the structure map was easy to use after given the appropriate PowerPoint training slides. At the end of the second interview session, all four students worked out the interview problem using the structure map. Unlike the questioning strategy, all four students' solutions improved from the previous interview. Only one student from the structure mapping interviews recognized the problems were paired. He found the pairings useful for comparing question answers between the two problems. Other students, when asked if there might be a reason why the problems are paired responded,

SS2: "I think there is a reason, I just don't know what it is."

SS3: "I haven't really thought about that, umm, prolly (sic) because they are similar problems but your given different information, I think it would be just as helpful if they were on separate sheets."

CONCLUSIONS

Overall, our results indicate that students believe these strategies are helpful in giving them good problem visualization and facilitating their ability to identify important information from the problem. Students from the questioning strategy group believe the questioning strategy is similar to their own problem solving techniques, providing well structured questions that attempt to draw important information from the problem statement. Students from the structure mapping interviews believe the structure maps are not comparable to their own problem solving techniques, but still feel the strategy is an effective tactic for representing problem information.

All eight students agreed that the purpose of the strategies was to help them work out problems, though the intended purpose of some of the questions from the questioning strategy was not clear to the students. When asked to implement the strategies to solve a problem, students showed difficulty expressing differences between concepts and equations, providing equations that fit some or all of the quantities provided in the problem as an appropriate means of defining problem concepts. Finally, the results of this study suggest the structure mapping and the questioning strategy, interlaced with case-reuse, were well-liked and user-friendly.

LIMITATIONS & FUTURE WORK

The goal of this study was *not* to assess the effectiveness of the strategies, since the effectiveness of any strategy can only be gauged in the long term. The goal here was to examine how students perceived these strategies. The next phase of our research is to adapt the strategies based on our results and implement and assess these strategies in class over the long term.

Based on our results, the questioning strategy will require several adaptations prior to implementation. The intended meaning of the questions was sometimes misaligned with student interpretation. Questions asking students to identify concepts, theories or laws were ultimately answered with equations containing quantities identified in the problem. These questions will be reworded in the future such that students may not phrase their answer in terms of an equation. Following these adaptations we will include long term quantitative and qualitative investigations of students' performance on these strategies.

ACKNOWLEDGMENTS

This work is supported in part by the National Science Foundation under grant DUE-06185459. Opinions expressed are those of the authors and not necessarily those of the Foundation.

REFERENCES

1. L. Hsu, E, Brewe, T.M. Foster and K.A. Harper, *American Journal of Physics* **72**(9), 1147-1156 (2004).
2. D.H. Jonassen, *Educational Technology and Research and Development* **48**(4), 63-85 (2000).
3. J. Hattie, J. Biggs and N. Purdie, *Review of Educational Research* **66**(2), 99-136 (1996).
4. M. Pressley and C.C. Block, "Comprehension Strategies Instruction: A Turn-of-the-Century Status Report.," in *Comprehension Instruction: Research-Based Best Practices*, edited by C.C. Block, Guilford Publications, Inc, New York, 2002, pp. 11-27.
5. K.A. Ericsson, R.T. Krampe, and C. Tesch-Romer, *Psychological Review* **100**(3), 363-406 (1993).
6. J.L. Kolodner, *American Psychologist* **52**(1), 57-66 (1997).
7. D.H. Jonassen, *Technology, Instruction, Cognition and Learning* **3**(1-2), (2006)
8. A.C. Graesser, W. Baggett, and K. Williams, *Applied Cognitive Psychology* **10**(7), 17-31 (1996).
9. D. Gentner, *Cognitive Science* **7**(2), 155-170 (1983).
10. J.D. Novak, B.D. Gowin, and G.T. Johansen, Science Education **67**(5), 625-645 (1983).
11. Y. Anzai, "Learning and Use of Representations for Physics Expertise," in *Toward a General Theory of Expertise*, edited by K.A.A.J. Smith, Cambridge University Press, New York, 1991, pp. 64-92.

Investigating Students' Ideas about Wavefront Aberrometry

Dyan McBride and Dean Zollman

Department of Physics, Kansas State University, Manhattan KS, 66506-2601

Abstract. We describe a qualitative study of student understanding of the functions of the human eye and the resources used in understanding wavefront aberrometry, a relatively new method of diagnosing vision defects. Twelve students enrolled in an introductory physics class participated in a semi-structured clinical interview in which the functions of the eye, traditional diagnosis methods such as the eye chart, and wavefront aberrometry were discussed. Results from this study indicate that students do not initially understand the subjective nature of traditional diagnosis techniques and that the use of physical models of the eye and aberrometer can facilitate the transfer of prior knowledge to these concepts.

Keywords: physics education, transfer of learning, human eye, wavefront aberrometry
PACS: 01.40.Fk

INTRODUCTION

Our group has undertaken several studies to investigate how students transfer their learning from a typical physics course and/or everyday life to contexts that they have not previously seen. These studies help us understand what reasoning and knowledge the students use appropriately or inappropriately as they learn physics. The goal of this component of the study is to investigate the ways in which students transfer prior learning to understand the physics related to a relatively new vision diagnostic tool, wavefront aberrometry. Rather than beginning with the aberrometry techniques, however, we first examined students' understanding of how the human eye works.

The main research question guiding this study is: *How do students use their existing knowledge to understand wavefront aberrometry methods of diagnosing vision defects and what resources do they use in constructing their understanding?*

LITERATURE REVIEW

Wavefront aberrometry is a relatively new method of diagnosing vision defects in the human eye. By shining light into the eye and measuring the properties of the reflected light, an aberrometer utilizes physical properties of light instead of subjective judgments of the patient for identifying aberrations within the eye [3-5]. Such methods are becoming increasingly important as the use of surgery (LASIK) becomes significant in correcting vision difficulties. It is also likely to become a common method for determining corrective lens prescriptions.

Aberration of wavefronts of light due to defects in optical instruments is not commonly taught in introductory physics courses. Thus it is an appropriate topic to use when studying how students transfer their learning to a new context. Transfer is defined as the application of knowledge from one context to another [6] or as the mediated association of information between contexts [7]. A useful approach to investigating student transfer is to identify and analyze the resources which they utilize when attempting to understand physics in a novel context. Resources can be thought of as the fragments of information, knowledge, and experience that individuals bring to a new situation or context. An overview of resources and their use in physics can be found in reference 8.

METHODOLOGY

To address our research question, we conducted formal, semi-structured interviews with 12 students (3 females, 9 males). All of the students were enrolled in a calculus-based introductory level physics course. All students were interviewed before

CP951, *2007 Physics Education Research Conference*, edited by L. Hsu, C. Henderson, and L. McCullough
© 2007 American Institute of Physics 978-0-7354-0465-6/07/$23.00

they had instruction about mirrors/lenses, but while they were in the process of learning about the electromagnetic properties of light.

Each participant was interviewed for approximately 45 minutes and all were encouraged to think-aloud as they responded to the questions. To place the interview in a context of diagnosis, students were first given a copy of a typical eye chart. This introduction led to a discussion of how light travels and how we are able to see. Following this, a model of the eye was used for clarification and often times prompted further discussion. A photo of the model is shown in Fig. 1. One convenient feature of this model is the pliable "lens" that is attached to a syringe system. By varying the amount of liquid in the lens, students change the radius of curvature. Students were also provided with paper and many made sketches to help illustrate their answers.

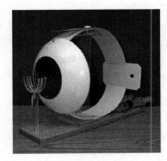

Figure 1. Model used during interviews [9].

The final part of the interview involved the aberrometer. We adapted the method of modeling an aberrometer used by Colicchia and Wiesner [10]. In this adaptation, the model shown in Fig. 1 was used along with an array of small lenses, an LED light source, and a paper screen. The lens array was placed in the slot (visible in the above model) in front of the "pupil." After a brief discussion with the participants about light being reflected from surfaces, students agreed that the light source could be clipped to the "retina" of the model to simulate a reflection of light being shone into the eye. The combination of the light source and lens array provided the formation of a grid pattern on the paper screen. This grid pattern is representative of the grid pattern obtained by wavefront aberrometry diagnosis techniques. Figure 2 shows the grid pattern formed by the model used by Collicchia and Wiesner, which is very similar to the pattern seen in our setup.

After the aberrometer was set up, the participants were asked to describe what was being modeled and then to predict what would happen if the lens of the model were somehow aberrated. The pliable nature of the lens enabled us to aberrate the lens by pushing or deforming it and reacting to the differences being created in the grid pattern.

Figure 2. Grid pattern formed with Collicchia and Wiesner model [10] from normal lens (left) and aberrated lens (right).

Finally, students were asked to describe how a system such as the one being modeled could be used by a doctor in order to diagnose vision defects. This also led to a discussion of the advantages and disadvantages of a wavefront system as compared to a more traditional method of diagnosis. When the interview had ended, students were given the opportunity to ask any questions about the eye, defects, and the aberrometry setup.

The interviews were video and audio recorded and afterwards transcribed. The transcriptions, student sketches, and field notes from the interviewer served as the data sources for this study.

A phenomenographic approach was taken during data analysis in order to illicit variations in student ideas instead of researcher conceptions [11]. Student responses were then examined in an effort to identify any resources that were being used. We considered the resources used by a single participant as well as all participants in order to extract possible themes.

RESULTS AND DISCUSSION

In the analysis we paid particular attention to the resources that students used along with their reasoning patterns. As the long term goal of the project is to create teaching and learning materials about wavefront aberrometry, these resources will be beneficial for creating learning materials to help the students in their understanding.

Resources

A list of some of the resources used by participants is included as Table 1. The listed resources were chosen because of their use by multiple participants and applicability to understanding wavefront aberrometry.

The first resource, *light can be represented by a line*, was extracted partially from statements from participants, but mostly from their sketches. Of the 12 participants, nine made sketches of ray diagrams,

TABLE 1. Selected Student Resources

	S1	S2	S3	S4	S5	S6	S7	S8	S9	S10	S11	S12
Light and Lenses												
Light can be represented by a line	X	X	X		X	X	X		X	X	X	
Light is a wave	X	X		X						X		
Concavity/thickness/curvature of a lens changes the focus	X			X	X	X	X			X	X	
Aberrometry												
Light entering a lens differently will focus differently	X		X	X	X	X	X	X	X	X		X
An aberration is an anomaly					X			X		X		X
Size of change in grid reflects size of aberration	X				X				X			X
Symmetry has value						X				X	X	
Can only measure one thing at a time					X			X				
Objectivity												
"Objective" means no human opinion/interpretation			X	X	X		X					X
"Objective" means consistent (always same for everyone)		X		X	X		X			X		

(though to varying levels of correctness) and each clearly represented light as a straight line coming from a source. Fewer mentioned the wave properties of light, but it should be noted that students were not directly asked about wave properties of light.

Students utilized many resources that can be applied appropriately to wavefront aberrometry. For instance, the fact that *light entering a lens differently will focus differently* is a very important concept for understanding aberrometry. Three students commented on the symmetry of a grid from a perfect eye and the lack of symmetry from an aberrated eye. Using the resource of looking at patterns and symmetry is also useful in understanding how the grid patterns resulting from wavefront aberrometry are interpreted.

However, some of the above resources may not necessarily be appropriate for understanding wavefront aberrometry. Alone, the resource that *light can be represented by a straight line* is not an inappropriate resource – however, if participants believe that light only travels as a straight line, this could hinder their understanding of the altered wavefronts that result from aberrations. Many participants noted that *a big change in the grid represents a big aberration*. This could be considered to be a phenomenological primitive (p-prim) as described by diSessa [12]. However, this is not entirely true; the type of aberration is determined by the properties of the resulting grid pattern, but the "size" of the aberration is determined in terms of severity and not spatial size as participants seemed to suggest. Two students also brought up the fact that *controlled experiments only measure one thing at a time*. A closer look into this resource is required to determine if it could hinder the understanding that an

aberrometer measures the function of the eye as a whole and not as individual components.

Transfer of Prior Knowledge

Most participants had a great amount of prior knowledge that they clearly used to describe how the eye works, including naming parts (iris, cornea, retina, as well as rods and cones) and that the image produced is upside down and must be "flipped" by the brain. However, it was found that when students had relatively little prior knowledge about how the eye works, it was far more difficult to get them to talk about the aberrometry model and techniques.

When it came to wavefront aberrometry, students had significantly less prior information to transfer. Most of the transfer came in the understanding of the two-lens system created by the eye lens and the lenses in the array. As one student put it, "the [eye] lens focuses light onto the area of the array, and then the [array] lenses are breaking up light … and focusing it to their own point."

Students also seemed to believe that there was an "ideal" grid, though different ideas existed of what that ideal might be. Some indicated symmetry as discussed above, while others thought that a specified intensity or the size of the dots should be known. Perhaps this ideal reading concept is transferred from ideal vision, e.g. 20/20.

The issue of subjectivity in measurement is one that we purposefully raised during discussion of both detection instruments. Most students (8 of 12) did not initially realize that any subjectivity was involved during diagnosis with an eye chart. In fact, five participants clearly stated that the eye chart was an objective diagnosis tool because it was exactly the same for every patient. This result indicates that the students' view of objectivity only may have a component of fairness. In any case where the issue of objectivity was not directly addressed by the student, they were prompted with questions such as "Did you ever try to guess at a letter you couldn't really see?" or "Did you ever have trouble telling the doctor how much clearer one line was than the next?" This scaffolding was in all cases adequate to get participants thinking along the lines of subjectivity. Interestingly, one student justified this type of guessing and subjectivity with the assertion that all people probably guess, so the results average. It should also be noted that no differences were noted between students who had glasses or contact lenses and those who did not.

After discussing aberrometry, students were asked what the advantages and disadvantages of that type of system could be. The issue of subjectivity was raised by nine of 12 students. Based on these responses, the idea of objectivity now included a component of "not open to human interpretation."

Reflection and Refraction

The words 'refraction' and 'reflection' were used improperly by nine of 12 participants (with one student using 'diffraction' as well). Common statements include light is: "refracted off a lens," "reflected through a lens," and "reflected into the eye."

We found no previous studies about this word usage. Further investigation will determine if this confusion is related to vocabulary or learning issues.

CONCLUSIONS AND FUTURE WORK

This study is one component of a larger project to understand transfer of physics learning to novel contexts and to design teaching and learning materials on the application of physics to contemporary medicine, including wavefront aberrometry diagnosis and techniques. The results of this study indicate that while most students have a large body of prior knowledge about the human eye and basic optics, much scaffolding will be needed in order to facilitate the transfer of that knowledge to wavefront aberrometry techniques. Students have a significant body of resources that they use to understand aberrometry – some appropriately and some inappropriately. These resources need to be considered carefully as we move forward in the design of teaching and learning materials. This study also indicates that while students do not immediately recognize the subjective nature of traditional

diagnosis, once prompted they both acknowledge and appreciate the value of objective methods such as the aberrometer.

Reliability checks need to be performed on the data before continuing and we would like to further investigate the issue of reflection/refraction that has been raised. We have also uncovered some rather unusual (to us) views on the concept of objectivity. Further investigation is necessary to see if these views affect students' views on the nature of science. As the study progresses, we plan to continue to carefully analyze what resources students use and to examine what scaffolding is needed to assist in the transfer of previous knowledge to the diagnosis techniques of wavefront aberrometry.

ACKNOWLEDGMENTS

This work is supported by the National Science Foundation under grant DUE 04-26754. Any opinions, findings, and conclusions or recommendations expressed in this material are those of the authors and do not necessarily reflect the views of the National Science Foundation. We would also like to acknowledge Dr. Hartmut Wiesner and Dr. Giuseppe Colicchia from Ludwig Maximilian University in Munich, Germany.

REFERENCES

1. K.M. Hamed, PhD Dissertation, Kansas State University, Manhattan KS, 1999.
2. E. Corpuz, PhD Dissertation, Kansas State University, Manhattan KS, 2006.
3. R.A. Applegate, S. Marcos, & L.N. Thibos, *Optometry and Vision Science*, **80**(2), 85-86 (2003).
4. R. Krueger. *Ophthalmology*, **108**(4), 674-678 (2001).
5. L.N. Thibos, *Journal of Refractive Surgery*, 16(Sept/Oct), S563-S565 (2000).
6. J. Greeno, D.R. Smith, & J.L. Moore, "Transfer of Situated Learned," in *Transfer on Trial*, edited by D. Dellerman & R. Sternberg, Norwood, NJ: Ablex. (1993) pp. 99-167.
7. N.S. Rebello. "Dynamic Transfer: A Perspective from Physics Education Research," in *Transfer of Learning from a Modern Multidisciplinary Perspective*, edited by J. P. Mestre, Greenwich, CT (2005) Information Age Publishing. pp. 217-250.
8. D. Hammer, *American Journal of Physics Physics Education Research Supplement*, **68**(S1), S52-S59 (2000).
9. A similar model is available, for instance, from American 3b Scientific. Item Number: W16002
10. G. Colicchia & H. Wiesner, *Physics Education*, **41**(4), 307-310 (2006).
11. F. Marton. *Journal of Thought*, 21: 29-39 (1986).
12. A. diSessa, *Cognition and Instruction*, 10(2-3): 105-225 (1993).

Student Estimates of Probability and Uncertainty in Advanced Laboratory and Statistical Physics Courses

Donald B. Mountcastle[1], Brandon R. Bucy[1], and John R. Thompson[1,2]

[1]*Department of Physics and Astronomy and* [2]*Center for Science and Mathematics Education Research*
University of Maine, Orono, ME

Abstract. Equilibrium properties of macroscopic systems are highly predictable as n, the number of particles approaches and exceeds Avogadro's number; theories of statistical physics depend on these results. Typical pedagogical devices used in statistical physics textbooks to introduce entropy (S) and multiplicity (ω) (where $S = k \ln(\omega)$) include flipping coins and/or other equivalent binary events, repeated n times. Prior to instruction, our statistical mechanics students usually gave reasonable answers about the probabilities, but not the relative uncertainties, of the predicted outcomes of such events. However, they reliably predicted that the uncertainty in a measured continuous quantity (e.g., the amount of rainfall) does decrease as the number of measurements increases. Typical textbook presentations assume that students understand that the relative uncertainty of binary outcomes will similarly decrease as the number of events increases. This is at odds with our findings, even though most of our students had previously completed mathematics courses in statistics, as well as an advanced electronics laboratory course that included statistical analysis of distributions of dart scores as n increased.

Keywords: Thermal physics, statistical mechanics, statistics, multiplicity, entropy, Second Law of Thermodynamics, equilibrium properties, probability, uncertainty, coin flips, physics education research
PACS: 01.40.-d, 01.40.Fk, 05.70.Ce, 65.40.De, 65.40.Gr.

INTRODUCTION

At the University of Maine (UMaine), we are actively involved in a research study of the teaching and learning of thermal physics at the advanced undergraduate level [1,2]. One theme of our research is the extent to which student mathematical conceptual difficulties may affect understanding of physics concepts. Our department offers both a one-semester classical thermodynamics course (text: Carter [3]) and a separate one-semester course in statistical mechanics (text: Baierlein [4]). Here we report research on student understanding of probability and statistics in three successive semesters of *Statistical Mechanics* (Spring 2005-07), along with a statistics project in a junior level *Physical Electronics Laboratory*.

Previous work in PER has explored introductory student understanding of measurement error analysis and measurement uncertainty [5-7]. We are investigating the extent to which advanced students connect measurement uncertainty in a laboratory context to probability and uncertainty in statistical physics.

Twenty-seven *Statistical Mechanics* students are included in this data; most were senior physics majors,

two were mathematics majors and four were physics graduate students. Seventeen of the twenty-three undergraduates in *Statistical Mechanics* included in this study had previously participated in the *Physical Electronics Laboratory* course, either the previous semester or a full year earlier; sixteen had previously taken a mathematics course in statistics.

The theoretical foundation of the learning objectives reported here is the *fundamental postulate in statistical mechanics*: All accessible microstates are equally probable in an isolated equilibrium system. Mathematical foundations are the *Central Limit Theorem*, and the *Strong Law of Large Numbers*.

Boltzmann gave us the extraordinary insight that the world is highly predictable *due entirely to probability*, expressed in multiplicity *(ω)*, entropy *(S)* (where $S = k \ln(\omega)$), and the Second Law of Thermodynamics. Statistical mechanics textbooks [3,4,8] typically introduce the definitions and meanings of multiplicity and entropy by exploring the binomial coefficient describing independent binary outcomes of n events (such as coin flipping) as n increases. The advantage of this approach is that it offers concrete countable events that can be easily visualized, measured, calculated and displayed as

CP951, *2007 Physics Education Research Conference*, edited by L. Hsu, C. Henderson, and L. McCullough
© 2007 American Institute of Physics 978-0-7354-0465-6/07/$23.00

histograms of probability, shown to converge, *with overwhelming probability*, to a single predictable system macrostate; i.e., the inevitable equilibrium state. Convergence is seen as a transition from a discrete, broad histogram (small n) into a continuous, smooth, sharply-peaked function (large n), accompanied by a decreasing relative uncertainty ($\Delta n/n$) (Figure 1). Full understanding of that transition requires the conceptual ability to move seamlessly from a summation to an integral, from obvious to hidden degeneracy, and from discrete states to a (continuous) density of states. This conceptual ability is fundamentally important in learning statistical mechanics [9], yet at the same time can be exceptionally challenging for students.

Often textbooks thus use independent binary outcome events, but rather quickly extrapolate from small n to large n, emphasizing the narrowing sharpness of the macrostate (probability peak; Figure 1) as n increases toward the thermodynamic limit, all based on simple rules of combinations, and probability. The (vanishing) *uncertainty* associated with the most-probable macrostate is usually acknowledged only indirectly by phrases such as *overwhelming probability*, or *narrowing peak-width*.

An exercise during the first week of our *Physical Electronics Laboratory* is a class group project, labeled the "darts project," adapted from Squires (2nd edition) [10]. Each student throws a specified number of darts at the center of a one-dimensional target, reports their scores (based on horizontal location) to the class electronic conference, analyzes and graphs the entire data set ($n > 200$), and compares this result to a Gaussian distribution.

One of the learning objectives of the laboratory course is facility with measurement error analysis and propagation of errors. Thus, the primary reason for the statistics of the darts project is to familiarize the students with an easily visualized, actual numerical distribution that can be seen to grow into a near Gaussian distribution as n, the number of reported scores, increases from ~ 20 to > 200. Students' observations of how the standard deviation (σ) and standard deviation of the mean (σ_m) of the dart score distribution change (or not) with increasing n are relevant to our questions about uncertainty reported here.

QUESTIONS, RESULTS, AND ITERATIONS

We were interested in whether or not our *Statistical Mechanics* students had the inclination to follow and to adopt the typical textbook arguments, using binary event outcomes, that a *single* equilibrium macrostate,

with vanishing uncertainty, will emerge as n increases towards the thermodynamic limit. These arguments articulate the microscopic basis for maximum multiplicity, entropy and the Second Law. We designed the diagnostic question Q1 (Figure 2) to probe our students' reasoning in this context both pre- and post-instruction.

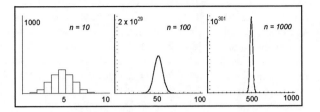

FIGURE 1. Typical textbook introduction to multiplicity, showing the convergence toward $n/2$ of the distribution of n independent binary events from a discrete histogram to a continuous smooth function for large n.

Estimate *all* of your answers as $x = \underline{\ a\ } \pm \underline{\ \Delta a\ }$. That is, give your best numerical *answers* for both (*a*) **_and_** numerical estimates of the *uncertainties* (the $\pm \Delta a$) that indicate your confidence level in the precision of your answers (*be realistic*).

If you flip one coin n times, estimate the value of and the uncertainty of the number of 'heads' that would occur for:

    a) $n = 4$

    b) $n = 100$

    c) $n = 1000$

    d) $n = N_A = 6.022 \times 10^{23}$

Explain the reasoning for your estimates in (a) - (d) above.

FIGURE 2. Question Q1, given in *Statistical Mechanics* both pre- and post-instruction about multiplicity and probability (pretest version shown).

It is important to point out that we are looking for the qualitative trend of *relative* uncertainties ($\Delta n/n$), as n increases, rather than precise numerical answers. Qualitatively, a prediction of $n/2 \pm n/2$ is *not unreasonable* for $n = 4$ coin flips, but is dubious for $n = 10$, extremely unreasonable for $n = 100$, and totally unreasonable for $n = 1000$ and $n = N_A$. On the pretest, essentially all students as expected correctly gave $n/2$ as their predicted cumulative number of heads. However, their responses for the "realistic uncertainty" were surprising. Fewer than half of the students seemed to believe with confidence in a decreasing trend in relative uncertainty ($\Delta n/n$). A substantial number ($\sim 20\%$) specifically stated uncertainties of $\pm n/2$, for parts (a) – (d), as if all n (or zero) heads is a realistic possibility (Table 1). Although not as severely unreasonable, another 13% predicted the same relative uncertainty (e.g., $\Delta n/n$ of 15%) in outcome for n=100 and n=1000. (Often they gave smaller relative uncertainties for part (d).) Another 17% refused to commit to any model for the n-dependence of uncertainty by omitting it in their responses to Q1. Even those students who did exemplary work previously on the darts project did not always estimate a decreasing uncertainty with increasing n, which was one of the intended learning outcomes of that exercise. On similar post-test questions about coin flips as functions of n (2006-07, N=12), essentially all students gave realistic estimates for both the values *and* uncertainties for the number of heads, although generally with superficial explanations of their reasoning (Table 1).

TABLE 1. Results from Q1 for relative uncertainties.

Response	Pre-instruction (N=24)	Post-instruction (N=12)
Decreases with n (correct)	11	11
Covers entire range	5	--
Same for n=100, n=1000	3	--
No uncertainty	4	--
Other	--	1

From the laboratory class and elsewhere, we know that students quite often accept, even take for granted, that taking multiple readings and averaging those values reduces uncertainties in measured quantities. So, in class the next two years we again gave students Q1, collected their written answers, and then immediately gave them Q2 (Figure 3). That collection sequence was followed in order to prevent students from changing their Q1 answers *after* reading Q2.

The groundskeeper at the Penobscot Valley Country Club wants to know the total amount of rain (in kg H_2O) that falls on the golf course during a predicted downpour. Before the storm, four atmospheric science majors report to the golf course with timepieces and rain gauges in hand; Andy brings one rain gauge, Betsy brings four, Charlie brings 40, and Debby brings 400. You may assume all k now how to use them.

a) Which student do you expect will be able to determine a value closest to the actual total amount of rainfall? Please explain your reasoning.

b) Rank the uncertainties (as a percentage of the total amount of rainfall) associated with each student's determined value. If any uncertainties are equal, state that explicitly. Explain.

FIGURE 3. Question Q2, given in Statistical Mechanics (2006, -07) as a pretest only.

All students (N = 13) answered both parts of Q2 as expected, indicating that the relative uncertainty in the measurement of rainfall should decrease with increasing n.

We have noted the richness of statistical information available in the darts project tracking dart scores with increasing n, and yet the poor learning outcomes evident in the later *Statistical Mechanics* pretest. Consequently, we changed the darts exercise protocol in Fall 2006 (Figure 4) to add explicit individual *predictions* (based on their own scores only) for several features of the dart score distribution, emphasizing what *should change* (e.g., smoother histograms and σ_m decreasing) and what *should not change* (e.g., the distribution σ) as n (the number or reported scores) increases.

For each of the following, provide your predictions, and a brief explanation of how you made them, for:

(*i*) n = 24 (your own data); (*ii*) n = 220; (*iii*) n = 500.

a) the *average* (mean; $< X >$) of the n scores

b) the *standard deviation* (σ) of the n scores

c) the *standard deviation of the mean* (σ_m) of the n scores

---

FIGURE 4. Question Q3, given in *Physical Electronics Laboratory* (2006); these three questions are a subset of a considerably longer pretest.

TABLE 2. Results from Q3 for predicted standard deviations (σ) of dart scores for *Physical Electronics Laboratory* students (Fall 2006; N = 13)

Response	N
No change in σ as n increases (correct)	5
σ increases as n increases	4
σ decreases as n increases	4

The Fall 2006 results contain evidence of student confusion with statistics in their predictions from the darts project (Table 2). Of 13 students, over 60% predicted (incorrectly) that the standard deviation (σ) of the distribution *should change* with increasing n. Of those, half predicted σ would *increase* with n, while the other half predicted a *decrease* in σ. Both groups gave written explanations that they found satisfying; e.g., σ will increase with n, since "with more throws, there will be a greater spread in values…"; or σ will decrease with n "since outliers will have less and less impact…" Some of the students who predicted a decrease in σ with increasing n may be confusing σ of the distribution with σ_m, the standard deviation of the mean of the distribution (related to the relative uncertainty in Q1). However, regardless of what students indicated as a trend for σ (increasing, decreasing, or staying the same) in part (b) of Q3, they typically determined values for σ_{m} for part (c) by dividing their answers to part (b) by $\sqrt{n}$.

DISCUSSION AND CONCLUSIONS

Our findings so far include several that were expected, as well as some surprises on both ends of the performance spectrum. We point out some of each here, and briefly outline future plans.

Prior to instruction in statistical mechanics, essentially all of our students exhibited attributes of concepts that are consistent with appropriate models for various applications of statistics, probability and uncertainty. For example, (1) all students predicted the cumulative outcome for the number of 'heads' recorded for n coin flips to most likely be the expected value of $n/2$ [Q1], and (2) in a qualitative ranking, students agreed that the uncertainties in measurement of a continuous macroscopic quantity will decrease with n, the number of measurements made [Q2(b)].

The most serious conceptual deficiency identified here is that most of our students do *not* predict the *convergence* towards the $n/2$ (binary) macrostate with increasing probability as n increases. As seen in Q1, approximately 20% of our students start with a reasonable model for $n = 4$ coin flips (# of heads = $n/2 \pm n/2$) but remain stuck there, regardless of the magnitude of n; another 13% give smaller, but still constant, percentages of n as the uncertainty for intermediate values of n.

However, results from Q2 indicate that all our students share the widespread qualitative belief that macroscopic measurement uncertainty is indeed reduced with increasing n. Apparently, something is quite different in student understanding of the statistics describing discrete binary events [Q1] and uncertainties of measurements of continuous variables [Q2]. Even though they all have at least qualitatively appropriate models for uncertainties in a continuum context (rainfall), most fail to make a transition from discrete to continuous models as an increasing n should indicate.

Finally, the results from parts (b) and (c) of Q3 indicate that many students know the formula $\sigma_{\mathrm{m}} = \sigma/\sqrt{n}$, yet lack the conceptual understanding of distinctions between σ and σ_{m}.

As noted earlier, the typical textbook presentation of the microscopic basis for predictable macroscopic equilibrium states via entropy and the Second Law is the dramatic convergence with increasing n to a single, well-defined, most probable macrostate of $n/2$, as seen in Figure 1. We have not yet offered curricular content that differs from such a textbook presentation in any significant way, and appropriate post-instruction student predictions for both outcomes ($n/2$) and associated uncertainties indicate that the traditional approach may be somewhat successful (Table 1). However, based on the quality and depth of the reasoning accompanying the post-instruction responses, we do not believe the conceptual difficulties with uncertainties of discrete outcomes seen with the pretests have been resolved by current instruction. Many students give correct answers supported only by terse descriptions of reasoning that are less than convincing. Since fluency with concepts of a smooth transition from the discrete to the continuous with increasing n is such a widespread and fundamental need throughout the physics curriculum, we plan to develop curricular materials to address this need, although specific instructional strategies have not yet been developed. Individual student interviews based on the written questions shown here will help us confirm common models, as well as find if students are even aware of the need for such concept development.

Informal conversations with individual students indicate one reason for incompatible rather than complementary models (discrete vs. continuous) may be that *each* binary coin flip event has zero associated uncertainty, while *each and every* measurement of a continuous variable *always* has inherent associated uncertainties. Interviews may explore further this line of reasoning.

ACKNOWLEDGMENTS

We thank an anonymous referee for useful comments. Supported in part by NSF Grant #PHY-0406764.

REFERENCES

1. J.R. Thompson, B.R. Bucy, and D.B. Mountcastle, *2005 Phys. Educ. Res. Conf.,* edited by P. Heron *et al.,* AIP Conf. Proceedings **818**, 77-80 (2006).
2. B.R. Bucy, J.R. Thompson, and D.B. Mountcastle, *2006 Phys. Educ. Res. Conf.,* edited by L. McCullough *et al.,* AIP Conf. Proceedings **883**, 157-160 (2007).
3. A.H. Carter, *Classical and Statistical Thermodynamics,* Upper Saddle River, NJ: Prentice-Hall, Inc., 2001.
4. R. Baierlein, *Thermal Physics,* Cambridge: Cambridge University Press, 1999.
5. D.S. Abbott, unpublished doctoral dissertation, North Carolina State University (2003).
6. S. Allie, et. al., *Phys. Teach.* **41**, 394-401 (2003).
7. R.L. Kung, *Am. J. Phys.* **73**, 771-777 (2005).
8. D.V. Schroeder, *An Introduction to Thermal Physics,* San Fransisco: Addison Wesley Longman, 2000.
9. A. Compagner, *Am. J. Phys.* **57**, 106-117 (1989).
10. G.L. Squires, *Practical Physics,* 2nd. ed., Cambridge: Cambridge University Press, 1976.

Peer-assessment of Homework Using Rubrics

Sahana Murthy

Massachusetts Institute of Technology, 77 Mass. Ave., Cambridge, MA 02139

Abstract. I have implemented a peer-assessment system in an introductory physics course, where students assess each other's homework. Students are provided with descriptive rubrics to guide them through the process. In this paper I describe the implementation of the peer-assessment process, discuss the role of rubrics, present data of agreement of students' assessment with the instructor's, and show evidence of student improvement in evaluation abilities.

Keywords: Rubrics, assessment, peer assessment, evaluation.
PACS: 01.40.Fk, 01.30.lb, 01.40.gb

INTRODUCTION

An assessment system that supports learning involves student participation in productive activity, distribution of student effort evenly across topics and weeks, communication of clear expectations by the instructor, and detailed, frequent and quick feedback that reinforces learning goals [1]. One way to meet the above goals is to have students systematically and consistently assess the work of their peers. Peer-assessment also engages students in the process of evaluating scientific information, which is often an important part of students' future professional careers [2]. In addition, peer-assessment could create a sense of community in the classroom and a shared ownership of the learning process. Instances of peer-assessment have previously been reported in middle school classrooms [3] and college-level courses [1] with favorable results.

In this paper I describe instructional and research efforts to implement a peer-assessment system in an introductory physics class in which students assess each other's homework. Students were provided with assessment rubrics to guide them through this process, and to incorporate descriptive, criterion-based feedback. In this preliminary study, I address two questions: how consistent is students' assessment with the instructor's (author of this paper), and do they develop and improve upon evaluation abilities?

COURSE DETAILS

The peer-assessment process was implemented in a two-semester calculus-based introductory physics course. The course is part of MIT's Experimental Study Group [4], an alternate academic program that offers highly interactive, small group learning in the core first-year subjects within a community-based setting. There were two sections, each of about 10 students, consisting mainly of engineering majors. Each section met for 4 hours a week. Instruction included in-class collaborative problem solving, some ISLE-style observational and testing experiments [5], visualizations [6], and class discussions. Throughout the course, there was a focus on developing and using scientific abilities such as multiple representation, experimental testing and evaluation [7].

Homework details

Homework was assigned on a weekly basis. Students wrote their solutions and submitted them in class a week later. Homework contained various PER-based questions, such as multiple-representation tasks [8], ranking tasks [9], convince-your-friend tasks [8], and evaluation tasks [10]. Longer analytic questions had parts that focused on scientific abilities. A sample question from kinematics is shown below:

Example. A car and a motorcycle travel on a straight highway. The motorcycle is initially at a distance d behind the car and moves at the same velocity as the car. The motorcycle starts to pass the car by speeding up at a constant acceleration a_m. When it is side by side with the car, it stops accelerating and is travels at twice the velocity of the car. Your goal is to determine the distance traveled by the motorcycle while it is accelerating.

a) Construct a position-time and a velocity-time graph for the two vehicles.

CP951, *2007 Physics Education Research Conference*, edited by L. Hsu, C. Henderson, and L. McCullough
© 2007 American Institute of Physics 978-0-7354-0465-6/07/$23.00

b) Determine the distance the motorcycle traveled while it was accelerating.

c) Perform at least two checks to see if your result is reasonable.

ELEMENTS OF THE PEER-ASSESSMENT PROCESS

Rubrics

The backbone of the peer-assessment process was the assessment rubrics. The rubrics contained detailed but general descriptors of criteria to assess problems. These criteria include: physics content, relevant representations, modeling the situation, problem-solving strategy and reasonableness of answer. The criteria were chosen based on problem-solving literature in physics and science education [such as 12, 13]. For each criterion, the rubrics contained a scoring scheme on a scale of 0-3 for different levels of performance, 3 being the desired level. The format of the rubrics is based on those described in [7]. Students used the same assessment rubrics throughout the year. A portion of the rubrics is shown in Table 1.

In addition, students were also provided with a taxonomy that described what the criteria in the general rubrics were, for a specific problem. For example, the taxonomy for the problem solving strategy in the example above includes choosing a coordinate system, identifying physical quantities with respect to the coordinate system, and equating the position of the two vehicles at the time of their meeting.

Timeline of the assessment process

Following students' submission of their homework, there was a 10-minute class discussion of assessment criteria for selected questions. Then each student was randomly assigned another student's homework for assessment. Students also took home solutions to the homework questions, rubrics, taxonomies and a score-sheet. Fig. 1 shows a sample score-sheet filled out by a student-grader. The score-sheet contained a blank grid in which students wrote the rubric score and comments for relevant criteria for every homework question. If they assigned a less than perfect score, they had to write a comment as to why the particular score was assigned. For example, see the entry under problem 1, 'Relevant representations' in Fig. 1. The student-grader has assigned a score of 2, and has commented that labels on forces are missing.

Students returned the assessed homework in 3-4 days to the instructor who then checked student assessment and made necessary corrections on the score-sheet. See for example, the circled entries (initialed by the instructor) under problem 4 in the score-sheet in Fig. 1. If a score-sheet contained too many errors, the instructor met with the student-grader and discussed the problems. Such meetings occurred more frequently and with more students in the initial weeks, but as students' assessment became more consistent, the need for these discussions reduced. Students got their homework with rubric scores and comments in 1-2 days after they handed in filled out score-sheets. The entire process from the time students handed in their homework to the time they got back their assessed homework took no longer than a week.

TABLE 1. Portion of assessment rubrics showing two criteria.

CRITERIA	0: MISSING	1: INADEQUATE	2: NEEDS IMPROVEMENT	3: ADEQUATE
Modeling the situation Is able to construct a useful model of the situation in the problem	No attempt is made to model the situation.	An attempt is made to model the situation but details are missing or unclear.	An attempt is made to model the situation. Some assumptions are stated, but there is an error or one important aspect missing.	The model fits the given situation. All assumptions about objects and processes are clearly stated.
Reasonableness of answer Is able to evaluate if the answer obtained is reasonable.	The answer is not reasonable and no attempt is made to evaluate its reasonableness.	The answer is reasonable, but student makes no attempt to evaluate why it is so. OR: The answer may not be reasonable and student recognizes that, but does not attempt to analyze what makes it so.	The answer is reasonable, and student makes an attempt to evaluate it, but the evaluation is incorrect or incomplete. OR: The answer may not be reasonable but student recognizes that and tries to analyze it. But the analysis is flawed or incomplete.	The answer is reasonable, and student evaluates why it is so (e.g. by limiting cases or dimensional analysis). OR: The answer may not be reasonable. Student correctly analyzes what makes it so.

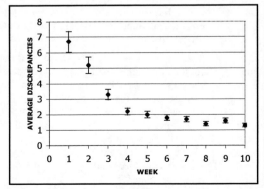

FIGURE 1. Score-sheet filled out by student-grader

Connection between peer-assessment and grades

The grade a student received on a homework was computed by adding the checked rubric scores from the score-sheet. The homework grades were worth 20% of the course grade. I considered reserving a portion of the course grade for peer-assessment, but I did not encounter many problems in terms of students not taking peer-assessment seriously. In the end, no extra points were assigned for the peer-assessment part of the homework, but a student had to return a classmate's assessed homework in order to receive credit for his/her own homework.

RESULTS

Consistency between students' and instructor's assessment

When students returned the assessed homework and score-sheet, the instructor checked their entries for rubric scores. Essentially, a student's homework was scored again, using the same rubrics. Discrepancies between student-assigned scores and instructor-assigned scores were noted. The left graph in Fig. 2 shows first semester (mechanics) data for the average number of such discrepancies per week. The total number of rubric scores in a given week was between 30 and 40, depending on the number of homework questions. We see that by week 4, the average number of discrepancies have stabilized to about 2 per week (about 7%). In the right graph in Fig. 2 we see the percentage of students whose assessments had more than 10% discrepancy with the instructor's assessment. Again we see that this percentage steadily decreases initially, and then stabilizes around week 4. A similar documentation (graph not shown) of assessment discrepancies from the second semester (electricity and magnetism) shows that students' assessment was consistent with the instructor's.

Development of evaluation abilities

Evaluation ability is defined as making a judgment about information based on specific criteria [7] and strategies. Since the process of peer-assessment involves judging scientific information based on specific criteria in the rubrics, a reasonable question to ask is if students' evaluation abilities change during the course of the semester. That is, can students learn strategies used by practicing physicists, such as dimensional analysis and special case analysis to evaluate a solution to a problem? To explore this, exam questions were assigned in which students had to evaluate a proposed solution to a given problem. The solution had errors in physics content, incorrect assumptions, numerical errors and so on. Students had to evaluate the proposed solution using different strategies, discuss whether the solution satisfied certain criteria and identify errors.

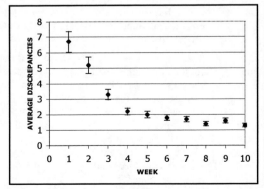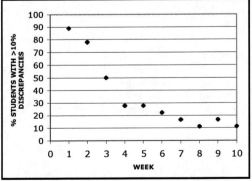

FIGURE 2. LEFT: Average number of assessment discrepancies per student per week. RIGHT: Percentage of students with 10% or more assessment discrepancies per week

Students' responses were scored using scientific abilities rubrics developed by the Rutgers PAER group [11]. (These are not the same as the rubrics students used for peer-assessment). Students' average scores on three evaluation strategies are shown in Table 2.

Table 2. Students' average scores (std. dev) on strategies used to evaluate solutions to problems			
Strategy	Exam 1	Exam 2	Final
Use of dimensional analysis	2.5 (0.95)	2.8 (0.62)	2.9 (0.51)
Use of special-case analysis	1.4 (1.04)	1.8 (0.98)	2.6 (0.65)
Identification of errors in modeling	1.6 (1.18)	2.0 (1.10)	2.2 (1.09)

Students quickly mastered the strategy of dimensional analysis to evaluate a solution. The strategy of using special case analysis took longer to be learned. Students struggled the most with identifying errors in assumptions and modeling a situation (such as, is it constant velocity or constant acceleration motion), but they improved upon this as the semester progressed.

DISCUSSION AND SUMMARY

Preliminary data from the first semester show that students do learn to systematically assess physics problem solutions using given criteria. In our study, students' assessment could be considered to be consistent with the instructor's after about four weeks. Students' evaluation abilities improved over the course of the semester. Some factors which contributed to the success of the peer-assessment process are: the use of rubrics which helped clearly communicate learning and performance goals and incorporated criterion-based descriptive feedback; a weekly schedule that encouraged quick feedback; dedicated class time to discuss details of assessment; and an effort by the instructor to develop among students a sense of the importance of evaluation abilities.

Implementing the peer-assessment process was not without challenges. The initial investment of time in writing the rubrics was large. Substantial effort was required in training students for the first 3-5 weeks. The overall time spent by the instructor was comparable to what one would spend grading homework traditionally. In terms of attitudes, most students accepted peer-assessment as part of the course. In class, the instructor repeatedly discussed the rationale behind peer-assessment and invited students to share their reactions. On the positive side, some students thought that peer-assessment helped them realize what they did or did not understand. On the other hand, one student, who mainly focused on the grading aspect, repeatedly expressed her discomfort. She commented on the mid-term evaluation that it was the teacher's job to grade, not the students'.

In terms of further work, one question to study is if there students' assessment work in a given content area is related to their conceptual understanding. Another open question is to understand what frame students are in when they are involved in assessment tasks: do they see it as a grading task, do they reflect on their understanding or are they looking to learn something new? A practical problem that needs to be addressed is how to scale up the peer-assessment process to larger classes. These are the subjects of present and future studies. Currently, results from this study are being used to develop and implement a peer-assessment process in two other classes at the Experimental Study Group.

ACKNOWLEDGMENTS

I would like to thank all members of the Experimental Study Group at MIT.

REFERENCES

1. Gibbs, G. and Simpson, C. "Does your assessment support your students' learning?" *Journal of Learning and Teaching in Higher Education*, **1**, 1 (2003).
2. Boud, D, "The role of self-assessment in student grading", *Assessment and Evaluation in Higher Education*, **14**, 20–30 (1989).
3. P. Sadler and E. Good, "The impact of self- and peer-assessment of student learning", *Educational Assessment* **11**(1), 1-31 (2006).
4. http://web.mit.edu/esg/www/
5. E. Etkina, A. Van Heuvelen, D. T. Brookes, and D. Mills, *Physics Teacher* **40**, 351 (2002).
6. http://web.mit.edu/8.02t/www/802TEAL3D/
7. E. Etkina, A. Van Heuvelen, S. White-Brahmia, D. Brookes, M. Gentile, S. Murthy, D. Rosengrant, and A. Warren. "Scientific abilities and their assessment." *Phys. Rev. ST Phys. Educ. Res.* **2 (2)**, 020103 (2006).
8. A. Van Heuvelen and E. Etkina, *The Physics Active Learning Guide*, San Francisco: Pearson Education, 2006.
9. T. L. O'Kuma, D. P. Maloney and C. J. Hieggelke, *Ranking task exercises in physics*, Benjamin Cummings, 2003.
10. For a description of these tasks, see http://paer.rutgers.edu/ScientificAbilities/Formative+Assessment+Tasks/default.aspx
11. http://paer.rutgers.edu/ScientificAbilities/Rubrics/default.aspx, specifically Rubric I.
12. R.E. Mayer, *Thinking, Problem solving, Cognition* (2nd ed.), New York: Freeman & Co., 1992.
13. A. Van Heuvelen. "Learning to think like a physicist: A review of research-based strategies," *Am. J. Phys.* **59**, 891-897 (1991).

Learning to Think Like Scientists with the PET Curriculum

Valerie K. Otero and Kara E. Gray

University of Colorado at Boulder
249 UCB Boulder CO 80309-0249

Abstract. Instructional techniques based on research in cognitive science and physics education have been used in physics courses to enhance student learning. While dramatic increases in conceptual understanding have been observed, students enrolled in these courses tend to shift away from scientist-like views of the discipline (and views of learning within the discipline) and toward novice-like views. Shifts toward scientist-like views are found when course materials and instruction explicitly address epistemology, the nature of science, and the nature of learning. The Physics and Everyday Thinking (PET) curriculum has specific goals for helping non-science majors explicitly reflect on the nature of science and the nature of science learning. We show that in PET courses with small and large enrollments, shifts toward scientist-like thinking ranged from +4% to +16.5% on the Colorado Learning Attitudes about Science Survey. These results are compared to results from other studies using a variety of similar assessment instruments.

Keywords: physics education, curriculum, student beliefs, teacher preparation
PACS: 01.40.Di, 01.40.Ha, 01.40.jc

INTRODUCTION

National documents have made clear the need for increased scientific literacy among the general public [1] [2]. Among the issues raised in these types of reports is the need for undergraduate science courses that not only address fundamental content goals, but also explicitly address the nature of scientific knowledge, science as a human endeavor, and the unifying concepts and processes of science. Science education researchers and curriculum developers have responded to this call by developing inquiry-based, physical science curricula especially for the post-secondary, non-science major population. Such curricula include Physics By Inquiry [3], Powerful Ideas in Physical Science [4], Operation Primary Physical Science [5], Physics and Everyday Thinking [6], and Physical Science and Everyday Thinking [7]. In most cases, large conceptual gains have been found to be associated with these specialized curricula [5][8][9]. However, less work has been done to determine how students' beliefs about learning science and their understanding of the nature of scientific knowledge are impacted by such courses.

Research has consistently demonstrated that K-12 teachers and their students do not develop desired understandings of the nature of science and the nature of science learning [10]. Abd-El-Khalick showed that pre-service elementary teachers developed a better understanding of certain aspects of the nature of science when nature of science instruction was embedded within content instruction [10]. Unfortunately, few science teacher education courses focus on content and few science courses focus on the nature of science and the nature of learning science.

Research has found that not only does instruction on the nature of science and the nature of learning need to be embedded in content, it also needs to be *explicit* in order to be effective [11]. Instruction that explicitly addresses issues about the nature of science and learning is much more effective in improving students' views and attitudes about aspects of the nature of science and science learning than implicit approaches that use inquiry-based science activities but lack explicit references to the nature of science knowledge and learning [11]. For example, Elby [12] demonstrated that a curriculum that was "epistemologically-focused" had greater success in helping students develop sophisticated beliefs about the nature of science knowledge than curricula focused on content only. Elby argued that "even the best curricula aimed at conceptual development *but not aimed explicitly at epistemological development* do not produce comparable epistemological results [12]."

Several instruments have been developed to assess students' views of the nature of learning science and their views of the nature of science knowledge. These include the Maryland Expectations Survey (MPEX) [13], the Epistemological Beliefs Assessment for

CP951, *2007 Physics Education Research Conference*, edited by L. Hsu, C. Henderson, and L. McCullough
© 2007 American Institute of Physics 978-0-7354-0465-6/07/$23.00

Physical Science (EBAPS) [14], the Colorado Learning Attitudes about Science Survey (CLASS) [15], and the Views of the Nature of Science (VNOS) survey [10]. While each of these instruments measures slightly different aspects of students' expectations, beliefs, and understandings of the nature of science and the nature of learning science, studies that use these instruments show remarkable similarities—students of all ages have difficulty learning how science knowledge is constructed and in most cases regress in sophistication over a semester-long science course. For example, Redish, Saul, and Steinberg [13] found that at six different schools, students' overall MPEX scores deteriorated over one semester of introductory physics rather than becoming more sophisticated. Although students enrolled in courses that use research-based curricula show significant conceptual gains as measured by conceptual instruments such as the Force and Motion Conceptual Evaluation, they do not show significant increases in beliefs about the nature of science or the nature of science learning as measured by the MPEX or the CLASS [16]. These results suggest that conceptual learning is not necessarily associated with expert-like beliefs about the nature of learning science. Hrepic et al. [5] studied a curriculum designed specifically for pre-service elementary teachers, Operation Primary Physical Science Curriculum (OPPS), from both the conceptual and attitudinal perspective. While significant conceptual gains were found using a conceptual instrument designed by the authors of the curriculum, only very small positive shifts in attitudes about science and the nature of science knowledge and learning were measured by the CLASS and in one course CLASS scores deteriorated [5]. Although the OPPS curriculum was developed for pre-service elementary teachers, it does not appear to explicitly address issues about learning science and the nature of science knowledge.

THE PET AND PSET CURRICULA

Physics and Everyday Thinking (PET-formerly Physics for Elementary Teachers) and Physical Science and Everyday Thinking (PSET) are inquiry-based curricula designed to meet the needs of elementary teachers. PET course content focuses on the themes of interactions, energy, forces, and fields. PSET is based on the PET curriculum but integrates significant chemistry content by focusing on the same themes plus atomic-molecular theory to account for physical and chemical changes, conservation of matter, and gas behaviors. Each curriculum consists of carefully sequenced sets of activities intended to help students develop physical science ideas through guided experimentation and questioning with extensive small group and whole class discussion. The curricula also include a series of *Learning About Learning* activities, in which students are *explicitly* asked to reflect on their own learning, the learning of other students, and the learning of scientists. The *Learning about Learning* activities are embedded throughout the curricula, often occurring between two content-focused activities. For example, after PET students develop a domain-like model of magnetism through experimentation and consensus discussions, they read about the historical development of the domain model of magnetism and investigate how and why the scientifically accepted model of magnetism changed over time. PET and PSET students study their own learning by reflecting on how their ideas about a particular concept changed over time and what classroom activities, discourse, or experiments influenced these changes. Finally, PET and PSET students investigate the learning of others by watching short video clips of elementary children struggling with scientific issues similar (but age appropriate) to those found in the PET/PSET curricula. The PET and PSET *Learning about Learning* activities are similar in format to the content activities and follow a guided-inquiry format. Each activity has three parts: (1) Initial Ideas, (2) Collecting and Interpreting Evidence, and (3) Summarizing Questions.

PET and PSET *explicitly* address issues about learning science and the nature of science knowledge and these activities are embedded within the content instruction. Thus, it would be expected that students who participate in the PET and PSET curricula would be more likely to move toward expert-like thinking about the nature of science and the nature of learning science than students who participated in physics curricula that either address the nature of science and science learning *implicitly* or do not attend to them at all. We investigated the hypothesis that students in PET and PSET courses would move toward expert-like thinking about the nature of scientific knowledge throughout one semester of taking PET or PSET by administering the CLASS instrument in multiple PET/PSET classroom settings.

Previous work using the EBAPS at one institution showed that PET and PSET students did indeed have higher positive shifts toward expert-like thinking than students from traditional courses [17]. The EBAPS was administered pre/post to 228 students enrolled in traditional lecture format earth science and general physics courses and to 173 students enrolled in PET and PSET courses at a single university. Composite scores indicate a +7% to +18% change toward "expert" thinking about the nature of knowledge and learning for students enrolled in the PET and PSET courses, and a change of -2% to -9% for students enrolled in the traditional courses. Furthermore, Post-test composite scores were 14-25% higher for the students enrolled in the PET and PSET courses than for the students enrolled in the traditional courses. In other words, the traditional students became more novice-like in their thinking about the nature of knowledge and learning during a semester of traditional instruction but the PET/PSET students became more expert-like in their thinking.

In the following section we discuss the administration and analysis of the CLASS survey in PET/PSET courses. The atypical shifts found among students enrolled in PET/PSET may be the result of the intensive, explicit focus on issues of Learning about Learning embedded throughout the curricula.

METHOD AND DATA ANALYSIS

We administered pre and post CLASS surveys to 360 students enrolled in nine different PET and PSET courses at seven universities. These PET and PSET courses had enrollments ranging from 13 to 100 students. The CLASS was chosen for this study because it is worded to be applicable to a wide variety of physics courses and to be meaningful to students who have not taken physics [15].

A total of 288 pre-surveys and 265 post-surveys were returned. Of the surveys returned, 182 were used for the analysis because they met the following criteria: the student submitted both pre and post-surveys, the student correctly responded to question #31 which is intended to catch students who are not reading the survey, the student did not select the same answer to almost all the statements, and the student answered all or almost all of the statements.

Characteristics of each of these universities and courses are shown in Table 1. One university offered three courses taught by two different instructors.

The CLASS consists of 42 statements for which students are asked to agree or disagree to using a 5-point Likert scale. Of these 42 statements, 36 are scored by comparing a students' response to the expert response. An overall favorable score is then calculated for each student by comparing the percentage of answers in which their response matches the expert response. An average score for each course is then found by averaging all student scores.

RESULTS

Pre and Post-CLASS scores for each participating course are shown in Table 2 along with the shifts for each course.

TABLE 2. Overall Favorable Scores in Pre/Post CLASS Survey (standard error of the means in parentheses)

Course	Pre-Test	Post-Test	Shift
1	69.9 (2.6)	73.8 (4.2)	3.9 (3.6)
2	53.6 (4.3)	67.0 (2.6)	13.3 (3.1)*
3	51.6 (2.0)	58.3 (2.2)	6.7 (2.0)*
4	49.5 (3.0)	59.0 (3.7)	9.5 (3.6)*
5	49.8 (3.7)	59.6 (3.8)	9.8 (2.8)
6	51.8 (2.7)	68.3 (3.3)	16.5 (3.8)*
7	51.6 (4.5)	62.1 (3.7)	10.6 (4.3)
8	64.2 (3.0)	70.4 (3.8)	6.3 (2.5)
9	55.0 (3.6)	63.1 (4.0)	8.1 (2.7)*
All Students	53.8 (1.15)	62.6 (1.18)	8.8 (1.1)*

* p-value < 0.05

Shifts toward favorable increased from +4% to +16.5%, which is similar to the shifts found by Allen and Kruse on the EBAPS. Aggregate results from the CLASS study show an average shift of 8.8% in PET and PSET courses compared to average shifts of -6.1% to +1.8% found in other physical science courses (OPPS) designed especially for elementary teachers with enrollments ranging from 14 to 22 students [5].

Table 3 shows comparisons of CLASS pre/post scores and shifts for different curricula and different instructional settings. As shown in table 3, the average CLASS pre-test score is 54 which is low in comparison to typical scores found in calculus-based courses. Traditional lecture style courses have been found to have shifts on the CLASS of -8.2% to +1.5% in calculus-based courses with enrollments of 40 to 300 students in each course section.

TABLE 1. Characteristics of Schools and Courses

Course	School	Instructor	School Type	Region	Curriculum	Enrollment	Surveys Scored
1	A	S	Community College	Mid-Atlantic	PET	13	6
2	B	T	Community College	South	PET	30*	18
3	C	U	Research University	MidWest	PET	100	54
4	D	V	Regional University	South	PSET	32	18
5	D	W	Regional University	South	PSET	32	18
6	D	W	Regional University	South	PET	25	10
7	E	X	Technical University	South	PET	30	17
8	F	Y	Regional University	Midwest	PET	48*	20
9	G	Z	Regional University	Midwest	PET	50*	21

*Students in these courses were divided into two different sections.

TABLE 3. Comparison of overall CLASS Scores for different courses.

Course	N	Pre	Post	Shift
PET/PSET	189	54%	63%	8.8%
OPPS Fa04*	14			1.8%
OPPS Sp05*	20			-6.1%
OPPS Fa05*	16			2.9%
Non-Sci Fa03	76	57%	58%	1.0%
Alg Fa03	35	63%	53%	-9.8%
Calc Fa03	168	65%	67%	1.5%
Calc Sp04	398	68%	70%	1.5%
Calc Fa03	38	65%	57%	-8.2%

* Pre and post scores were not reported for these courses

Shifts of 1.0% have been found in courses for non-science majors, and -9.8% in algebra-based courses for pre-med students [18]. Of the courses shown in table 3, only PET and PSET embeds activities that explicitly address the nature of science and the nature of learning science. Recent research however, has demonstrated significant positive shifts on CLASS scores in Physics by Inquiry courses which have a strong implicit focus on the nature of science and the nature of learning science but do not explicitly address these issues [19].

CONCLUSIONS

The PET and PSET curricula showed large positive shifts on the CLASS compared to other courses. Akerson et al. [11] and Elby [12] have argued that explicit instruction on the nature of science and learning is necessary if we expect to see students make gains in these areas. Abd-El-Khalick [10] has argued that this type of instruction must be embedded in content-focused instruction. Our data support these arguments although further work is needed to distinguish the effects of the "explicit" variable and the "embedded within content" variable. Future and practicing teachers must develop positive views of science since their views will greatly influence the views of their students. Our results suggest that the PET and PSET curricula helped non-science majors become more expert-like in their thinking about the nature of science and learning science.

REFERENCES

1. NRC, *National Science Education Standards.* Washington D.C.: National Academy Press, 1996.
2. AAAS *Benchmarks for Scientific Literacy.* New York: Oxford University Press, 1993.
3. L.C. McDermott, *Physics by Inquiry*, Vols. I and III, Wiley, New York, 1996.
4. *Powerful Ideas in Physical Science.* College Park, MD: AAPT, 1995.
5. Z. Hrepic, P.Adams, J. Zeller, N. Talbott, G. Taggart, L. Young, "Developing an inquiry-based physical science course for preservice elementary teachers" In *Physics Education Research-2005*, edited by P. Heron, L. McCollough, and J. Marx, AIP Conference Proceedings 818, American Institute of Physics, Melville NY, 2006 pp. 121-124.
6. F. Goldberg, S. Robinson, and V. Otero, *Physics for Elementary Teachers.* It's About Time, Armonk, NY, 2006.
7. F. Goldberg, S. Robinson, R. Kruse, N. Thompson, and V. Otero, *Physical Science and Everyday Thinking.* It's About Time, Armonk, NY., In press.
8. L.C. McDermott, *Am. J. Phys.* **59**, 301-315, 1991.
9. M. Jenness and P. Miller, Mallinson Institute for Science Education, Western Michigan, 2005.
10. F. Abd-El-Khalick, *J. Sci. Teacher Educ.* **12**(3), 215-233, 2001.
11. V. Ackerson, F. Abd-El-Khalick, and N. Lederman, J. Res. in Sci. Teaching, 37 (4), 295-317, 2000.
12. A. Elby, Phys. Educ. Res., Am. J. Phys. Suppl. **69**(7), S54-S64, 2001.
13. E. Redish, J. Saul, and R. Steinberg, Am J. Phys. **66**(3), 212-224, 1998.
14. A. Elby , J. Fredriksen, C. Schwarz, and B. White, *Epistemological Beliefs Assessment for Physical Science*, http://www2.physics.umd.edu/~elby/EBAPS/home.htm (2001).
15. W. Adams, K. Perkins, N. Podolefsky, M. Dubson, N. Finkelstein and C. Wieman. *Phys. Rev ST: Phys. Educ. Res.* **2**(1) 010101, 2006.
16. S. Pollock, "Transferring transformations: Learning gains, student attitudes, and the impact of multiple instructors in large lecture classes," in *Physics Education Research -2005*, edited by P. Heron, L. McCollough, and J. Marx, AIP Conference Proceedings 818, American Institute of Physics, Melville NY, 2006, pp.141-144.
17. Allain, R. and Kruse, R. Unpublished results.
18. K. Perkins, W. Adams, S. Pollock, N. Finkelstein, and C. Wieman, "Correlating student attitudes with student learning using the Colorado Learning Attitudes about Science Survey," in *Physics Education Research-2004*, edited by J. Marx, P. Heron, and S. Franklin, AIP Conference Proceedings 790, American Institute of Physics, Melville, NY, 2005, pp. 61-64.
19. L. Hsu, K. Cummings, and J. Taylor, "Assessing adaptations of Physics by Inquiry: Students attitudes," presented at the bi-annual meeting of the American Association of Physics Teachers, Greensboro, NC, 2007.

Salience of Representations and Analogies in Physics

Noah S. Podolefsky and Noah D. Finkelstein

University of Colorado at Boulder, 390 UCB, Boulder, CO 80309-0390

Abstract. This paper focuses on the dynamics as students reason using analogies. We describe analogical scaffolding, a model of cognitive processes by which students can use prior knowledge to learn new material, and apply this model to demonstrate its utility in describing the dynamics of student reasoning about EM waves in an interview. The present fine-grained analysis confirms prior large-scale findings, that representations play a key role in student use of analogy.

Keywords: analogy, representation, electromagnetic waves
PACS: 01.40.Fk

INTRODUCTION

This paper focuses on a fine-grained analysis of a student interview using the analogical scaffolding model [1] as an analytic tool. We describe this model and apply it to describe the dynamics of student reasoning about EM waves using wave on a string and sound wave analogies. The present fine-grained analysis confirms prior large scale findings, that (external) representations play a key role in student use of analogy. [1] [2] In this paper, we argue that an approach that treats representations as *part of* concepts can be extremely productive for understanding dynamic student reasoning. Though controversial, this view is articulated and supported elsewhere. [3]

A standard view posits an analogy as a mapping from a familiar conceptual domain, *base*, to an unfamiliar conceptual domain, *target*. Analogies in this sense can be productive when generated by the user, but students may not generate and/or use analogies productively. This general finding is paradoxical: how does the student know what mappings to make in an analogy if they do not have sufficient knowledge of the target *a priori*? (Not to mention the base.)[1]

This theoretical paradox is manifest empirically. Student interviews, such as the example analyzed in this paper, reveal that simply suggesting an analogy to students (e.g., stating "an EM wave is like a wave on a string") does not generally enhance these students' reasoning in any apparent way about the target. We have, however, found effective ways of promoting the

productive use of analogy by students. These findings called for an explanatory model, and based on student interviews, classroom observations, and large-scale (N>100) studies, we developed the analogical scaffolding model.

Prior empirical studies demonstrated the utility of analogical scaffolding. Students demonstrated significantly greater learning gains when taught about EM waves with multiple (vs. no) analogies [1] and with multiple (vs. single) representations. [2] Here, we extend these large-scale results, using a fine-grained approach to study the dynamics more directly.

ANALOGICAL SCAFFOLDING

A more detailed account of analogical scaffolding is provided in a prior paper. [1] We represent the relationship between a signifier, *sign*, the thing the sign refers to, *referent*, and a knowledge structure mediating the sign-referent relationship, *schema* (Figure 1). [4] In the case of a sound wave, the sign could be a *sine wave*, the referent *sound*, and the schema would include the elements *longitudinal* and *three dimensional (3D)*. Sign-referent-schema spaces can *blend*, producing new schemata and new sign-schema

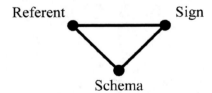

Figure 1. Sign-referent-schema diagram.

---

[1] This paradox is reminiscent of a problem elaborated by Plato. Namely, if one is to learn something that they do not already know, how does that person know what to learn?

CP951, *2007 Physics Education Research Conference*, edited by L. Hsu, C. Henderson, and L. McCullough
© 2007 American Institute of Physics 978-0-7354-0465-6/07/$23.00

associations. [5] For instance, a relatively abstract sign for an EM wave, a sine wave, can take on concrete features of a wave on a string, which uses the same sign (representation). Abstract ideas, such as an EM wave represented by a sine wave, gain particular meaning as result of a series of layered blends. [6] The interview analysis below provides a specific example of layering, building on string and sound to compile meaning into the canonical representation of EM waves.

REPRESENTATIONS AND SALIENCE

According to our framing of analogy, students presented with an unfamiliar (and perhaps challenging) problem can draw on existing mental structures to solve that problem by analogy. These mental structures may be cued by the student recognizing some similarity between the problem and prior experience (which may include another problem the student has solved previously). Two possible mechanisms for recognizing similarity, and hence making productive use of an analogy, are the following. If the two domains contain surface features (e.g., signs) that are similar, this can cue the student to make an analogical comparison. This mechanism explains why students may consistently solve problems that include inclined planes using kinematic equations (even for problems where the optimal solution method uses conservation of energy). [7] Note that in this case, students *do* use an analogy, albeit the base domain may be inappropriate. A second possible mechanism is that similarity is based on mental structures that transcend the surface features of a problem representation. This explains why physics experts can productively apply different solution methods to different problems which, say, all involve blocks on inclined planes. Still, *something* contained in the problem representation cues physicists to use one method or another, so the representation and mental structure cannot be completely separated.

A productive dimension for analyzing analogy use is *salience*, the strength of associations between sign features and schemata. [8] We adopt this notion of salience to argue that it is not the surface features of a sign that are salient *per se*, but the associations made with those surface features. Salience depends on the individual and context. A student presented with a sine wave may associate that sign with a material object (e.g., a wave on a string or a water wave). A physicist may associate the same sine wave with a graph (e.g., of electric field strength oscillating in time). Both the student and physicist cue on surface features – where the sine wave goes up, *something* goes up – but *what goes up* is very different.

How can the salient associations that students already use (often quite readily) lead to these students making associations that are salient to physicists? We seek a solution by proposing analogical scaffolding, focusing on representation use. This model suggests productive ways of scaffolding students' use of analogy by capitalizing on associations that are already salient for students and layering these associations towards more expert-like ideas.

DYNAMICS IN AN INTERVIEW

We have interviewed numerous introductory physics students in our studies of analogy. These think-aloud interviews generally involve students using curricular materials. Here, we focus on an interview with student "S". We describe several segments of this interview briefly and then apply analogical scaffolding in detail to a segment involving sound waves. At the time of the interview, S was enrolled in the first semester of a calculus-based introductory physics course. At this point in the semester, S told us the class was "doing oscillations and starting pressure."

In the interview, S was presented with the EM wave concept question in Figure 2 and asked to give his best answer.[2] S said he had "never seen anything like this before" and cycled through several answers over the next five minutes, finally settling on answer choice B (3>2>1=4). His reasoning was that "3 would be the first" and that "1 and 4 are on the same position on the two different waves." At one point, S stated that "x would…be the time."

We can already identify blends in S's reasoning. S appears to blend this picture, where *x* is position, with a graph in which the horizontal axis is *time* – hence 3 is "first". This idea may have been cued by the word "time" in the problem statement. In this case, a *position*

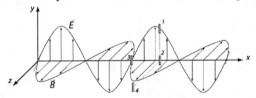

An electromagnetic *plane* wave propagates to the right in the figure above. Four antennas are labeled 1-4. The antennas are oriented vertically. Antennas 1, 2, and 3 lie in the *x-y* plane. Antennas 1, 2, and 4 have the same x-coordinate, but antenna 4 is located further out in the z-direction.

Which choice below is the best ranking of the *time averaged signals* received by each of the antennas. (*Hint*: the time averaged signal is the signal averaged over several cycles of the wave.)

A) 1=2=3>4   B) 3>2>1=4   C) 1=2=4>3   D) 1=2=3=4   E) 3>1=2=4

FIGURE 2. EM wave concept question. D is correct.

---

[2] Answering this question correctly requires a blend in which the EM wave sign is coupled to a schema including *3D* and *traveling (moving in time)* schema elements.

is time blend is salient for S. However, the peaks in the wave also cue a salient blend, along the lines of *a peak is a position* (S does not distinguish between the *E*- and *B-field* waves). Note that these salient blends are coupled to signs – the *x* axis, the two sine waves, and the word "time".

Next, the interviewer suggested analogies *verbally* and asked S if this helped with the EM wave concept question. First, the interviewer stated that an EM wave was like a wave on a string, but with no elaboration on how to use the analogy. S said that the string would "follow a similar pattern" to the EM wave, but that this would not help him answer the concept question. The interviewer then suggested that an EM wave was like a sound wave, again with no elaboration. S said immediately that, for making sense of the concept question, the sound wave analogy "wouldn't change too much." Note that in both cases, the sign is the verbal statement of the analogical comparison, but in these cases productive blends are not salient for S. We emphasize the distinction between not *salient* and not *existing*. As we will see, S does use several productive ideas about strings and sound, but these ideas become salient only under different conditions.

S was next presented with two new signs printed on a sheet of paper: one a sine wave and the other pictorial, depicting a hand at one end of a realistic representation of a string. He was told both represented a wave on a string. S stated that one representation was "a physical form of the other," implying a blend *sine wave is physical object*. S applied this blend to the EM wave, stating "2 is not really on... I guess none of them, other than 3, are on it." S also stated "as the hand moves it would follow the up and down with the hand," implying a blend *sine wave is moving object*. Note that the *moving object* element is not explicitly contained in the sign (a static picture), but becomes salient for S after he sees the two signs of a string.

Here, we reserve the detailed application of analogical scaffolding for sound waves. Selected transcript segments accompany a schematic representation of analogical scaffolding in Figure 3. Following the string discussion, S was presented with two different signs, a sine wave and a picture of a loud speaker with the "arrangement of air particles", and told these both represented a sound wave. These signs are shown in the topmost boxes of Figure 3. S first stated that the particles sign is a "physical representation of the sine", implying the blend *sine wave height is particle density*. In Figure 3, we place an additional node between sign and schema to indicate a prior blend contributing to a new sign-schema association. [1] S related quantities of the sine wave (e.g., "negative") with arrangements of particles (e.g., "grouping") and descriptive terms like "strong" and "weak" signal. S then applied these ideas to the EM

wave at time 17:37. Previously, S had said antennas 1 and 4 are less than 3 (answer B), but with the sound wave he said 1 and 4 are greater than 3 because they are where there are "more particles".

Now S has a new idea about antenna 2. Using the sound wave blend, S suggests that 2 is the same as 1 and 4, since the particles are "all the way down" (pointing at the dense region of particles in the pictorial sound representation, 17:59). He has applied ideas from the sound wave, now part of the *sine wave height is particle density* blend, to the EM wave.

S next tried to decide whether the particles extend in the *z* direction (in the picture, they are drawn only in the *x-y* plane). He created a new blend, drawing on experiential knowledge of sound – i.e., it is *3D* (18:37). Finally, S used a blend – *sine wave is 3D particle density* – to reason that 1, 2, and 4 are equal.

At this point, S proceeded to reason about point 3 by using the *sine wave is moving object* blend from the string. He could not deduce from the materials in front of him whether the wave at antenna 3 is stationary or moves as the wave propagates, (essentially whether the EM wave is a standing wave or a traveling wave) and he wavered between answers C and D in Figure 2. Nonetheless, S explicitly voiced the idea that a static picture of a string represents a moving object, and that an EM wave exhibits a similar property.

CONCLUSION

What does it mean to *know* a concept? S did not articulate significant knowledge of string or sound waves nor did he apply theses analogs productively to EM waves when these analogies were cued verbally. Presented with pictorial signs he did both. Did these signs cue schemata that S already had but simply did not articulate at first, or did S create new schemata during the blending process? We take a pragmatic position, framing concepts as observable through the reification of students' talk, gesture, and interactions with the environment. Our observations suggest that S's string, sound, and EM wave concepts changed dramatically under different conditions. The analogical scaffolding model captures this observed coupling between sign, schema, and referent. For S, concepts appear to depend strongly on the salience of sign-schema associations – for instance, he did not know to associate a sine wave, sound, and 3D, nor did he know to apply a sound wave blend to EM waves, without the signs in Figure 3. These associations became salient as S layered blend upon blend. We therefore argue that including signs as *part of* concepts can be extremely productive for understanding the dynamics of student learning. Furthermore, analogical scaffolding can be a useful tool for studying these dynamics.

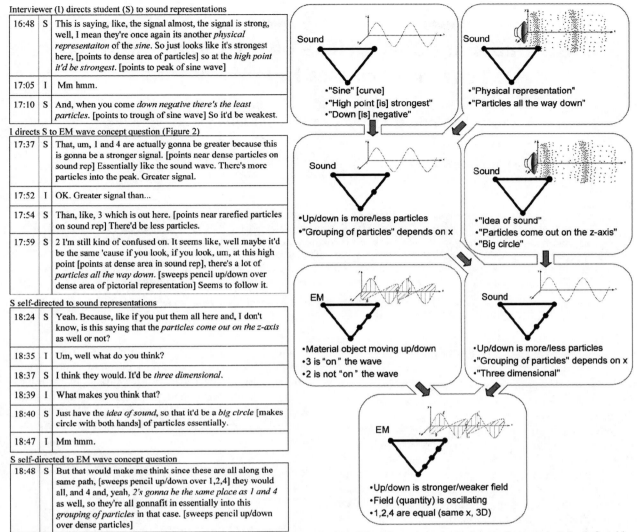

Interviewer (I) directs student (S) to sound representations

16:48	S	This is saying, like, the signal almost, the signal is strong, well, I mean they're once again its another *physical representaiton of the sine*. So just looks like it's strongest here, [points to dense area of particles] so at the *high point it'd be strongest*. [points to peak of sine wave]
17:05	I	Mm hmm.
17:10	S	And, when you come *down negative there's the least particles*. [points to trough of sine wave] So it'd be weakest.

I directs S to EM wave concept question (Figure 2)

17:37	S	That, um, 1 and 4 are actually gonna be greater because this is gonna be a stronger signal. [points near dense particles on sound rep] Essentially like the sound wave. There's more particles into the peak. Greater signal.
17:52	I	OK. Greater signal than...
17:54	S	Than, like, 3 which is out here. [points near rarefied particles on sound rep] There'd be less particles.
17:59	S	2 I'm still kind of confused on. It seems like, well maybe it'd be the same 'cause if you look, if you look, um, at this high point [points at dense area in sound rep], there's a lot of *particles all the way down*. [sweeps pencil up/down over dense area of pictorial representation] Seems to follow it.

S self-directed to sound representations

18:24	S	Yeah. Because, like if you put them all here and, I don't know, is this saying that the *particles come out on the z-axis* as well or not?
18:35	I	Um, well what do you think?
18:37	S	I think they would. It'd be *three dimensional*.
18:39	I	What makes you think that?
18:40	S	Just have the *idea of sound*, so that it'd be a *big circle* [makes circle with both hands] of particles essentially.
18:47	I	Mm hmm.

S self-directed to EM wave concept question

| 18:48 | S | But that would make me think since these are all along the same path, [sweeps pencil up/down over 1,2,4] they would all, and 4 and, yeah, *2's gonna be the same place as 1 and 4* as well, so they're all gonnafit in essentially into this *grouping of particles* in that case. [sweeps pencil up/down over dense particles] |

FIGURE 3. Transcript (left) and analogical scaffolding analysis (right) for selected portions of transcript. In left table, timestamps in column 1; interviewer (I) and student (S) in column 2. Italics in column 3 indicate quotations used to code schema elements (repeated in quotations in the diagram on the right). For instance, the two sound spaces on the top (with sine wave and air particles signs) blend as shown by the arrows. A blend space can then become one of two inputs for another layered blend. The EM wave input space (3rd row from top, left) comes from a blend with a wave on a string earlier in the interview.

ACKNOWLEDGMENTS

This work has been supported by the National Science Foundation (DUE-CCLI 0410744 and REC CAREER# 0448176), the AAPT/AIP/APS (Colorado PhysTEC program), and the University of Colorado. We wish to thank the PER at Colorado group for significant contributions to this work.

REFERENCES

1. N. S. Podolefsky and N. D. Finkelstein, *Phys. Rev. ST Phys. Educ. Res* 3, 010109 (2007)
2. N. S. Podolefsky and N. D. Finkelstein, in review.
3. D. Givry & W. M. Roth, *J. of Research in Science Teaching*, **43**(10):1086-1109 (2006)
4. W. M. Roth, & G.M. Bowen, *Learning and Instruction* **9**, 235 (1999)
5. G. Fauconnier & M. Turner, *"The Way We Think: Conceptual Blending and the Mind's Hidden Complexities"*, Basic Books (2003)
6. G. Lakoff & R. Nunez, *"Where Mathematics Comes From: How the Embodied Mind Brings Mathematics into Being"*, Basic Books (2001)
7. M. T. H. Chi, P. J. Feltovich, & R. Glaser, *Cognitive Science* **5**, 121-152 (1981)
8. S. Vosniadou, *"Analogical reasoning as a mechanism in knowledge acquisition: a developmental perspective"* in *Similarity and analogical reasoning*, edited by S. Vosniadou and A. Ortony, Cambridge: Cambridge University Press, 1989, pp. 413-437

Student Understanding Of The Physics And Mathematics Of Process Variables In *P-V* Diagrams

Evan B. Pollock,[1] John R. Thompson,[1,2] and Donald B. Mountcastle[1]

[1]*Department of Physics and Astronomy and* [2]*Center for Science and Mathematics Education Research*
University of Maine, Orono, ME

Abstract. Students in an upper-level thermal physics course were asked to compare quantities related to the First Law of Thermodynamics along with similar mathematical questions devoid of all physical context. We report on a comparison of student responses to physics questions involving interpretation of ideal gas processes on *P-V* diagrams and to analogous mathematical qualitative questions about the signs of and comparisons between the magnitudes of various integrals. Student performance on individual questions combined with performance on the paired questions shows evidence of isolated understanding of physics and mathematics. Some difficulties are addressed by instruction.

Keywords: Physics education research, thermodynamics, mathematics, upper-level, *P-V* diagrams, integrals.
PACS: 01.30.Cc, 01.40.–d, 01.40.Fk, 05.70.-a.

INTRODUCTION

As part of ongoing research at the University of Maine (UMaine) we are exploring student learning in upper-level thermal physics courses, primarily taken by physics majors. There exists a limited body of research on this topic [1-6]. Prior work has suggested that particular difficulties experienced at the introductory level are also evident at the advanced undergraduate level [1]. Previous results on the teaching and learning of the First Law have documented several common difficulties, such as indiscriminate application of the concept of a state function [2,3].

One theme of our research is the extent to which student mathematical conceptual difficulties may affect understanding of physics concepts in thermodynamics. We have presented results of this nature for partial derivatives and their relationship to material properties and the Maxwell relations [4,6]. Our purpose for this study is to determine if student difficulties with mathematical concepts are impacting students' abilities to solve problems related to the First Law of Thermodynamics. In this paper we describe studies on energy transfers in and out of a system.

Mathematics is a vital part of solving many physics problems. It is a universal language that can often condense a complicated conceptual problem into a simple relationship between variables. Many convenient representational tools exist in physics, e.g.,

equations, graphs and free-body diagrams, that simplify analysis of a complex problem. Appropriate interpretation of these representations requires recognition of the connections between the physics and the mathematics built into the representation and subsequent application of the related mathematical concepts [7].

In thermodynamics, two- or three-dimensional graphical representations of physical processes are especially useful in helping to understand these processes. These diagrams can contain information about the thermodynamic "path" followed in a process, regions of different phase, and critical behavior. *P-V* diagrams are used extensively as representations of physical processes as well as of the corresponding mathematical models. Information can quickly and easily be obtained from this representation. We focus on student understanding of *P-V* diagrams in the context of determining work related to the First Law of Thermodynamics.

Meltzer developed a set of questions probing student understanding of the First Law and related quantities based on processes shown on a *P-V* diagram [3]. In addition to replicating Meltzer's experiment in our upper-level thermodynamics course, we wanted to know if some of the conceptual difficulties might originate in the mathematics. We designed questions that are analogous to Meltzer's questions, but purely mathematical in nature – all physics is removed from the setting. The math context involves qualitative

CP951, *2007 Physics Education Research Conference*, edited by L. Hsu, C. Henderson, and L. McCullough
© 2007 American Institute of Physics 978-0-7354-0465-6/07/$23.00

questions regarding comparisons and determinations of magnitudes and signs of integrals.

We present results based on three years of data from UMaine's upper-level *Physical Thermodynamics* course, taught in the fall semesters of 2004-2006 (by DBM). UMaine's introductory calculus-based physics sequence does not include thermal physics; for some students this was their first exposure to the topic. All students in the data set had completed three semesters of calculus, including multivariable calculus, and one or more courses in ordinary differential equations.

PHYSICS QUESTION: WORK COMPARISON

Meltzer asked students to compare energy transfers and changes (work, heat transfer, internal energy changes) for two ideal gas processes (#1 and #2), shown on a *P-V* diagram, that both begin and end in the same states (A and B) (Fig. 1). The first problem, P1, asks the students to compare the works done by the gas in the two processes. The correct response for P1 can be obtained by recognizing that the work done by the gas is $\int P dV$, and that this integral is represented by the area under the curve for that process.

Meltzer found that between 20% and 30% of students incorrectly stated that the works done by the gas over both paths were equal. Common student reasoning used the idea that work was path independent, and/or that the work only depended on the endpoints in the diagram. Meltzer attributed this to one of two reasons: first, a (reasonable) overgeneralization of work in the context of conservative forces in mechanics to thermodynamics, and second, a conceptual difficulty distinguishing between heat, work, and internal energy. Meltzer's conclusion was that students were inappropriately attributing state function properties to work.

We asked these physics questions prior to explicit instruction on the First Law and *P-V* diagrams in 2004 and 2005 (*N*=15), but after instruction in 2006 (*N*=6). The pretest performance on P1 at UMaine was generally consistent with, although worse than, Meltzer's in his post-instruction introductory courses: of 15 students, 5 (33%) gave the correct comparison, while 8 (53%) stated that the works were equal. Students with correct responses used "area under the curve" or "$\int P dV$" in their explanations. Students justified equal works using mostly state function-like reasoning: "Work done is independent of path"; "Both processes start and end in same place"; "Both processes have same beginning and end points."

Students did much better when given the question as a post-test: 5 of 6 students gave correct responses, and none said that the works were equal. Correct

explanations included explicit references to the integral, to area under the curve, and to the path dependence of work.

This Pressure-Volume *(P-V)* diagram represents a system consisting of a fixed amount of ideal gas that undergoes two *different* processes in going from state A to state B:

[In these questions, *W* represents the work done *by* the system during a process; *Q* represents the heat transfer *to* the system during a process.]

P1.   Is *W* for Process #1 *greater than*, *less than*, or *equal to* that for Process #2? Explain.

FIGURE 1. Question P1. (From Ref. 3.)

MATHEMATICS QUESTION: INTEGRAL COMPARISON

As mentioned, Meltzer's work question (P1) was paired with an analogous mathematics question (M1) designed to elicit any underlying mathematical difficulties in this context. In M1, we asked students about the magnitudes of two integrals that have identical beginning and end points (M1). The two processes in P1 are replaced with two different curves. We asked the math question before the physics question, usually in the first week of class.

Two versions of M1 were given, to different students (see Figs. 2 and 3). In Version 1 (V1), the curves are distinguished with labels *Path 1* and *Path 2*. The integrals are expressed using the same variable, *z*, for both paths. Because the two curves were represented using the same variable, this version is a mathematical isomorph of P1, in which the two processes are both considered graphs of pressure vs. volume (*P* vs. *V*). In Version 2 (V2), the curves are labeled as two different functions, *f(y)* and *g(y)*; the integrals also reflect this distinction. In V2 the notation of the two curves is more mathematically rigorous; the question is analogous to P1 rather than isomorphic.

The results for M1 (see Table 1) lend insight into student conceptual difficulties related to integration. Approximately one-quarter of the students (5/21) stated that the two integrals were equal in magnitude, regardless of which version they were given. The reasoning given for these responses were often identical to that given for P1: "The function does not depend on the path, but only on the end points"; "same endpoints"; "both have same end points and equal functions." These explanations of integral path independence are consistent with treatment of both integrals as antiderivatives rather than as Riemann sums.

Furthermore, the data from the two versions of question M1 suggest that students treat the versions differently. A much higher fraction of students (77% vs. 38%) gave correct responses in V2, when the different functions were labeled as such, rather than in V1, where they were labeled as the same variable, with only the path to distinguish them.

Two paths have been traced out on the *z-y* graph shown at right, and are labeled Path *1* and Path *2*. Both paths start at point *a* and end at point *b*.

Consider the integrals $I_1 \equiv \int_a^b z dy$ and $I_2 \equiv \int_a^b z dy$

taken over Path *1*, and I_2 is taken over Path *2*.

Is the absolute value of the integral I_1 **greater than**, **less than**, or **equal** to the absolute value of the integral I_2, or is there **not enough information to decide**? Please explain.

FIGURE 2. Question M1 Version 1 (V1), a mathematical analog of P1.

Two functions have been graphed on the *z-y* graph shown at right, and are labeled *f(y)* and *g(y)*. Both functions start at point *a* and end at point *b*.

Consider the integrals

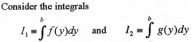 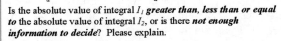

$$I_1 \equiv \int_a^b f(y) dy \quad \text{and} \quad I_2 \equiv \int_a^b g(y) dy$$

Is the absolute value of integral I_1 **greater than**, **less than or equal** to the absolute value of integral I_2, or is there **not enough information to decide**? Please explain.

FIGURE 3. Question M1 Version 2 (V2), a mathematical analog of P1.

TABLE 1. Student responses to M1, overall and by version.

I_1 vs. I_2	All (N=21)	V1(N=8)	V2 (N=13)
Greater Than	**13 (62%)**	**3 (38%)**	**10 (77%)**
Not Enough Information	1 (5%)	1 (12%)	---
Equal	5 (24%)	3 (38%)	2 (15%)
Blank	1 (5%)	---	1 (8%)
Less Than	1 (5%)	1 (12%)	---

ANALYSIS OF MATHEMATICS-PHYSICS RESPONSE PAIRS

A comparison of individual student responses to the question pair of P1 and M1 (Table 2) provides a stronger indication of the mathematical or physical nature of the student difficulty. On the pretest, one third answered both questions correctly, and one third stated that both the works and integrals were equal. On the post-test, 5 of 6 were correct on both questions. We enumerate notable points that the data uncover.

1. All of the students who answered the physics question correctly also answered the math question correctly.
2. All of the students who answered the math question incorrectly also answered the physics question incorrectly.
3. Of the students who said the works were equal, more than half said the integrals were also equal.
4. None of the students in the *other* category answered either question correctly.

TABLE 2. Student paired responses to M1 (combined V1 and V2) and P1. Two classes answered the physics questions before instruction, one class after instruction.

M1	P1	Physics Pre (N=15)	Physics Post (N=6)
$I_1 > I_2$	**$W_1 > W_2$**	**5 (33%)**	**5 (83%)**
$I_1 > I_2$	$W_1 = W_2$	3 (20%)	—
$I_1 = I_2$	$W_1 = W_2$	5 (33%)	—
other		2 (13%)	1 (17%)

Analysis of the data suggests the following interpretations. Students who answered that the math integrals were equal and that the works were equal demonstrate, to us, a math difficulty as well as a possible physics difficulty. The all-or-nothing nature of the responses (points 1, 2, 4 above) suggests that correct understanding of the mathematics is necessary in order to answer the physics question correctly.

As point 3 implies, less than half of the erroneous work comparisons were paired with correct integral comparisons; these students had genuine conceptual difficulties with the physics while possessing an understanding of the math. Two of the three students in the M1-correct/works-equal category (second row of Table 2) used *area under the curve* reasoning to compare integrals in M1 but then used *path independence* or *state function* reasoning to equate works in P1. These students are inappropriately invoking *state function* reasoning in the physical context of work, as described by Meltzer and by Loverude et al. [2,3]; however, these students do not use this reasoning in the pure mathematical context.

The high post-test performance implies that instruction may successfully address this issue.

DISCUSSION AND CONCLUSIONS

We have three components to this research that we summarize and discuss here. First, we asked questions comparing work done based on paths on a *P-V* diagram. Our pretest results on upper-level students were consistent with the original results obtained after instruction at the introductory or intermediate level [3], namely that a significant fraction of students are attributing the property of path independence to thermodynamic work. We also see much better results obtained after instruction at the upper level, implying that this issue may be a matter of familiarity with the material.

Second, we asked two different versions of a mathematics question analogous to the work comparison described above. The results show that it is not trivial for students to compare integrals of functions with the same endpoints but that follow different paths (curves). The notation used to describe, and to distinguish between, the functions seems to affect student performance: labeling two different curves using the same variable leads to more students equating integrals.

This second result is directly relevant to thermodynamics, in which one variable often represents multiple functions (e.g., $P(V)$ over two different paths). It is arguably necessary to use less mathematically rigorous notation in physics to keep the physical meaning. However, what might be considered a trivial distinction in notation to physicists seems to make a substantial impact on even advanced students' interpretations of the diagrams.

Third, we checked individual student responses on the paired physics and math questions. We found that some of the difficulties that arise when comparing thermodynamic work based on a P-V diagram – e.g., treating work as path independent – may be attributed to difficulties with the *mathematical* aspect of the diagram, in particular with the correct application of an understanding of integrals, rather than simply physics conceptual difficulties (e.g., treating work as a state function). These results suggest that some students aren't attributing state function properties to work so much as failing to recognize the same variable as two different functions during integration.

In related work, we have also administered and analyzed paired questions for Meltzer's internal energy comparison and a mathematical analog (i.e., integrating a path-*independent* quantity). (Space constraints prevent a full description.) Performance on the physics question is very high (>90%); students typically invoke the state function property of internal energy. The math question performance is much worse. The difference in outcomes on these questions shows that some students lack an understanding of the mathematical properties of state functions, and only rely on the physics and/or the physical aspects of the system in question. These results suggest that while students have a better understanding of *state function* than of *process function* in physics, the reverse is true for the mathematical distinction between these two concepts. The extent to which this may be related to

the calculus concepts of *antiderivative* and *Riemann sum*, or to the larger grain-sized distinction between function and variable, is yet to be determined.

Finally, while this analysis does not endorse any theoretical framework, an analysis of the data through the lens of resource activation may be enlightening. The idea of "compelling visual attribute" [8] could provide alternate explanations of two of our results. First, for P1, the endpoints of the P-V diagrams may tend to make students focus only on the two states represented, rather than the process followed between those states. Second, M1 contains a diagram and symbolic expressions of the integrals; for V1, these symbolic expressions differ only in the path label. Perhaps when presented with both symbolic versions of integrals and a graphical representation of the functions, some students may preferentially cue on the symbolic version to decide that integrals of identical variable expressions, with identical limits, but over different paths, are equal.

Future work

The data set contains additional question pairs that add to this work. We have also collected homework sets and exams from this population. In addition, a more in-depth probing of student understanding of the state function concept is warranted, including the relationship to calculus concepts.

ACKNOWLEDGMENTS

Brandon R. Bucy made substantial contributions to the initial phase of this work. We appreciate discussions with David E. Meltzer. Work supported in part by NSF Grant PHY-0406764.

REFERENCES

1. D.E. Meltzer, *2004 Phys. Educ. Res. Conf.*, edited by J. Marx *et al.*, AIP Conf. Proc. **790**, 31-34 (2005).
2. M.E. Loverude, C.H. Kautz, and P.R.L. Heron, *Am. J. Phys.* **70**, 137-148 (2002).
3. D.E. Meltzer, *Am. J. Phys.* **72**, 1432-1446 (2004).
4. J.R. Thompson, B.R. Bucy, D.B. Mountcastle, *2005 Phys. Educ. Res. Conf.*, edited by P. Heron *et al.*, AIP Conf. Proc. **818**, 77-80 (2006).
5. M.J. Cochran and P.R.L. Heron, *Am. J. Phys.* **74**, 734-741 (2006).
6. B.R. Bucy, J.R. Thompson and D.B. Mountcastle, *2006 Phys. Educ. Res. Conf.*, edited by L. McCullough *et al.*, AIP Conf. Proc. **883**, 157-160 (2007).
7. E.F. Redish, in Proceedings of *World View on Physics Education in 2005: Focusing on Change*, Delhi, 2005 (2005). Available at http://www.physics.umd.edu/perg/papers/redish/IndiaMath.pdf.
8. A. Elby, *J. Math. Behav.*, **19**, 481-502 (2000).

A Longitudinal Study of the Impact of Curriculum on Conceptual Understanding in E&M

S. J. Pollock

Department of Physics, University of Colorado, Boulder, CO 80309-0390

Abstract. We have collected extensive data on upper-division Electricity and Magnetism (E&M) student performance at CU Boulder since we introduced the University of Washington's *Tutorials in Introductory Physics* in 2004 as part of our freshman curriculum. In the earliest semesters, all upper-division students had themselves taken a non-Tutorial introductory Physics, providing a baseline at this upper-division level surprisingly close to post-scores in our reformed introductory course. More recently, the population in the upper-division is mixed with respect to freshman experience, with over half having been taught with Tutorials as freshmen. We track those students and find that on average, their individual BEMA scores do not change significantly over time. However, we do find a significantly stronger performance at the upper division level for students who went through Tutorials compared to those who had other introductory experiences, and stronger scores still for students who taught in the introductory sequence as Learning Assistants, indicating a long-term positive impact of Tutorials on conceptual understanding.

Keywords: course transformation, course assessment, longitudinal, upper-division electricity and magnetism.
PACS: 01.40.-d. 01.40.Fk, 01.40.gb, 01.40.G-

INTRODUCTION

In Fall 2004, we introduced the University of Washington's (UW) *Tutorials in Introductory Physics*[1] into Physics II (Electricity and Magnetism, E&M) at the University of Colorado (CU). There is considerable published evidence regarding the efficacy of Tutorials[2-4], and we use the BEMA (Brief E&M Assessment)[5] as one of a set of measures of student learning. Since relatively small amounts of BEMA data have been published elsewhere, we simultaneously collected data from upper-division students to provide a CU baseline. We assumed upper division physics majors could provide a reasonable target goal for learning in our large introductory class.

We find several interesting results regarding the development of student understanding of E&M concepts, as measured by the BEMA. We do *not* observe much change in performance for individual students over the 2+ year period between when they complete Physics 1120 and when they complete either first or second semester upper-division E&M. We *do* see evidence that the addition of Tutorials has had a significant impact (statistically and pedagogically) on student performance on this instrument at the upper division. We also see further evidence of the benefits of undergraduates serving as teachers (the CU Learning Assistant, or LA, model [6]).

STUDENT POPULATIONS

Physics 1120 is the introductory Calculus-based second term course on E&M at CU. The class serves engineers (who dominate the class by number), physics majors, and a variety of other hard-science majors. Class size ranges from ~325 to ~475. The basic course structure and content has not changed in many years, but in Fall 2004, traditional recitations were replaced by UW Tutorials[1]. A more detailed description of the transformed classes at CU can be found in our earlier work [3-4]. Phys 1120 classes are taught by different faculty every term, but almost all these faculty regularly used *Peer Instruction*[7] in the large lectures (with clickers), computer-based homeworks[8], and a staffed help-room. The primary curricular switch which occured in Fall '04 was the addition of Tutorials with trained undergraduate Learning Assistants [6].

Physics 3310 and 3320 are CU's upper-division physics majors' E&M sequence, with an average of ~30 (in 3320) to ~40 (in 3310) students. A few students in these classes are e.g. astrophysics or engineering majors, but the large majority are junior physics majors. The textbook for many years has been Griffiths[9]. The course is taught in a very traditional physics lecture style by a variety of different faculty.

CP951, *2007 Physics Education Research Conference*, edited by L. Hsu, C. Henderson, and L. McCullough
© 2007 American Institute of Physics 978-0-7354-0465-6/07/$23.00

MEASUREMENTS

The BEMA[5] instrument has been given at the start and end of Physics 1120 every semester since we began using UW Tutorials, on paper, during the first and last weeks of the term. We obtain matched, valid (e.g. most questions attempted, not all answers the same, etc) pre/post scores for ~70% of students. In the upper division courses, students were asked only at the *end* of the semester by their instructor to take an online version of the BEMA. This voluntary approach resulted in a smaller fraction of returns, approximately 45%. However, average course grades for the students who took the BEMA online are statistically indistinguishable from the remaining students.

DATA AND RESULTS

The Physics 1120 data were collected primarily to evaluate the impact of our curricular reforms, with results in progress and reported elsewhere [3,4]. Here we are interested in the longitudinal aspect, an evaluation of the difference in BEMA results between introductory and upper-division students, and the change over time of students' scores on this instrument. We present some of the principal results, first for the two classes separately, then examining longitudinal effects more directly.

Introductory Physics 1120

The average BEMA *pre*score in Phys 1120 has been stable for six terms at 26+/-1% (with stand. dev. ~10%). Posttest scores range from 50-60%, (s.d. ~15%). The average for all students in the first three terms of data collection was 56% (and it is a small fraction of *these* students who fed into the upper-division courses for which we have data, as described below.) The correlation coefficient (r) of BEMA post-score to course grade ranges between .55 and .65 for different terms. We are not aware of many published BEMA results to calibrate these results, but the BEMA is difficult - our incoming graduate students average just over 80% on it. The authors of the BEMA[5] indicate that these prescores are typical, and that our postscores are well above what might be expected from a purely traditional introductory lecture course.

Upper Division Phys 3310 and 3320.

For the first three terms (Fa04 to Fa05) that we gave the BEMA to upper-division students, none had themselves gone through an introductory course with Tutorials. Thus, we have baseline data for how upper division physics majors with a mix of introductory

courses (but none with Tutorials) performed on this conceptual exam *after* upper-division E&M. Average results were stable from term to term; there do not appear to be any systematic instructor effects. Average upper-division BEMA scores are 54+/-2% (s.d. = 19%) for N=76 unique students over those 3 terms, representing ~45% of all enrolled students. This is our "control" group, shown on the left in Fig 1. Note that this average score is marginally *below* the postscore we obtained after our freshman level course with Tutorials, a result we found somewhat surprising, and discuss below.

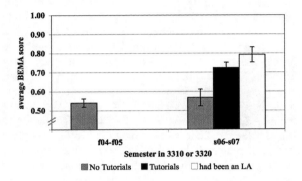

FIGURE 1. BEMA scores *after* upper division E&M, separated into terms in which no one had been through freshman Tutorials (grey, far left), and terms (right) in which some students had not been through freshman Tutorials (grey), but others had (black). The last (white) bin shows students who had been LAs for Phys 1120. Error bars are standard error of the mean.

Starting in Sp06, the upper-division population changed - some students came up through the usual sequence from a freshman Tutorial experience, while others did not. This latter group included e.g. students who passed out of freshman physics due to AP credit, transfer students, and students who took extra time to go through the sequence. Because of this diverse population, the first question we wanted to answer was whether the subpopulation of upper division students without Tutorial experience had a different BEMA score than the control group from earlier terms. The answer was essentially no. The average BEMA for Sp06-Sp07 for the "non-Tutorial experience" students was 57+/-4% (s.d. = 21%) (N=23), see Fig 1. Recall the control group of upper-division students from earlier terms had an average BEMA of 54%. So, although the "non-Tutorial" subgroup demographics differs over time, their BEMA scores are not statistically significantly different.

We now consider the scores for upper-division students who *did* come through freshman Tutorials. The results are also shown in Fig 1. These students (middle bar on the right of Fig 1) have an average

BEMA score of 72+/-3% (s.d.=15%) (N=33), statistically significantly higher than their non-Tutorial compatriots in the same courses, and/or the control group of *all* upper-division students from earlier terms. This is one of the central observations of our study - it appears that the Tutorial experience leads, after several years, to more than a 15 point difference on the BEMA, comparing to either students in the same classes, or to the earlier upper-division control group who never had Tutorials as freshmen. This difference is statistically significant (p<.01, 2-tailed t-test), and pedagogically significant (effect size > 0.75)

The final bin in Fig 1 shows BEMA scores for students who taught as Learning Assistants in Phys 1120 at some point between taking freshman physics and taking Phys 3310. Their BEMA in 3310/20 is higher still: 79+/-4%, (s.d.=12%), but the small number (N=7) is subject to larger fluctuations - indeed, one poor student had a strong impact on this group, most had extremely high BEMA scores. We do not include the LAs in our pool of "Tutorial" students in Fig 1, due to their extraordinary additional contact with the material, but these data provide additional evidence for the impact of teaching experiences for undergraduates [6]

A final question on the online BEMA asks students to tell us how hard they tried. Counting *only* students who said they tried "very hard" (just under half the students who took it), the results are consistently shifted up by 5-10 points, but this does not change the qualitative results or *differences* between the groups discussed above. LAs all indicated they had tried very hard, perhaps an interesting observation itself.

Some of our more traditional faculty have explicit course goals in 3310/20 which are far from the sort of elementary conceptual understanding addressed by the BEMA. They often have a highly mathematical and quantified problem-solving focus. One can therefore ask about course grades in the upper-division courses as determined by these faculty. The average course grade is quite consistent over time (close to 3.0, "B-centered"). For the students in the control group (everyone in both 3310 and 3320 from F04-F05), the average course grade was 3.1+/-0.1 (out of 4.0=A) (s.d.=0.8). For students in the "No-Tutorials" subgroup in S06 through S07, the average course grade was 3.0+/-0.1 (s.d. =0.7), further evidence that this subgroup is not so different from the earlier control group. All "No-Tutorials" students combined (both courses, all six terms of the study) have an average grade of 3.04+/-0.07 (s.d.=0.8, N=124). For the "freshman Tutorial" subgroup, the average course grade was 3.3+/-0.10 (s.d=0.7, N=40), statistically significantly higher than students without Tutorials (p<.05, 2-tailed z-test) The average grade in 3310 and 3320 for the 7 students who had been LAs was 3.2.

The students with Tutorial experience thus had marginally higher course grades than the cohort in the *same* semesters who did not have a Tutorial background, and also higher than the control group of all students from earlier terms, but the difference is not great. Nonetheless, one certainly cannot argue from this data that our increased focus on conceptual development at the freshman level is *harming* our upper division majors by any of these measures, as at least a few traditional faculty seem to fear.

Direct Longitudinal Comparisons.

The data above indicates several long-term benefits of having taken Tutorials as freshmen, both in BEMA scores and possibly course grades. But what is the direct impact of Phys 3310 or 3320 on student conceptual understanding?

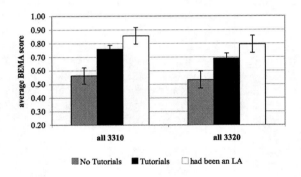

FIGURE 2. BEMA scores in upper division courses, separated into Phys 3310 (E&M I) and 3320 (E&M II) Errors are standard error of mean.

In Fig 2, we separate the results from Physics 3310 and 3320. The leftmost bin for each class combines students from early terms (none of whom had Tutorial experience) with the sub-population in later terms who also had no Tutorial experience (since we saw above there was little difference in these groups' scores or grades). We see a very slight, but not statistically significant, decline in BEMA scores from 3310 to 3320. One might wonder about "exam fatigue", but only 39 students took the BEMA in *both* 3310 and 3320, and for these students, their average *shift* in BEMA score from 3310 to 3320 was 0.0% (s.d.=15%). It appears that Phys 3320 does not have any measurable positive impact on conceptual understanding of freshman-level material. Fig 2 shows that the difference of Tutorial and non-Tutorial students persists even after 2 semesters of upper division physics, and that the LAs remain stronger as well.

Table 1 (below) allows a comparison of average scores on "clusters" (3-8 questions per category) of BEMA questions taken by upper-division students at the end of the course. Even when looking at different conceptual categories, the UW Tutorials appear to have a significant positive residual impact on students.

TABLE 1. Comparison of performance of upper-division students on groups of BEMA questions.

Rough category of BEMA questions	No 1120 Tutorial	Had 1120 Tutorial (and prior BEMA score in Phys 1120)
• E-fields, Gauss, and Coulomb	71%	86% (84%)
• Fields in Matter	50%	72% (72%)
• Current and Voltage	56%	72% (74%)
• Circuits	40%	60% (75%)
• Magnetostatics	62%	83% (80%)
• Faraday's law	28%	41% (45%)

The "no Tutorial" column consists of N=98 individuals, combining the control group with any later students who had never taken freshman Tutorials. Standard errors of these entries vary, but are typically +/-2%. We have no 1120 BEMA data for this population. The "Tutorial" group (last column in Table 1) has N=33 students. Standard errors vary by category, but are typically +/-3%. As in Fig 1, we do not count scores a second time if a student takes the BEMA in Phys 3310 and again in 3320, nor do we include students who had been a Learning Assistant.

We were able to track some individuals from Phys 1120 through upper division. The numbers in parentheses in the 2nd column of Table I are older data from when the students had originally taken the BEMA at the end of Physics 1120, for the (N=29) students for whom we have these matched data. There is little change over a 2-year period, on average, with only one category (circuits, the BEMA topic perhaps least likely to be revisited in upper-division) showing a statistically significant drop.

Recall that the overall average BEMA score in Phys 1120 in the semesters we're considering was 56%, but the average score for the 29 students who would ultimately take Phys 3310 and 3320 (later in their careers) was 74%. (Their score *in* 3310/3320 would become 75%) It appears, then, that our future physics majors are already doing well above average in their introductory course as freshmen, and retaining their skills years later.

CONCLUSIONS

Upper division courses *by themselves* do not appear to have much impact on BEMA scores, evidenced by direct measurement of shifts (or lack thereof) for individuals taking the BEMA before and after either upper division course. However, we do measure a significantly (>15%) higher BEMA score in 3310/20 for the student population who went through freshman Tutorials, along with a marginally higher course grade.

It is interesting to speculate about the origin of these strong differences in the upper division populations. UW Tutorials do not directly address all of the particular topics and questions on the BEMA, although there is considerable overlap, nor are they typically quantitative. Tutorials focus on conceptual understanding, sense-making, and explanations, which appears to manifest itself in improved performance on this conceptually focused exam. Apparently some of the qualitative understanding built at the introductory level persists over time, and benefits students at the upper division level, as evidenced by improved BEMA scores and (marginally) improved grades. We also see continuing evidence that the LA experience is beneficial. We believe such data are of value to PER researchers trying to understand mechanisms and outcomes of reformed curricula such as the Tutorials, and also to traditional faculty who are trying to decide on the value (and perceived costs) of such curricula.

ACKNOWLEDGMENTS

Thanks to PhysTEC (APS/AIP/AAPT), NSF CCLI (DUE0410744) and NSF LA-TEST (DRL0554616) for support of our transformed classes. Enormous thanks to the CU Physics Department, upper division faculty who helped facilitate this research, the University of Washington's Physics Education Group, and the PER at Colorado group.

REFERENCES

1. McDermott, L., Shaffer, P., and the PEG, (2002). *"Tutorials in Introductory Physics"*, Prentice Hall.
2. see McDermott, L, Redish, E., *Am. J. Phys* 67 (1999) p. 755 for many original references from the UW group.
3. Pollock, S., *2004 PERC Proc* 790, p. 137, Pollock, S., *2005 PERC Proc.* 818, 141, Pollock, S., and Finkelstein, N. *2006 PERC Proc. 883, p. 109.*
4. Finkelstein, N. Pollock, S., (2005). *Phys. Rev. ST Phys. Educ. Res.* 1, 010101.
5. Ding, L et al, (2006). *Phys Rev ST: PER*, 2, 010105, see www.ncsu.edu/per/TestInfo.html, and private communication with the authors. We supplement the BEMA with three questions from the ECCE instrument of Thornton and Sokoloff, see physics.dickinson.edu
6. Otero, V. et al, Science 28 July 2006: 445-446.
7. Mazur, E. (1997). *Peer Instruction*, Prentice Hall
8. CAPA homework system: see www.lon-capa.org, and Mastering Physics: www.masteringphysics.com
9. Griffiths, *Intro. to Electrodynamics*, Prent. Hall (1999)

FCI-based Multiple Choice Test for Investigating Students' Representational Coherence

Antti Savinainen, Pasi Nieminen, Jouni Viiri, Jukka Korkea-aho and Aku Talikka

Department of Teacher Education, P.O. Box 35, FI-40014 University of Jyväskylä, Finland

Abstract. We present the Representation Test derived from the FCI for evaluating students' representational coherence on some aspects of gravitation and Newton's third law. The test consists of 23 questions addressing verbal, graphical, bar chart, and vectorial representations. Matched high school student data (n = 54) on the pre- and post-test are analyzed in terms of representational coherence and scientific correctness.

Keywords: multiple representations, representational coherence, the Force Concept Inventory
PACS: 01.40.Fk

INTRODUCTION

Many studies have been published on multiple representations (such as texts, pictures, diagrams, graphs, or mathematical ones) in physics education (e.g., [1], [2], [3], [4]). Van Heuvelen and Zou [1] offer several reasons why multiple representations are useful in physics education: they foster students' understanding of physics problems, build a bridge between verbal and mathematical representations and help students develop images that give meaning to mathematical symbols. These researchers also argue that one important goal of physics education is helping students to learn to construct multiple representations of physical processes, and to learn to move in any direction between these representations. Furthermore, it has been pointed out that in order to thoroughly understand a physics concept, the ability to recognize and manipulate that concept in a variety of representations is essential [2].

Several research-based multiple choice tests have been developed for evaluating students' conceptual understanding in the domain of introductory mechanics, the most widely used being perhaps the Force Concept Inventory (FCI) [5]. The FCI addresses several representations in a variety of contexts but it does not provide a systematic evaluation of students' ability to use multiple representations when a context is fixed. It is important to note that both the context and the representation affect on students' responses: the student might be able to apply a concept in a familiar context using a certain representation but fail when the context or the representation is changed [6].

The existing tests are limited in that they do not permit comprehensive evaluation of students' skills in using multiple representations. This is why we have developed a multiple-choice test – the Representation Test - to evaluate students' representational coherence: i.e., their ability to use different representations consistently when the context is kept as constant as possible.

In this paper we present the rationale and structure of the Representation Test. High school students' pre- and post-test data on the Representation Test are analyzed from the point of view of representational coherence. We also analyze students' written responses explaining their choices in the post-test to support the validity of the Representation Test.

THE REPRESENTATION TEST

The Representation Test is based on the Force Concept Inventory [5]. Firstly, FCI questions which might be easy to transform into different representations were identified. Secondly, after inspection, FCI questions 1, 4, 13, 28 and 30 were chosen: questions 1, 13 and 30 deal with gravity, and questions 4 and 28 deal with Newton's third law. We could not use all the FCI questions since the questionnaire would then have been much too long. Therefore, the Representational Test does not deal with all the aspects of the force concept addressed by the FCI.

The questions chosen for the Representational Test questions are in verbal form in the original FCI. The description of the context, the question and the

CP951, *2007 Physics Education Research Conference*, edited by L. Hsu, C. Henderson, and L. McCullough
© 2007 American Institute of Physics 978-0-7354-0465-6/07/$23.00

different alternative multiple choice items are verbal. Each question has five alternative answers of which the student must choose one. In developing the new questions we tried to keep the contexts and alternative forms of the answers as similar as possible to the original questions. For each of the selected five FCI questions, two to four new questions were formulated in different representations. We made some changes after the pilot phase [7] on the basis of the data gathered and feedback from an expert [8]: there are 23 questions altogether in the new version (one question has three parts).

We use the term *theme* for the question set developed from an FCI question and formulated in different representations (see Table 1). The representations are verbal, graphical, bar chart, and vectorial. In every question the description of the question situation is verbal, but the different multiple choice alternatives are described in different representations. For instance, the theme FCI4 corresponds to the question set developed from FCI question number 4; the alternatives in the theme FCI4 are in verbal, graphical, vectorial and bar chart representations (see Table 1). The questions of each theme were split across the Representation Test.

TABLE 1. Themes of the Representation Test, the concepts the items deal with, and the context of the theme.

Theme	Concept	Context	Representations
FCI1t	gravitation (time of falling)	falling ball	verbal, pictorial
FCI1f	gravitation (gravitation force)	falling ball	verbal, graphical, vectorial
FCI1a	gravitation (acceleration)	falling ball	verbal, bar chart
FCI4	Newton III	collision of cars	verbal, graphical, vectorial, bar chart
FCI13	gravitation (forces)	a steel ball is thrown vertically upwards	verbal, graphical, vectorial
FCI28	Newton III	students sitting on office chairs push each other off	verbal, graphical, vectorial, bar chart
FCI30	gravitation (forces)	a tennis ball passes through the air after being struck	verbal, bar chart, vectorial

A sample question in the vectorial representation is provided in Figure 1: it is derived from the FCI question 4 addressing a collision of a car and a truck (the original alternatives are in verbal representation). Additional information had to be included in the question to enable sensible formulations in multiple

representations. Naturally, these additions did not change the physics involved.

FIGURE 1. A sample question from the Representation Test.

DATA COLLECTION AND ANALYSIS

The data for this study were collected from Finnish high school students (aged 17). The students were taught mechanics in two groups (n = 54 altogether), following the same textbook with the same teacher (author AS). Hence, the groups are combined in the following analysis. The students had had an introduction to both kinematics and the force concept before the course in mechanics. The teaching of the mechanics course was geared to foster conceptual understanding and made use of multiple representations.

The Representation Test was administered before and after teaching kinematics and the force concept to the students. The students also had to provide a written explanation in the post-test of why they selected one of the multiple choice alternatives in every item. This was done to investigate the validity of the Representation Test, i.e. to detect possible false positives (choosing the correct answer for the wrong reason) and false negatives (choosing a wrong answer while understanding the idea in question correctly).

The coherence of a student's understanding in each of the themes was determined by studying how consistently he/she had answered the different questions belonging to each theme. To answer consistently the student had to choose the corresponding multiple choice alternatives in all questions belonging to the same theme. Students' representational coherence was evaluated from two points of view. Firstly, students' answers were investigated solely from the point of view of the representational coherence regardless of the scientific correctness: their answers were considered to exhibit representational coherence if they chose the corresponding alternatives in different representations within the themes, i.e. answered all the questions in a given theme in the same way. Secondly, students' answers were investigated from the point of view of how consistent they were between representations in a given theme and also whether their answers were scientifically correct.

Students' answers in a given theme were graded in the following way:

- two points, if they had answered every question in the theme in the same way.
- one point, if they had answered every question except one in the theme in the same way.
- zero points, if they had answered two or more questions in the theme differently.

We used the same criteria in analysing students' responses when both representational coherence and scientific correctness were required: in this analysis 'answering the same way' means only scientifically correct responses.

Student's points in seven themes (see Table 1) were added up so they got from zero to fourteen points in the whole test. On the basis of this grading system students' representational coherence was categorized into three coherence classes:

- Class I: 13–14 points indicate that representational thinking was *coherent*.
- Class II: 11–12 points indicate that representational thinking was *moderately coherent*.
- Class III: 0–10 points indicate that representational thinking was *incoherent*.

The first analysis involves representationally coherent classes with or without scientific correctness: these classes are *representationally correct*. The second analysis also demands scientific correctness: these classes are *scientifically correct*.

RESULTS

To provide a general view of the data the students' matched pre- and post-test results (averages and standard deviations) in the Representation Test are presented in Table 2. There are 23 questions in the Representation Test so the maximum score is 23: one point was awarded for each scientifically correct answer.

TABLE 2. Students' (n =54) results in the Representation Test. Standard deviations are given in the parentheses.

Pre-test (%)	Post-test (%)	Hake's gain	Cohen's d
52.7 (26.7)	78.3 (19.9)	0.54	0.77

Hake's gain (the average normalized gain) and Cohen's d (effect size) indicate that the students did improve their command of the concepts and representations addressed by the Representation Test reasonably well. There was almost no correlation ($r = -0.06$) between the pre-Representation Test scores and individual students' normalized gains. This indicates that the students' initial knowledge state regarding the gravitation and force concepts and their multiple representations did not affect students' conceptual gains.

Students' Representational Coherence

Students' representational coherence results in the pre-test and post-test are presented in Table 3 and in Table 4, respectively.

TABLE 3. Students' (n = 54) results in the matched pre- and post-tests in terms of representationally correct classes.

Representationally correct	I (%) coherent	II (%) moderately coherent	III (%) incoherent
Pre-test	20.4	42.6	37.0
Post-test	59.3	29.6	11.1

TABLE 4. Students' (n = 54) results in the matched pre- and post-tests in terms of scientifically correct classes.

Scientifically correct	I (%) coherent	II (%) moderately coherent	III (%) incoherent
Pre-test	7.4	5.6	87.0
Post-test	24.1	25.9	50.0

There is a clear shift from the pre- to post-test results towards representational coherence in both classes. Students' coherent use of representations increased quite substantially (from 20.4% to 59.3%). However, the change towards both representationally and scientifically correct understanding is not so prominent (from 7.4% to 24.1%).

Differences in Students' Ability to Use Various Representations

There were some statistically significant differences in students' ability to use different representations with scientific correctness. In the pretest in theme FCI28 (Newton's third law) students had fewer correct answers in graphical representation (31.5%) than in other representations (verbal 66.7%, vectorial 63.0% and diagrammatic 64.8%). The differences were statistically significant (McNemar's test: $p < 0.001$). Also in theme FCI13 (gravitation, forces) students had fewer correct answers in graphical representation (9.3%) than in other representations (verbal 27.8% and vectorial 20.4%). However, only the difference between graphical and verbal representations was statistically significant ($p = 0.013$).

In the post-test there was only one theme in which differences in students' ability to use different representations with scientific correctness were statistically significant. In theme FCI13 83.3% of students chose the scientifically correct answer in the verbal question whereas 61.1% of students did so in the graphical question ($p = 0.04$).

Validity of the Representation Test

We were particularly interested in finding out how well the students could justify their answers. For this purpose, the written responses for each multiple choice were examined. The results indicate that 91% of correct answers were accompanied by correct explanations (7% had partially correct explanations). Hence, the number of clear false positives is very small (2% of all correct answers). However, the number of false negatives is surprisingly high (14% of all incorrect answers). One possible reason for some false negatives might be that some students made mistakes in writing down the answers on the answer sheets; this seems likely in some cases where the verbal explanation was perfect and did not match the chosen answer at all.

CONCLUSIONS

The analysis of students' representational coherence in the post-test revealed that over half of the students could consistently answer all questions in each theme in the same way after the teaching, i.e. exhibited representationally correct coherence, either scientifically correct or not. These results suggest that students can use representations quite consistently even if they do not understand the underlying physics correctly. The scientifically correct coherence class was, however, significantly smaller: about a quarter of the students mastered both representations and the concepts addressed by the post-test. It should be noted that the results provide genuinely new information which could not be deduced from the students' averages in the Representation Test.

Our pre-test results show that there were statistically significant differences between the representations in the three themes, graphical representation being harder for the students than other representations. Our findings on the pre-test results provide evidence that students' performance may vary with the representation even if the context of a problem is very similar. This is well in line with the findings of Meltzer [2] and Kohl & Finkelstein [3]. However, in the post test only one theme had statistically significant differences between the representations: the verbal question was easier for the students than the corresponding graphical question. This suggests that teaching can decrease students' difficulties with multiple representations at least to some extent. The post-test results also suggest that the students' command of physics was more constrained in terms of scientific correctness than in terms of multiple representations.

The written responses provided support for the validity of the test: they show that the test scores do reflect the students' understanding very well. This suggests that the Representation Test is a useful instrument for both researchers and teachers in determining students' ability to use multiple representations in the context of gravitation and Newton's third law.

REFERENCES

1. A. Van Heuvelen and X. Zou. "Multiple Representations of Work-energy Processes", *Am. J. Phys.* **69**(2). 184-194 (2001).
2. D.E. Meltzer. "Relation between students' problem-solving performance and representational format", *Am. J. Phys.* **73**(12). 463-478 (2005).
3. P.B. Kohl and N.D. Finkelstein. "Student representational competence and self-assessment when solving physics problems", *Phys.Rev.ST PER* **1**, 010104 (2005).
4. P.B. Kohl and N.D. Finkelstein. "Effect of instructional environment on physics students' representational skills", *Phys.Rev.STPER* **2**, 010102 (2006).
5. I. Halloun, R.R. Hake, E. Mosca and D. Hestenes. Force Concept Inventory (Revised 1995). Password protected at <http://modeling.asu.edu/R&E/Research.html>.
6. A. Savinainen. "High school students' conceptual coherence of qualitative knowledge in the case of the force concept". Ph.D thesis, University of Joensuu, 2004.
7. P. Nieminen, A. Savinainen and J. Viiri. "Investigating Students' Representational Coherence: Developing a Multiple Choice Test". Submitted for publication.
8. D.E. Meltzer (private communication).

The Effect of Field Representation on Student Responses to Magnetic Force Questions

Thomas M. Scaife and Andrew F. Heckler

The Ohio State University, Department of Physics, 191 West Woodruff Avenue, Columbus, Ohio 43210

Abstract. We examine student understanding of the magnetic force exerted on a charged particle and report three findings from a series of tests administered to introductory physics students. First, we expand on previous findings that many students believe in "charged" magnetic poles and find that although students may answer according to a model where a positive charge is attracted to a south pole and repulsed by a north, these students may not believe that the poles are charged. Additional models produce identical answer schemes, the primary being magnetic force parallel to magnetic field. Second, the representation format affects responses: students answer differently when the magnetic field is portrayed by a field source vs. by field lines. Third, after traditional instruction improvement in student performance is greater on questions portraying field lines than for questions portraying field sources.

Keywords: Physics Education Research, Magnetic Fields, Student Models, Lorentz Force
PACS: 01.40.Fk, 01.40.gf, 01.40.Ha, 01.50.Kw

INTRODUCTION

There are many similarities between the electric force and the magnetic force on a charged particle. Ultimately, the concept of field, which could be represented by a source (e.g. charged plates, magnetic poles, etc.), by field lines, or both, must applied to properly determine the force for any given situation. In addition, both fields have two types of sources, positive/negative charge for an electric field or north/south pole for a magnetic field. Because of these similarities, it may be natural for the novice to assume that there is effectively no difference between determining the electric and magnetic force on a particle. This ambiguity may at least in part be responsible for the common misconception, first studied by Maloney, that a magnetic pole exerts a force on a charged particle, regardless of its velocity. In other words, Maloney concluded that some students interpret magnetic poles as being charged, thus affecting electric charge [1].

Maloney's finding that many students answer according to a "charged pole" model raises several questions. First, the concept of charge and the concept of field are distinct; since the concept of field is an ultimate goal and necessary for eventual complete understanding, to what extent are students thinking about charges and/or fields when asked these force questions? Second, how sensitive are the student responses to representation? For example, if instead of being presented with poles, the students were presented with a representation of the field only, would the students still apply the "charged pole" model? If so, are they envisioning charges as sources of the fields and then making conclusions based on the presence of charges, or are they simply attending to the field alone and answering directly from knowledge of the field [2-4]? The effects of representation on student performance are the subject of many other studies (e.g. Meltzer in ref. 5). Maloney's study does not examine the distinction between student understanding of charge and field, nor do the questions posed represent field lines.

To gain more insight into student understanding of magnetic forces, we expand upon Maloney's findings by examining the effect of representation, i.e. poles or field lines, on student responses to magnetic force questions. We also include both a pre and post assessment to determine the effects of traditional instruction.

EXPERIMENT 1

The students in this experiment (N=122) were enrolled in the second of a three-quarter introductory, calculus-based physics sequence at The Ohio State University, a large research university. Instruction in the course was traditional. Short, multiple choice quizzes were administered during weekly lab meetings

CP951, *2007 Physics Education Research Conference*, edited by L. Hsu, C. Henderson, and L. McCullough
© 2007 American Institute of Physics 978-0-7354-0465-6/07/$23.00

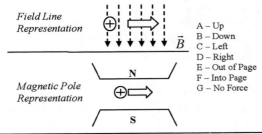

A charged particle is placed in a magnetic field, as shown in the arrangements below. In some cases, the particle is given an initial velocity v in the direction indicated by the large arrow. For each case, circle the direction of the force, if any, experienced by the charged particle. Ignore gravity.

Field Line Representation

$\vec{B}$

A – Up
B – Down
C – Left
D – Right
E – Out of Page
F – Into Page
G – No Force

Magnetic Pole Representation

N

S

FIGURE 1. Example question from pre-test. Each student was given either the *Field Line Representation, or the Magnetic Pole Representation.*

and graded for participation. All questions asked were similar to those of the Maloney study – students were given a positively charged particle either moving perpendicular to the magnetic field or with zero velocity [1]. However the questions included two significant differences. First, the answer choices included all three dimensions. In Maloney's study, answer choices were limited to directions collinear with the field. In order to be correct, students had to repeatedly choose "neither" for all questions. This lack of directional choices potentially inflated responses in these directions. Second, in order to test the effects of representation format, approximately half of students were given questions in which the magnetic field was represented by magnetic poles, and the other half were given questions in which the magnetic field was represented by field lines. Examples of each question format are shown in Figure 1.

The pre-test was administered in the fifth week of a ten-week course, directly following instruction in electrostatics but previous to instruction in magnetostatics. Figure 2 presents students' answering patterns on both a non-zero velocity question and a zero velocity question. Both figures indicate that students in the field line representation condition are more likely to answer incorrectly, choosing "F Parallel to B" or "F Anti-Parallel to B". However, only the zero velocity question produces a statistically reliable difference between the conditions ($t=2.25$, $dF=120$, $p<0.05$), compared to the non-zero velocity condition ($t=0.99$, $dF=120$, $p=0.32$).

The post-test was administered in the 10th (final) week of instruction (N=124), about one week after the material was covered in class. There is a clear effect of instruction: 13% of the students obtained a correct answer in the pre-test compared to 45% in the post-test on the two reported questions. These results are somewhat different than Maloney's, who found that even after instruction approximately 60% of students answered according to the "charged pole" model [1]. Overall, students in the field line representation obtained higher scores than students in the pole representation on the 12-question quiz (field lines: $\bar{x}_f=0.53$, magnetic poles: $\bar{x}_m=0.39$, $t=3.04$, $dF=122$, $p<0.01$).

Figure 3 shows post-test results of questions virtually identical to the pre-test questions. On the post-test the performance of students in the two conditions reversed from the pre-test; students in the field line representation answer correctly significantly more than students in the pole representation in both the zero velocity ($t=2.56$, $dF=122$, $p=0.01$) and the non-zero velocity ($t=2.69$, $dF=122$, $p<0.01$) questions. Notably, students in the pole representation chose the answer consistent with the "charged pole" model more often than the student in the field line condition.

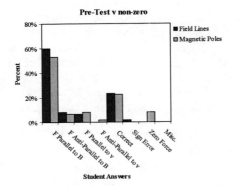

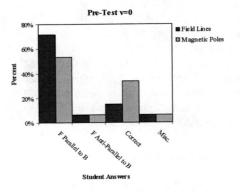

FIGURE 2. Plots of student responses to the non-zero and zero velocity pre-test questions (Field Line Representation: N=60, Magnetic Pole Representation: N=62).

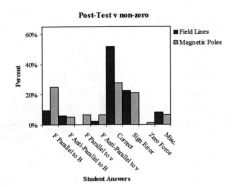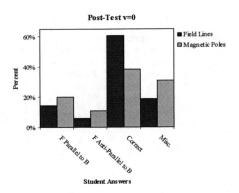

FIGURE 3. Plots of student responses to the non-zero and zero velocity post-test questions (Field Line Representation: N=69, Magnetic Pole Representation: N=55).

Discussion

The pre-test results suggest that when given a question represented with magnetic field lines, students with little prior knowledge of magnetic forces tend to answer that the magnetic force is in the direction of the field, similar to the rules for electric field lines, which they just learned in an earlier section in the course. Note that this answer scheme is not necessarily the same as that of the "charged pole" model, since there is no explicit representation of a pole. When students are given a question represented with magnetic poles, they tend to answer in a way consistent with a "charged pole" model. Although not directly tested in this experiment, students may be first imagining field lines from the represented poles and then using the field lines to answer. It is not unreasonable to infer from this data that in fact these two different representations are cueing two completely different models: either the "charged pole" model or the "force-in-direction-of-field" model.

The post-test results present an interesting shift in the dependence of representation: when students are presented with the field representation they appear to have abandoned the "force-in-direction-of-field" model and learned the correct model, whereas the when cued with the pole representation, many students still answer according to the "charged pole" model, even after traditional instruction (although less than before instruction). Nonetheless, these results are only suggestive; it is still not clear whether students are using any particular model.

EXPERIMENT 2

Experiment 2 was administered to a separate but similar population of students (N=239) in a different quarter. Experiment 1 indicated that when compared

between students in different conditions, student answering depends on the representation format of the question, namely field lines or poles. In Experiment 2, we administer a within-student test to determine whether the same student will answer differently when presented with the two representations. In order to account for effects of order of presentation of the two representations, one half of the class received one representation first and the second half received the other first.

Results indicate that while student performance did not depend significantly on question order (field line question: $\chi^2=4.47$, $dF=2$, $p=0.11$; magnetic pole question: $\chi^2=1.34$, $dF=2$, $p=0.51$), students did answer according to the "charged pole" model more frequently to the field line representation (67% compared to 57%) (McNemar's test, $\chi^2=6.7$, $dF=1$, $p=0.01$). Figure 4 shows student responses to the both the magnetic pole representation and the field line representation. The "Charged Pole" category includes all responses in the dimension of the magnetic field, including parallel and anti-parallel, while the "Correct" category includes all responses in the dimension of the correct answer.

The post-test was administered under somewhat different conditions than in Experiment 1, namely in the 9th week, which was during (i.e. essentially directly after) instruction of the Lorentz force and the right hand rule. In addition, this post-test was graded for credit. Again, question order did not affect student performance (field line question: $\chi^2=0.63$, $dF=2$, $p=0.73$; magnetic pole question: $\chi^2=0.67$, $dF=2$, $p=0.72$).

As in Experiment 1, students in the post-test answer according to the "charged pole" model slightly more frequently to the magnetic pole representation (8% compared to 5%) (McNemar's test, $\chi^2=2.6$,

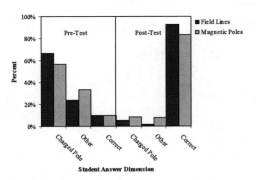

FIGURE 4. Plot of within student responses to the non-zero velocity pre-test questions (N=239) and post-test questions (N=225).

p=0.17), and answer correctly more frequently to the field line representation (93% compared to 84%) (McNemar's test, χ^2 =13.3, dF=1, p<0.001). Figure 4 shows student responses to the both the magnetic pole representation and the field line representation.

Discussion

The results from the Experiment 2 replicate the results of Experiment 1 in a within-subject design. Most notably, the same student will answer differently to magnetic pole and field line representations, and traditional instruction tends to affect the student responses to field line representation more than responses to magnetic pole representation.

CONCLUSIONS

There are two primary conclusions to be drawn from this study. First, we replicate and the findings of Maloney [1], which claim that many students answer in accordance with a "charged pole" model, and extend these results to include the field line representation. In the "charged pole" model, students believe that charged particles experience magnetic forces directed toward or away from poles and/or along field lines, much like the electrostatic force. While traditional instruction reduces the occurrence of this incorrect conception, our findings suggest that the difficulty persists in a significant number of students.

Second, our data demonstrate that the representation format of a question significantly affects students' correct determination of the direction of magnetic force on a charged particle. Following traditional instruction, students answer the magnetic field line questions more correctly than the magnetic pole questions, suggesting that the different representations tested cue different solution paths or models. One possible solution path, which is consistent with our data, is that when students are presented with the relatively abstract representation of field lines, they easily apply the right hand rule, which explicitly refers to the magnetic field, as taught in class. Conversely, when presented with magnetic poles students might take one of several different paths. The relatively concrete representation of the poles could lead many students to choose an answer consistent the "charged pole" model. Alternatively, students may first use the magnetic poles to draw the magnetic field and then apply the right hand rule to determine the direction of the force. This additional step of drawing the field introduces some degree of error. Many students did, in fact, spontaneously draw field lines when presented with the magnetic poles representation.

An additional explanation for the difference between student performances in differing representations is that in the course of traditional instruction, field line representation tends to be favored by the instructor, thus providing the students with a greater degree of practice and familiarity. Nonetheless, even if this explanation is the primary reason for the difference in performance, our results clearly demonstrate that instruction in one representation alone is not sufficient to ensure complete comprehension of a concept. Instead, these results reveal that students will resort to preconceived notions—or an altogether different model—whenever confronted by a different context.

ACKNOWLEDGMENTS

This research is supported by a grant from the Institute of Educational Sciences of the U.S. Department of Education (#R305H050125).

REFERENCES

1. D. P. Maloney, "Charged poles?" *Phys. Educ.* **20**, 310-316 (1985).
2. J. Guisasola, J. M. Almudi, and J. L. Zubimendi, "Difficulties in Learning the Introductory Magnetic Field Theory in the First Years of University," *Sci. Ed.* **88**, 443-464.
3. M. C. Pocovi, "The Effects of a History-Based Instructional Material on the Students' Understanding of Field Lines," *J. Res. Sci. Teach.* **44**, 107-132.
4. S. Tornkvist, K. A. Pettersson, and G. Transtromer, "Confusion by representation: On student's comprehension of the electric field concept," *Am. J. Phys.* **61**(4), 335-338.
5. D. E. Meltzer, "Relation between students' problem-solving performance and representational format," *Am. J. Phys.* **73**(5), 463-478.

Multiple Modes of Reasoning in Physics Problem Solving, with Implications for Instruction

David Schuster, Adriana Undreiu and Betty Adams

Physics Department and Mallinson Institute for Science Education
Western Michigan University, Kalamazoo, MI 49008, USA

Abstract. Problem-solving is an important part of physics teaching, learning and assessment. It is widely assumed that the way that experts solve problems, and students should, is by systematic application of basic physics principles. Model solutions are laid out this way, and teaching of problem-solving usually consists of 'going over' such solutions step by step. However, while this does represent the physics structure of the final solution, it does not adequately reflect how people actually think when tackling problems. Real cognition is complex. This study was prompted by students trying to 'map across' result features recalled from previous cases instead of working from basics. Since our instruction emphasizes the power and generality of basic principles, our first response was to re-emphasize principles, but we found that experts in fact draw extensively and effectively on rich compiled case knowledge. We investigated cognition in detail for geometrical optics. Research methods included analysis of written solutions, reflections on thinking, and interviews. Cognitive modes emerged from the initial research stages, and were then used to code individuals' problem-solving pathways. Learners and experts alike used multiple modes of cognition, significantly principle-based reasoning, case-based reasoning and experiential-intuitive reasoning. Case-based reasoning using pre-compiled knowledge played a pervasive role in conjunction with, and sometimes in conflict with, principle-based reasoning. The implications for instruction are that it should reflect what we know about cognition and expertise, and hence include teaching case-based as well as principle-based reasoning. We are doing this in optics, by using cases and variations, identifying topic knowledge schema 'sub-assemblies', and modeling their use in problems.

Keywords: Physics education research, cognition, physics problem-solving, case-based reasoning, optics.
PACS: 01.40.Fk, 01.40.Ha, 01.40.gb, 01.40.–d

INTRODUCTION

A goal and characteristic of science is the use of relatively few fundamental concepts and principles to explain a variety of situations, and accordingly our inquiry-based physics course for prospective teachers emphasizes a 'powerful ideas' theme throughout. We hope that students come to appreciate this perspective in both learning the physics fundamentals and applying them to solve problems.

It would appear to follow that we should teach problem solving by modeling how solutions are obtained by systematic application of concepts and principles. Thus teaching physics problem solving, if done explicitly at all, usually consists of 'going over' worked-out solutions as a stepwise application of principles. While this does represent the physics structure of the final solution, it turns out that it by no means reflects the richness of how people actually think when tackling problems. Hsu et. al.[1] recently provided a useful review of work on problem solving in physics, and Bodner [2] notes how typical problem instruction in chemistry does not reflect actual thought processes.

Observations during teaching of geometrical optics led us to believe that the cognitive processes and compiled knowledge involved in physics problem-solving might be far more complex and multifaceted than is generally realized, and not reflected in solutions conventionally presented in teaching and textbooks. Subsequently, we found that both learners and experts use multiple modes of reasoning in physics problem solving, three significant modes being principle-based reasoning, case-based reasoning and experiential-intuitive reasoning. By case-based reasoning we mean drawing quickly by association on outcome features of recalled similar cases, or retrieving previously compiled knowledge parts (schemata or subassemblies), and adapting to the new case, rather than constructing a principled solution from scratch. Kolodner [3] discusses case-based reasoning in several fields.

Elucidating the nature of cognition and expertise should have important implications for more effective instruction [4, 5]. Toward that end, we undertook a study of cognition in problem solving in optics, consisting of an observational exploratory stage and a designed experimental stage.

CP951, *2007 Physics Education Research Conference*, edited by L. Hsu, C. Henderson, and L. McCullough
© 2007 American Institute of Physics 978-0-7354-0465-6/07/$23.00

OBSERVATIONAL STUDY

The impetus for this study arose from observing unexpected student attempts to solve geometrical optics questions in our physics course for prospective teachers. It is clear that many entering students are not used to thinking and working via fundamentals. However, our emphasis on powerful ideas does have an effect during the course, and most students learn to work in principled fashion applying those ideas. Nevertheless we observed a mode of operating where students seemed to try 'mapping across' specific results recalled from similar-looking problems encountered earlier, instead of proceeding from principles.

This tendency first became apparent in the section on reflection and images. Students had investigated reflection, formulated the law, learned about image formation, and applied these ideas to locate virtual and real images by ray construction for various plane, convex and concave mirror cases. Given a new variant of a concave mirror problem, students produced all sorts of diagrams in their attempts.

Fig. 1 illustrates two concave mirror cases. The left-hand diagram, for a source close to the mirror, is a correct construction to find the (virtual) image.

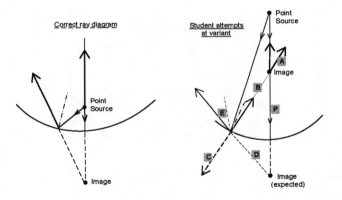

FIGURE 1. Ray constructions and image location for two curved mirror cases.

The right-hand diagram, for the case of a source further away, is a composite diagram to illustrate the main kinds of things that students did. These are labeled A–E. Most students reflected the perpendicular ray P correctly, but did all sorts of things for the other ray. Students who reflected it correctly using the basic law constructed rays ending at (A) and marked the image at the ray intersection. Some who initially reflected rays correctly erased them later (B), then did other things. Others made dotted ray extensions, diverging behind the mirror (C) and then gave up. Some drew dotted 'lines' converging (D) independently of rays. Some drew 'reflected' rays (E) actually

violating the law, with dotted extensions converging behind the mirror to an 'image' point.

Mystifying attempts like this had first been interpreted as displaying near complete lack of understanding. However, an incident cued us as to what might be happening. A distraught student approached an instructor in an exam with her attempt saying: "*I don't know what I'm doing wrong... I did it like we did before... and my rays sort of reflect up here somewhere... I know the image is supposed to be here, behind... but I just can't make the rays go there...*"

Despite her confusion, she actually had most of the declarative and procedural knowledge needed to solve image problems. It seemed that she was trying to balance her knowledge of principles and procedures with a conflicting expectation of where the image should be, arising from the 'mapping across' of result features from a previous case onto a new case – a simplistic form of case–based reasoning.

Learner modes of operating

Intrigued by this mode of operating we undertook a study of learners' problem solving in geometrical optics. We focused on reflection and refraction, each of which involved both a fundamental law and its geometric application to image formation.

Our simple 'mapping' hypothesis was subject to immediate test – if the tendency was not idiosyncratic but more widespread, further examples should be found in other students' existing responses. We revisited all previous responses to the concave mirror problem. Encouraging for the predictive power of the hypothesis was the fact that supporting examples were now recognized in other responses to the same question, once we knew what we were looking for and better able to interpret students' diagrams.

Furthermore, we made a second testable prediction: if the tendency was not specific to this problem, but more general, it should be found in solutions to other problems on the same exam. This second hypothesis was also borne out, in a quite different problem on plane mirrors.

In several problems subsequently used in the course it became clear that most students were at least able to reflect rays correctly and seek ray intersections. Nevertheless, some students tended not to go that way or not to persist in it, but tried to work backwards from the kind of result they thought they ought to get. The feature-mapping mode of operating was often strong enough to override working from scientific principles when results conflicted, so that students would even resort to distortions to resolve this and get the features they had induced from other cases.

Our main focus was case-based and principle-based thinking, but we also noticed that in situations related to everyday experience, students' responses could sometimes be interpreted as arising from experiential resources such as phenomenological primitives (p-prims) [6]. An example is believing that no image is possible if an object is off to the side of a mirror, based on personal experience with being able to see (or not see) one's *own* image in a mirror. We have called this mode experiential-intuitive reasoning. The tendency seems considerably less for optics than mechanics.

EXPERT MODES OF OPERATING

Our first response to the result-mapping tendency had been to discourage it and re-emphasize working from basic principles. However, we also undertook a preliminary study of *expert* cognition, in areas of mechanics and optics, in which it became clear that experts too used case-based reasoning in problem solving, extensively and effectively! From experience, they drew on a rich case knowledge, in conjunction with basic principles. However, experts were also aware of case applicability conditions and limitations. Reif and Allen [7] discuss the use of case-specific knowledge for the concept of acceleration.

EXPERIMENTAL STUDY OF LEARNER REASONING MODES

Recognition of the major role of case-based reasoning for both novices and experts indicated that instead of discouraging associative thinking in learners, we should investigate it more closely, and learn how best to include it in instruction. Therefore, we extended the study to a designed experimental phase. We produced an instrument consisting of five sets of reflection and refraction problem tasks. Each set contains three different case variants, for use in examples, homework, and tests. The variants have different result features but the same underlying principles, and are thus of known relation to each other. For each problem students had to describe and explain their reasoning, thinking paths and levels of confidence.

Qualitative and quantitative data sources were: • Students' written problem solutions. • Students' written (metacognitive) accounts and reflections on what they did and why. • Explanations of confidence level. • Individual interviews. • Group observations.

Cognitive coding scheme

Significant modes of reasoning were first identified in the observational stage and extended and refined in the experimental stage. Usually more than one mode occurred and a subject's solution process was analyzed into episodes. Useful mode subdivisions emerged for classifying and coding episodes along thinking paths. Case knowledge items invoked in episodes were also noted. An individual's thinking path could thus be

represented by a code sequence. Table 1 presents the codes and subdivisions devised:

Table 1. Cognitive coding scheme
Reasoning modes:
PBR– Principle-Based Reasoning:
F–Forward from principles.
B–Backward to principles.
PEC–Principles/Procedures Extracted from Case
CBR– Case-Based Reasoning:
RM–Result Mapping.
FM–Feature Mapping (partial or complete).
PPT–Principles & Procedures Transfer,
FMA–Feature Mapping Adjusted for case differences
EIR – Experiential-Intuitive Reasoning:
P-Prim–Phenomenological Primitive
Comparison and resolution:
C&R: Comparison & Resolution:
TP–Trusting Principles
BB–Bending the Basics (to fit result)
NA–No Answer reached, knows something is wrong
Correctness: Each episode coded, as well as overall solution
√ – used correctly X– used incorrectly
Confidence level:
VC-very confident, SC-somewhat confident, NC- not confident
Case Knowledge items accessed.
Specified for each episode.

Illustrative results

Qualitative and quantitative data analysis involved written, observational and interview data, coded as described. To illustrate briefly the nature of some quantitative results, Table 2 presents data from the Summer 2007 physics section (20 students).

Table 2. Results for the concave mirror image problem Figures are number of students out of 20 in the section.
Reasoning modes:
4/20   Reported only PBR
13/20 Used combination PRB and CBR
3/20   Confused; not enough information to classify
No apparent EIR evoked by this particular problem.
Comparison and resolution:
11/20 Compared PBR and CBR, of which:
3/11 PBR overrode CBR
2/11 CBR overrode PBR; student 'bent basics'
6/11 Unresolved by student, no solution.
Correctness of aspects:
8/20 Correct solution overall, of which:
4/8 Correct by PBR,
3/8 Correct by combination PBR & CBR
1/8 Case as a starting point to extract Procedure.
12/20 Incorrect overall, of which:
8/12 Potential to succeed if conflict resolved well
4/12 Unable to do it at all.

This is consistent with both the exploratory studies and our findings for other problems, though mode mix naturally depends on problem and context. The work is being extended to larger numbers in regular semesters.

FINDINGS AND DISCUSSION

We find that learners and experts alike invoke multiple modes of cognition in physics problem solving. Three important modes of reasoning emerging from this study in optics are principle-based reasoning, case-based reasoning and experiential-intuitive reasoning. Case-based reasoning, drawing extensively on pre-compiled knowledge, is pervasive during the process, though this may not be evident in final constructed solutions, and there is much interplay between case- and principle-based thinking. Experiential-intuitive thinking was little evident for these particular optics problems. Significantly, working linearly and systematically from basic principles is not usually the primary mode of initial thinking, of either learners or experts, although it is the way that we conventionally present model solutions, post-hoc, in teaching and textbooks.

Thinking by association seems to be a natural human tendency. It has pros and cons, in both everyday and scientific contexts. Working by association instead of via principles, although it did not serve some students well in these physics problems, is not broadly undesirable. It works well in everyday situations and is quick. Used correctly it can be very productive scientifically too. Tapping into previously compiled case knowledge is very efficient, compared to working everything out anew from scratch every time a new situation is encountered. Experts do this as part of their repertoire! They have a rich store of compiled knowledge and particular cases to draw upon. In fact this is one of the features of expertise, and indeed of structural understanding of a topic, and it is hard to imagine operating without it – everything would be cumbersome. But when drawing on case knowledge, experts are aware of what they are doing, know the status, applicability conditions and limitations of that knowledge, and are able to crosscheck against basics. They also recognize that various cases are just particular instances arising from the same underlying principles or mechanisms.

Returning to the mirror example, note that once one has solved one or two cases it is hard to imagine solving a variant without drawing on the mental picture already in place, which represents not just the result but also the method, and encapsulates one's previous construction of understanding of such situations. However this kind of compiled case knowledge, where the essence is understood as well, is different from the type of simplistic mapping of results features which we observe in some students, who do not perceive both similarities and differences in cases.

Note that in discussing how case features compiled from previous examples are used in tackling new problems we are not talking about the conventional use of example problems to teach principle-based reasoning.

Work on transfer is relevant to this study, recently that of Marton [8] on recognizing both sameness and difference in transfer. Of course if transfer between cases is to occur, instructors would prefer students to carry across deep features (underlying principles and procedures), rather than surface features of results. Students are also known to

categorize problems by surface features [9], though this is different from solving by mapping *result* features.

Associative thinking and previous examples play a pervasive role in problem solving, but novices do not use them very well without instruction. Students need to be aware of how science works, and also that scientific thinking is different from everyday thinking in certain ways [10]. They need to appreciate the status of both general principles and particular examples, and know that applicability and generalizability need to be kept in mind. Such epistemological knowledge, though often tacit, is as important as declarative and procedural knowledge of the subject matter.

IMPLICATIONS FOR INSTRUCTION

Implications for more effective instruction are that it should reflect what we know about real cognition and the nature of expertise. This includes teaching case-based as well as principle-based reasoning and making such thinking 'visible.' Simply re-teaching the physics without the cognition will not work effectively.

We are proceeding in optics in a number of ways: identifying knowledge 'sub-assemblies' for each topic: presenting case variations to assist discernment of both commonalities and differences [8], as formative assessment for learning [11]; promoting explicit reflection and metacognition; and modeling these aspects in teaching and the design of learning tasks.

REFERENCES

1. Hsu, L., Brewe, E., Foster, T. M., & Harper, K. A. (2004). Resource letter RPS-1: Research in problem solving. *American Journal of Physics*, 72(9), 1147-1156.

2. Bodner, G. (2003). Problem solving: The difference between what we do and what we tell students to do. *J. Chem.Ed.*, 7, 37-45. Bodner (1987), The role of algorithms in problem-solving *J Chem Ed*, 64, 513-514.

3. Kolodner, J. (1993). *Case-based reasoning*. San Mateo, CA: Morgan Kaufmann.

4. Redish, E. (1994). The implications of cognitive studies for teaching physics. American Journal of Physics, 62(6), 796-803.

5. Bransford, J., Brown, A., & C., R. (Eds.). (2000). *How people learn: Brain, mind, experience, and school.* Washington, D.C.: National Academy Press.

6. diSessa, A. A. (1993). Toward an epistemology of physics. *Cognition and Instruction*, 10(2&3), 105-225.

7. Reif, F. & Allen, S. (1992). Cognition for interpreting scientific concepts. *Cognition and Instruction*, 9(1), 1-44.

8. Marton, F. (2006). Sameness and difference in transfer. *The Journal of the Learning Sciences*, 15(4), 499-535.

9. Chi, M. T. H., Feltovich, P. J., & Glaser, R. (1981). Categorization and representation of physics problems by experts and novices. *Cognitive Science*, 5, 121-152.

10. Reif, F., & Larkin, J. (1991). Cognition in scientific and everyday domains: Comparison and learning implications. *Journal of Research in Science Teaching*, 28(9), 733-760.

11. Black, P., Christine, H., Clare, L., Bethan, M., & Dylan, W. (2003). *Assessment for learning: Putting it into practice.* New York: Open University Press.

Explicit Reflection in an Introductory Physics Course

Michael L. Scott, Tim Stelzer, and Gary Gladding

Department of Physics, University of Illinois at Urbana-Champaign, IL 61801

Abstract. This paper will explore a classroom implementation in which explicit reflective activities supplemented the problems students worked during class. This intervention spanned a 14 week period and was evaluated based on the relative performance between a control and treatment group. Instruments used in this study to assess performance included the Force Concept Inventory (FCI), a physics problem categorization test, and four class exams. We will discuss fully our implementation of the reflective exercises along with results from the accompanying measures.

Keywords: physics education research, explicit reflection, metacognition.
PACS: 01.40.Fk

INTRODUCTION

The field of physics education research (PER) is the scientific study of the teaching and learning of physics. There has been extensive research, both within and outside of PER, that examines the behavioral and cognitive differences between novices and experts when solving problems.[1,2,3,4,5,6,7] Research has shown that when compared to experts, novices tend to have an incoherent knowledge structure,[8,3,4] employ means-ends analysis,[1,4] tend to categorize problems by surface features,[3] and have difficulty transferring their knowledge to other contexts.[9] Experts, on the other hand, generally begin solving problems with a conceptual analysis, have a very organized knowledge structure,[3,4,8] categorize physics problems by fundamental principles,[3] and can transfer their knowledge when confronted with a new situation.

The goal of instruction is to help shift a novice to a more expert-like state in matters of knowledge, understanding, and thinking. One important behavior of expert thinking is time spent reflecting upon the meaning and structure of things learned and of tasks worked. Of interest here is how does the supplementation of physics problems with reflective exercises help facilitate the transformation of students from novices to experts. In the following research, we require students to engage in explicit reflective activities upon the completion of their worked physics problems. We then attempt to examine the effect this intervention has on students' understanding of physics.

METHOD

For this investigation, we used student subjects enrolled in the fall 2006 semester of an introductory calculus-based mechanics course at the University of Illinois, Urbana-Champaign (UIUC). The course is part of the reformed introductory physics sequence at Illinois [10] and had an enrollment of approximately 430 students. Its components included 2-75 minute lectures, a 2 hour discussion, and a 2 hour laboratory each week. This study operated within the context of that course's discussion environment, which spanned a period of 14 weeks. In this environment, students work in collaborative, small groups on a set of 4-6 physics problems, where each week the problem set focuses on one or two physical principles.[11] Nine of those weeks also included an in-class quiz which is used as a formative assessment instrument.[12]

This intervention was assessed by evaluating the relative performance of a treatment group to a control group. Four discussion classes out of twenty two were chosen for the study, with two discussion classes appointed as the treatment group and two as the control group. Both conditions had the same instructor; in addition, they also worked with identical problems and quizzes as were common to the course. However, students in the treatment group had reflective activities added to the end of most discussion problems, along with four of their quizzes including a reflective question which was substituted for a quantitative question found on the control group's quiz.

The reflective exercises accompanying each discussion problem were to engage the students in

CP951, *2007 Physics Education Research Conference*, edited by L. Hsu, C. Henderson, and L. McCullough
© 2007 American Institute of Physics 978-0-7354-0465-6/07/$23.00

finding meaning in the problems they were doing, were to help them see commonalities between problems in problem sets, and to assist them in knowledge integration. These exercises were placed at the bottom of each discussion problem's page and were made explicit by a prompt that said "Reflecting Back," which preceded each set. Some examples of the reflective activities include:

"Outline your strategy in solving this problem."

"Write down what you believe the intent of this problem is and/or what you learned from doing this problem."

"Compare this problem to the previous problem. How are they similar? How are they different?"

Students were to work on these activities in their groups and were to write their responses in the space provided on their problem sheet.

ASSESSMENTS

The implementation of explicit reflective exercises as an intervention was assessed by evaluating the relative performance of a treatment condition to a control condition on set of assessment instruments. Instruments used in this study included the Force Concept Inventory (FCI), a physics problem categorization test, and four in-class exams. This section will discuss each of these instruments, how they were used to assess performance, and the results from each assessment.

Force Concept Inventory

The FCI [13] was used as an assessment in this study for two reasons. First, through reflection students should begin recognizing common principles and concepts applied over a group of physics problems that are cast in quite different contexts. We should therefore expect to see a larger gain [14] in conceptual knowledge for the treatment group than with the control group. The second reason for its inclusion is that there are numerous, well-published papers on prior findings from its use that we can draw on to compare our findings. We will thus be able to gauge our intervention against other findings.[15]

The FCI was given at the beginning of the first week of the discussion class as a pretest. As a posttest, it was given as an in-class quiz at the end of the semester. There were 33 students in the control group and 22 students in the treatment group that took both the pre- and posttest.

Prior research has shown that traditionally taught courses have an average normalized gain of 0.2 and that of courses employing more interactive engagement components have mean gains around

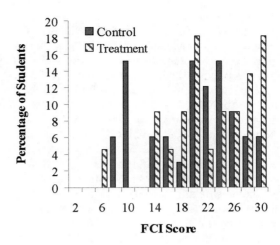

FIGURE 1. Histogram of FCI pretest scores for both groups.

0.4.[14,16] Findings from our study show that the treatment group had a normalized gain of 0.44 compared to the control group's gain of 0.41. Although the difference between the two group's normalized gains is not statistically significant, both groups had average pre- and posttest scores that were among the highest reported by universities in Hake's 1998 study. Also, the distribution of pretest scores, as shown in Fig. 1, reveals a ceiling effect. This effect limited the possible gain in scores for both groups, but this is especially true for the treatment group where one third of the group had pretest scores ≥ 28. This high level of performance by our students limited the resolution of the FCI for measuring student gains.

Problem Categorization Test

The categorization test was based on earlier research that investigated the differences between the classification of groups of physics problems by physics experts and novices.[17,18] These studies demonstrate that experts will classify physics problems based upon the principles used to solve them (deep structure similarity); whereas, many novices will identify surface feature similarities in their classifications of problems. Reflecting on problems during a problem set, where students are forced to think about the similarities between problems, should help students to categorize problems more along the spectrum of principles than on surface features. In the Hardiman et al study, which used a similar class of students, they reported that 40% of their subjects classified problems primarily by surface features. Given this percentage, the categorization test was included to see if students in the treatment group would categorize more like experts than novices.

Modeling the Hardiman study, the categorization test was a booklet with eight pages of problems. On each page there was a Model problem that appeared at the top of the page. Beneath it were two subsequent problems (Problems A and B). Students were asked if each of the subsequent problems would be solved similarly to the Model problem, and they were to provide written explanations for their answers. There were four distinct Model problems, each appearing twice. Among each of the four comparisons per Model problem, one sub-problem had only surface feature similarity (S), one had both surface feature and deep structure similarities (SD), one had only deep structure similarity (D), and one had neither surface feature nor deep structure similarities (N).

The test was administered during the last week of class and was voluntary. To encourage serious participation, students were offered extra credit. There were 27 students in the control group and 22 students in the treatment group who completed this test.

The test was scored two ways: 1) by looking at whether students said "yes" or "no" when asked if a pair of problems would be solved similarly, and 2) by coding their written explanations. Data for the correct responses of the yes or no questions is shown in Table 1. Among the four comparison types and on overall scores, there was no statistical difference between the control or treatment groups. Both groups' scores are slightly higher, but comparable, to the "novice" group in the Hardiman study.[19]

TABLE 1. Percentage of Correct Responses for the Control and Treatment Groups on the Four Comparison Types.

Comparison Type	Control Group	Treatment Group
S	32	25
D	60	66
SD	84	86
N	96	98
All	68	69

Note—S = surface features, D = deep structure, SD = both surface features and deep structure, N = neither surface features nor deep structure.

Student explanations were coded based whether the explanation focused on surface features, principles, equations, or physics terminology. These codes were not mutually exclusive per explanation. Subjects were then classified into three groups: (1) *surface-feature*, (2) *principle*, or (3) *mixed*. Classification into the surface-feature or principle group depended upon the subjects having those respective codes on half or more of their explanations. Pictured in Fig. 2 are the relative percentages of student types in both conditions. Juxtaposed against our data is data from Experiment 2 from the Hardiman study.[19] Both

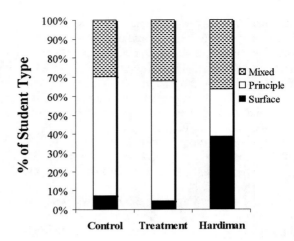

FIGURE 2. Percentages of the classification of subjects from the categorization test.

control and treatment conditions had statistically equivalent percentages of students in all three groups. An unexpected result was the large number of students in both conditions (63%) who were categorizing primarily by principle-based arguments compared to only 25% of subjects in the Hardiman study. Also surprising was that only 2 subjects in the control and 1 subject in the treatment conditions categorized primarily by surface features. Because the control subjects' performance exceeded our expectations, this categorization test was not sufficient at discriminating between the two conditions.

Class Exams

The students were also administered four multiple-choice exams throughout the semester as part of their regular course requirements: three midterms and a final. These exams are used in the course as summative assessments of the students' mastery of the material taught. Each midterm exam consisted of 25 questions, with the final exam having twice that number of questions. Question type varies from conceptual to quantitative, and from 3-choice to 5-choice option questions.[20]

Looking at the mean score on common questions between the two groups, the treatment group outperformed the control group at the 1.7 sigma level. Dividing those questions into conceptual and quantitative, the difference is greater for the conceptual questions with the treatment outperforming the control at the 2.3 sigma level; whereas, on the quantitative questions, the treatment group outperformed the control group at the 1.4 sigma level.

To interpret this result, we need to worry about systematic effects due to the selection of the control and treatment groups. In our study the treatment group had a higher FCI pretest average; therefore, we used the control group's FCI pretest scores to predict how the treatment group should have performed. Looking at the difference in attained versus predicted mean score on common conceptual questions, the treatment group exceeded its predicted average at the 1.4 sigma level. Similarly, on the quantitative questions they exceeded their predicted mean at the 1.1 sigma level.

DISCUSSION

It was expected that through reflection, students would recognize concepts and physical principles across different contexts more easily than without reflection. Unfortunately, two of our tests that were intended to test this, the FCI and a categorization test, were unable to discriminate between our control and treatment conditions. This inability to discriminate may have been due to the high level of performance on both measures by both groups. When analyzing the FCI data, the treatment group had little room to improve upon its pretest scores. Analysis of the categorization test showed that most subjects in the control group are already categorizing problems by principle-based arguments. This was unexpected given findings from earlier research. The high level of performance on both these tests is probably a function of the course itself, where the entire introductory sequence at the University of Illinois was reformed to include more interactive components.[10]

To test overall physics knowledge and performance, we looked to the four multiple-choice class exams. Only on this measure was the data suggestive of an effect favoring the treatment group, and this difference was marginal at best.

Instructor observations of student interactions with the reflective activities note that many of the students appreciated the intent of the activities and worked them without complaint. There were some students, however, who saw limited value in them and would skip them until the instructor approached their table and asked about them. Another observation was that the lower ability students, who tended to sit together in groups, had difficulty just getting to the reflective exercises. These students usually had trouble with the "tools" needed to solve the discussion problems, and in general they did not complete the discussion packet in the time allotted. Whereas, students at the other end of the spectrum would finish the packet with time to spare in class. In most cases, overall exposure to the reflective activities was less than 10% of class time.

Further research on this particular study will include more extensive analyses of our measures and exploring student responses to reflective questions found on their quizzes.

REFERENCES

1. R. Bhaskar and H. A. Simon, *Cog. Sci.* **1**, 193-215 (1977).
2. J. H. Larkin and F. Reif, *European Journal of Science Education* **1** (2), 191-203 (1979).
3. M. T. H. Chi, P. J. Feltovich, and R. Glaser, *Cog. Sci.* **5**, 121-152 (1981).
4. J. Larkin, J. McDermott, D. P. Simon, and H. A. Simon, *Science* **208**, 1335-1342 (1980).
5. J. H. Larkin, "The role of problem representation in physics," In *Mental Models*, D. Gentner and A. L. Stevens (Eds.), Lawrence Erlbaum, NJ, 1983, 75-98.
6. Y. Anzai and T. Yokoyama, *Cog. & Inst.* **1** (4), 397-450 (1984).
7. P. T. Hardiman, R. Dufresne, and J. P. Mestre, *Memory and Cognition* **17** (5), 627-638 (1989).
8. T. deJong and M. G. M. Ferguson-Hessler, *Journal of Educational Psychology* **78** (4), 279-288 (1986).
9. J. P. Mestre, Ed. *Transfer of Learning from a Modern Multidisciplinary Perspective*, Information Age Publishing Inc., (2005).
10. C. M. Elliot, D. K. Campbell, and G. E. Gladding. "Parallel parking an aircraft carrier: Revising the calculus-based introductory physics sequence at Illinois," Forum on Education of the American Physical Society, Summer 1997.
11. The discussion problems can be accessed on the Illinois Physics Education Research group's website at http://research.physics.uiuc.edu/PER/Materials.htm.
12. The quizzes are free-response and consist of one conceptual question and one quantitative question with multiple parts. Students are allotted the last 20 minutes of class for the quizzes. They are graded and returned to the students the following week.
13. D. Hestenes, M. Wells and G. Swackhamer, *Phys. Teach.* 30, 141-158 (1992).
14. R. R. Hake, *Am.J. Phys.*, **66** (1), 64-74 (1998).
15. The FCI had not been administered in this course at UIUC prior to this study.
16. E. F. Redish, *Teaching Physics with the Physics Suite*, John Wiley & Sons (2003).
17. M. T. H. Chi, P. J. Feltovich and R. Glaser, *Cognitive Science*, **5**, 121-152 (1981).
18. P. T. Hardiman, R. Dufresne and J. P. Mestre, *Memory & Cognition*, **17** (5), 627-638 (1989).
19. Comparative data comes from Experiment 2 of the Hardiman et al study. (See Ref. 7). That study involved eight Model problems, and thus had twice as many comparison types than our study. Because of the voluntary nature of our test and of time constraints, we used only four Model problems.
20. For more information on our exams, see the paper by M. Scott, T. Stelzer, and G. Gladding, *Phys. Rev. ST Phys. Educ. Res.*, **2**, 020102 (2006).

Introducing Ill-Structured Problems in Introductory Physics Recitations

Vazgen Shekoyan and Eugenia Etkina

Rutgers, The State University of New Jersey, Piscataway, NJ 08854

Abstract. One important aspect of physics instruction is helping students develop better problem solving expertise. Besides enhancing the content knowledge, problems help students develop different cognitive abilities and skills. This paper focuses on ill-structured problems. These problems are different from traditional "end of chapter" well-structured problems. They do not have one right answer and thus the student has to examine different possibilities, assumptions and evaluate the outcomes. To solve such problems one has to engage in a cognitive monitoring called epistemic cognition. It is an important part of thinking in real life. Physicists routinely use epistemic cognition when they solve problems. We present a scaffolding technique for introducing ill-structured problems in introductory physics recitations and describe preliminary results of an exploratory study of student problem solving of ill-structured problems.

Keywords: Student problem solving, Ill-structured problems, Cognition, Epistemic cognition, Metacognition.
PACS: 01.40.Fk

INTRODUCTION & MOTIVATION

Studies of workplace needs indicate that problem solving is one of the most important abilities students need to acquire. [1] Are the types of problems one encounters in the workplace similar to problems that students solve in a traditional educational setting? Do they invoke the same cognitive abilities and problem solving skills?

Most of real-life and professional problems are ill structured. However, in educational settings we polish problems and make them well-structured problems. [2] Shin *et al.* [3] compared the problem skills required for solving well-structured and ill-structured problems in the context of open-ended, multimedia environment in astronomy. They found that ill-structured problem-solving scores were significantly predicted by domain knowledge, justification skills, science attitudes, and regulation of cognition, whereas only the first two categories were significant predictors for well-structured problem-solving scores. So, a wider range of skills and cognitive abilities are required for being a good problem solver.

It is equally important to consider the benefits of using ill-structured problems for improving students' content knowledge of physics. Kim and Pak [4] found that there is little correlation between the number of traditional well-structured problems solved and conceptual understanding. Is it possible that solving ill-structured problems helps understand physics better?

Ill-structured problems have been used in different instructional settings in physics education research. Heller et al. [5] have designed context-rich problems (which can be considered as a subcategory of the ill-structured problem domain) and used them in recitations. Students can be engaged in ill-structured problem solving through science research projects [3, 6], as well as in non-traditional laboratory sessions in physics courses. [7] In this paper we describe a scaffolding technique for introducing ill-structured problems in general (not only context-rich problems) in physics recitations and focus specifically on their implications on students' epistemic cognitive skills.

EPISTEMIC COGNITION

There are several identifying characteristics of ill-structured problems. Typically, they fail to present one or more of problem elements, have vaguely defined goals and unstated constraints, possess multiple solutions and solution paths, posses multiple criteria for evaluating solutions, and represent uncertainty about which concepts, rules and principles are necessary for successful solution. [3] Examples of ill-structured problems are found in Appendix(# 1and 3).

Kitchener [8] proposed a three-level model of cognitive processing to categorize the thinking steps one makes when faced with an ill-structured problem (cognition, metacognition, epistemic cognition). At the first cognition level, individuals read, perceive the problem, perform calculations, etc. At the second metacognitive level, individuals monitor their progress and problem-solving steps performed in the first level. At the third epistemic cognition level, individuals reflect on the limits of knowing, the certainty of knowing, and the underlying assumptions they make. Epistemic cognition influences how individuals understand the nature of problems and decide what kinds of strategies are appropriate for solving them.

An example of first-level cognitive activity in solving a simple mechanics problem can be reading the problem, writing down Newton's equations, and solving for an unknown. An example of metacognitive level activity is monitoring first-level cognitive tasks, such as checking the math, making appropriate notations, choosing productive representations (e.g., drawing a free-body diagram), making time management decisions, etc. An example of epistemic cognitive activity is reflecting on the limits of knowing, the criteria of knowing, the assumptions, the types of strategies that should be chosen, limiting cases, reasonableness of the answer, etc.

ILL-STRUCTURED PROBLEMS IN AN INTRODUCTORY PHYSICS COURSE

We introduce ill-structured physics problems in cooperative group solving activities in recitations [5] using a scaffolding technique described below.

Heller et al. [5] have shown the effectiveness of group solving of context-rich problems (they usually are ill-structured). Working in groups helps students acquire complex problem-solving skills while easing the frustration students might have due to the difficulty of finding the correct approach to the problem.

We believe that one of the important features of scaffolding ill-structured problems is prompting students to ask themselves the following epistemic questions during every step of problem solving: "How do I know this? Am I making any assumption while doing these steps?" Schoenfeld used a similar technique to help students develop metacognitive skills in mathematical problem solving. [9] He asked students metacognitive questions during classroom problem solving: "What exactly are you doing?" "Why are you doing this?" "How does it help?"

One of the first steps of problem solving is constructing the problem space of the task. [10] The problem space consists of all possible actions the problem solver can take. Experts not only construct richer problem spaces, but also more productive and meaningful ones. Novices are not able to recognize

problem spaces as well as experts, as they pay more attention to surface characteristics of problems.

Once the problem space is constructed, the problem solver needs to reflect upon the alternative steps she/he can make. For ill-structured physics problems this often means reflecting upon what assumption to make and how reasonable the assumption is. Depending on the problem context, she/he can also try several of the possible solutions.

Then the problem solver should assess the validity of the different solutions by constructing arguments, justifications, and by improving domain knowledge.

The next step is monitoring the problem space and solution options. Although monitoring the problem-solving plan is a useful metacognitive ability for all problems, it becomes especially important when one is solving an ill-structured problem. In such an activity, one should monitor the validity of the different solutions in addition to the metacognitive comprehension monitoring used for well-structured problems. One should constantly ask oneself: "How do I know this?"

The last step of ill-structured problem solving is the adoption of a solution or solutions. If other solutions are identified, one can try them out and thus engage in an iterative process of monitoring and adapting the chosen solution based on feedback. [11]

Instructors can scaffold student work by prompting students to ask epistemic questions during the problem solving steps described above. [12]

DESCRIPTION OF THE STUDY

Setup: We conducted a pilot study in the second semester of a two-semester large-enrollment (225 students) algebra-based introductory physics course for science majors at Rutgers University. There were two 55-min lectures, one 80-min recitation and one 3-hr laboratory per week. The course followed the Investigative Science Learning Environment (*ISLE*) format [13]. During recitations (8 sections) students worked in groups of four (an optimal size for cooperative group problem solving [5]). At the end of each recitation students handed in their work which was graded for effort and clarity.

Intervention: After the first midterm, one of the recitation instructors (V. Shekoyan) replaced some of the assigned problems with ill-structured problems that covered the same content. He did this in two of his recitation sections while other sections (two taught by him and four by other instructors) were doing problems from the Physics Active Learning Guide. [14] Overall, his students worked on five ill-structured physics problems during the five weeks between the first and second midterms. As these problems replaced some of the recitation problems, experimental students solved fewer well-structured problems than their counterparts. During this period the course was covering waves and vibrations, magnetism, electromagnetic induction and mirrors. After the second midterm students did not solve ill-structured problems.

The instructor provided scaffolding while students were working on the ill-structured problems. The level of scaffolding depended on the difficulty of the problem. An example of such scaffolding for problem #3 in the Appendix is provided below. For example, a student solved the problem assuming that the negatively charged ball stays on the table and he did not mention that this was true only for a ball which was heavy enough, so that the gravitational force exerted on the ball was bigger than the Coulomb's attractive force due to the other charged ball. Then the instructor could ask him to write the condition under which the outcome would be true. Once the student wrote the condition, the instructor would ask for a solution when the condition he wrote was violated. Alternatively, the instructor could just ask the student why he thought the ball stays on the table. Basically the instructor devised prompts to help students learn to ask themselves: How do I know this? Am I making any assumption while doing these steps?

Student sample: The experimental and control groups had 55 and 155 students respectively. On the first midterm prior to the intervention, the experimental and control groups were indistinguishable based on the exam grades. There was no pre-test in the course.

Data collection: The midterms consisted of 10 multiple-choice and 2 open-ended questions; the final exam had 18 multiple choice and 6 open-ended questions. On the second midterm and on the final one of the open-ended problems was ill-structured.

We collected exam data for three problems. Two were ill-structured problems: on the second midterm exam and on the final. These problems were based on the content of the recitations where ill structured problems replaced five of the well-structured problems in the experimental sections. The third problem was a second midterm open-ended problem that was not ill structured but was based on the above content.

We graded these exam problems based on the evidence of content knowledge and correctness of chosen solutions. *Thus, students who did not consider multiple possibilities for ill-structured problems, but gave correct answers for one of the possibilities, still received high grades.* For example, if a student solved problem #3 in the Appendix only for the case of the charge staying motionless on the table without writing any constraints on the unknown mass of the charge, she/he still received full credit.

PRELIMINARY RESULTS AND DISCUSSION

Our general research hypothesis was that engaging students in solving ill-structured physics problems through the classroom and instructional settings described above should help students develop epistemic cognition and master the physics concepts that were covered by the problems.

We performed two-tailed unequal-variance t-test on the grades of the second midterm problems. Tables 1 and 2 show that at the confidence level $p = 0.05$, the difference in the grades is statistically significant.

TABLE 1. Ill-structured problem (second midterm)

T-test -> 0.01	Control grp	Exp. grp
Effect size = 0.4	N = 155	N = 55
Average grade	12/20	13.6/20
St. deviation	4.4	3.8

TABLE 2. Open-ended problem (second midterm)

T-test -> 0.04	Control grp	Exp. grp
Effect size = 0.3	N=155	N=55
Average grade	14.6/20	16/20
St. deviation	5.4	3.6

Students' grades on the final exam ill-structured problem were indistinguishable (16 and 15.8 for the experimental and control groups respectively). As the grades reflected only the understanding of physics, we can say that students who spent time solving ill-structured problems in recitation mastered the content at the same level or better than those who worked on well defined problems.

Additionally, after the end of the semester, we coded students' written solutions for the two ill-structured problems for evidence of epistemic cognition and noted the students who considered multiple possibilities in their solutions. We found that students in the experimental group considered multiple possibilities much more often than in control group, although the total numbers were small in both groups. For the midterm problem, 11% of students from the experimental group considered multiple possibilities, as opposed to 1.3% in control group. For the final exam problem, 24% of the experimental group and 18% of the control group considered multiple possibilities. Chi-square test showed that the differences between two groups were statistically significant ($p=0.01$ and $p=0.03$ respectively).

We believe that these preliminary results show a positive trend and warrant further investigation. It is likely that the difference would have been greater if students solved more ill-structured problems in recitations.

This was an exploratory pilot study. We plan to increase the number of ill-structured problems in recitations this year and closely monitor student progress. Also, we would like to explore how our approach might affect students' inclination towards building more globally coherent knowledge structures. [15]

ACKNOWLEDGMENTS

We would like to thank Alan Van Heuvelen, Maria Ruibal Villasenor, David Rosengrant, Anna Karelina, and Michael Gentile for help with implementation of the tasks, inter-rater checks, and for feedback.

APPENDIX

Problem #1 The ill-structured problem for the second mid-term examination: A uniform block with a significant fracture through its middle is attached to a spring that is initially compressed to the left to position $-A$ (see the figure below). When released, the block starts moving right toward the equilibrium position as it begins to vibrate horizontally on a frictionless surface. The vibration frequency with the complete block and spring is 2.0 Hz. Sometime during the first period of vibration, one half of the block breaks loose at the fracture and leaves the remaining half attached to the spring, which continues to vibrate.

a) What is the frequency of vibration of the system with the half block? *Explain*.

b) What could happen to the amplitude of vibration for the half block system after the other half falls off? *Explain* your reasoning.

Hint: Think about the energy of the vibrating system as well as the frequency of that vibration. Remember that the energy when the block passes equilibrium is all kinetic energy and when at the extreme positions farthest from equilibrium is all elastic potential energy.

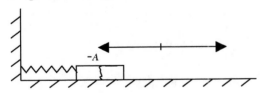

Problem #2 The open-ended problem for the second mid-term The following description is included in an electric gadget catalog: *"Squeeze No Battery Flashlight"*: No batteries or power plug will ever be needed!! An environmental-friendly flashlight, it saves energy without producing pollution to the environment. As long as you continually squeeze the handle in and out, the light works.

a) Devise an explanation for how this flashlight might work. Your explanation should allow someone else to build a model of this device.

b) Describe how you would test your explanation about how the flashlight works—without opening it.

Problem # 3 The final exam ill-structured problem: A positively charged small object with mass m = 10 g and charge q_1= +3 x 10^{-6} C hangs from a nylon string attached to the ceiling. The object's distance from the surface of a table is h = 20 cm. Imagine that you place another small object 2 with a small unknown mass and with negative charge q_2 = -3 x 10^{-6} C on the surface of the table.

a) Determine the force that the string exerts on the hanging object 1 immediately after you place object 2 on the table and before you remove your hand.

b) Does the force that the string exerts on object 1 stay the same after you place object 2 on the table and shortly after you remove your hand? If it does not stay the same, how qualitatively would the magnitude of the force that the string exerts on object 1 change a short time after you remove your hand from object 2? What assumptions did you make? Explain.

REFERENCES

1. R. Czujko, in *The changing role of Physics Departments in Modern Universities*, edited by E.F. Redish and J.S. Rigden [AIP Conf. Proc. **399**, 213-224, 1997].
2. K. A. Harper, R. F. Freuler, and J. T. Demel, in 2006 PERC proceedings, edited by L. McCullough, L. Hsu, and P. Heron, AIP conference proceedings, V. 883, 141-145, 2007.
3. N. Shin, D.H. Jonassen, and S McGee, *J. Res. Sci. Teach.* **40**, 6-33, 2003.
4. E. Kim, and S.-J. Pak, *Am. J. Phys.*, **70**, 759-765, 2002.
5. P. Heller, R. Keith, and S. Anderson, *Am. J. Phys.* **60**(7), 627-636, 1992; and P. Heller, and M. Hollabaugh, *Am. J. Phys.* **60**(7), 637-644, 1992 .
6. E. Etkina, T. Matilsky, and M. Lawrence, *J. Res. Sci. Teach.* **40**, 958-985, 2003.
7. E. Etkina, S. Murthy, and X. Zou, *Am. J. Phys.*, **74**, 979-986, 2006.
8. M. K. S. Kitchener, *Hum. Dev.* **26**, 222-232 (1983).
9. A.H. Schoenfeld, *"Mathematical Problem Solving"*, (Academic, San Diego, CA, 1982).
10. M.T.H. Chi, P.J. Feltovich, and R. Glaser, *Cog. Sci.* **5**, 121-152, 1981.
11. D.H. Jonassen, *ETR&D* **1**, 1042-1629, 1997.
12. X. Ge, and S.M. Land, *ETR&D* **52**(2), 1042-1629, 2004.
13. E. Etkina, and A.Van Heuvelen, in the *Proceedings of the 2001 Physics education Research Conference*, pp. 17-20, (PERC, Rochester, NY, 2001).
14. A.Van Heuvelen E. Etkina, *Physics Active Learning Guide* (Pearson Education, San Francisco, 2006).
15. M. Sabella and E. F. Redish, to be published in *Am. J. Phys.*

Effect of Misconception on Transfer in Problem Solving

Chandralekha Singh

Department of Physics and Astronomy, University of Pittsburgh, Pittsburgh, PA, 15260

Abstract. We examine the effect of misconceptions about friction on students' ability to solve problems and transfer from one context to another. We analyze written responses to paired isomorphic problems given to introductory physics students and discussions with a subset of students. Misconceptions associated with friction in problems were sometimes so robust that pairing them with isomorphic problems not involving friction did not help students fully discern their underlying similarities.

INTRODUCTION

Here, we explore the use of isomorphic problem pairs (IPPs) to assess introductory physics students' ability to solve problems and successfully transfer from one context to another in mechanics. We call the paired problems isomorphic because they require the same physics principle to solve them. We examine the effect of misconceptions about friction as a potential barrier for problem solving and analyze the performance of students on the IPPs from the perspective of "transfer". [1, 2, 3, 4] Transfer in physics is particularly challenging because there are only a few principles and concepts that are condensed into a compact mathematical form. Learning requires unpacking them and understanding their applicability in a variety of contexts that share deep features, e.g., the same law of physics may apply in different contexts. Cognitive theory suggests that transfer can be difficult especially if the "source" (from which transfer is intended) and the "target" (to which transfer is intended) do not share surface features. This difficulty arises because knowledge is encoded in memory with the context in which it was acquired and solving the source problem does not automatically manifest its "deep" similarity with the target problem. [1] Ability to transfer improves with expertise because an expert's knowledge is hierarchically organized and represented at a more abstract level in memory, which facilitates categorization and recognition based upon deep features. [2, 3, 4]

In this investigation, students in college calculus-based introductory physics courses were given IPPs in the multiple-choice format with one problem in each pair involving friction to examine the effect of misconceptions about friction on problem solving and transfer. One common misconception about the static frictional force is that it is always at its maximum value because students have difficulty with the mathematical inequality that re-lates the magnitude of the static frictional force with the normal force. [5] Students also have difficulty determining the direction of frictional force. If students are given IPPs in which one problem has distracting features such as common misconceptions related to friction, students may have difficulty transferring from the problem that did not have the distracting features to the one with distracting features. We analyze why it is difficult for students to fully discern the deep similarity of the problems in such IPPs and apply the strategies they successfully used in one problem to its pair.

In selecting and developing questions for the IPPs, we used our prior experience; one of the questions of each IPP was a question that had been found difficult by introductory physics students previously. Some students were given both problems of an IPP while others were given only one of the two problems. Students who were given both problems of an IPP were <u>not</u> told explicitly that the problems given were isomorphic. In some cases, depending upon the consent of the course instructor (due to time constraint for a class), students were asked to explain their reasoning in each case to obtain full credit. The problems contributed to students' grades in all courses. In some of the courses, we discussed the responses individually with a subset of students.

DISCUSSION

We find that students have misconceptions about friction [5] that prevented those who were given both problems of the IPPs from taking advantage of the problem not involving friction (which turned out to be easier for them) to solve the paired problem with friction. Questions (1)-(4) are related to an IPP about a car in equilibrium on an incline which are equilibrium applications of Newton's second law. Students have to realize that the car is at rest on the incline in each case,

CP951, *2007 Physics Education Research Conference*, edited by L. Hsu, C. Henderson, and L. McCullough
© 2007 American Institute of Physics 978-0-7354-0465-6/07/$23.00

so the net force on the car is zero. Also, since the weight of the car and the normal force exerted on the car by the inclined surface are the same in both problems, the only other force acting on the car (which is the tension force in one problem and the static frictional force in the other problem) must be the same. The correct answer (the magnitudes of the tension and frictional forces) in the problems is 7,500 N, equal to the component of the weight of the car acting down the incline. The correct responses in the questions below have been italicized:

Note: These trigonometric results might be useful in the next four questions: $\sin 30^0 = 0.5$, $\cos 30^0 = 0.866$.

• **Setup for the next two questions**

A car which weighs 15,000 N is at rest on a frictionless 30^0 incline, as shown below. The car is held in place by a light strong cable parallel to the incline.

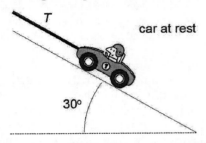

(1) Which one of the following is the correct free body diagram of the car (with correct directions)? N, mg, and T are the magnitudes of the normal force, the weight of the car and the tension force, respectively.

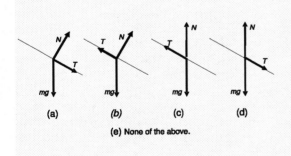

(2) Find the magnitude of tension force $\vec{T}$ in the cable.
(a) 7,500 N
(b) 10,400 N
(c) 11,700 N
(d) 13,000 N
(e) 15,000 N

• **Setup for the next two questions**

A car which weighs 15,000 N is at rest on a 30^0 incline, as shown below. The coefficient of static friction between the car's tires and the road is 0.90, and the coefficient of kinetic friction is 0.80.

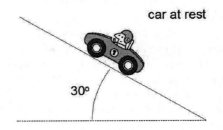

(3) Which one of the following is the correct free body diagram of the car (with correct directions)? N, mg, and f_s are the magnitudes of the normal force, the weight of the car and the static frictional force, respectively.

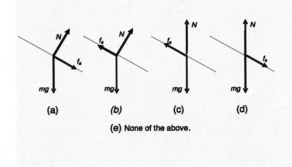

(4) Find the magnitude of the frictional force on the car.
(a) 7,500 N
(b) 10,400 N
(c) 11,700 N
(d) 13,000 N
(e) 15,000 N

One common misconception about the static frictional force is that it must be at its maximum value $f_s^{max} = \mu_s N$ where μ_s is the coefficient of static friction and N is the magnitude of the normal force. We wanted to investigate if student performance was significantly different for the cases in which students worked *only* on the problem with friction vs. when that problem is preceded by the problem involving tension in the cable. We wanted to explore if students in the latter group observed the underlying similarity of the IPPs and realized that the static frictional force was less than its maximum value and exactly identical to the tension in the cable in the problem pair.

First, 115 students were only asked question (4) and a different 272 students were asked both questions (2) and (4), given in that order. Questions (1) and (3) about the free body diagrams were not given to either of these groups. Out of the 115 students, 16% provided a correct response when only question (4) was asked. Out of the 272 students, 61% provided a correct response for question (2) with the tension force holding the car at rest but only 17% provided a correct response for the case with friction. The two most common incorrect responses in question (4) were $\mu_s N = 11,700$ N ($\sim 40\%$) and $\mu_k N = 10,400$ N ($\sim 20\%$). Thus, giving both problems in the IPP did not improve student performance on the problem with friction. In other words, the misconception about friction dominated and most students who were given both problems did not discern the underlying similarity of the two problems.

In order to help students discern the similarity between questions (2) and (4), we later introduced two additional questions that asked students to identify the correct free body diagrams for each of the questions. We wanted to assess whether forcing students to think about the free body diagram in each case helps them focus on the similarity of the problems. Forty-five students were given questions (1)-(4). We find that 89% and 69% of the students identified the correct free body diagrams in questions (1) and (3), respectively. The most common incorrect response ($\sim 25\%$) in question (3) was choice (a) because these students believed that the frictional force should be pointing down the incline. Unfortunately, the percentage of correct responses for questions (2) and (4) were 69% and 14%, respectively, not statistically different from the corresponding percentages without the corresponding free body diagram questions. Thus, although 89% and 69% of the students identified the correct free body diagrams in questions (1) and (3), it did not help them see the similarity of questions (2) and (4) or challenge their misconception about the frictional force (again $\sim 40\%$ believed that friction had a magnitude $\mu_s N$ and $\sim 30\%$ believed it was $\mu_k N$). In individual interviews, students often noted that the problem with friction must be solved differently from the problem involving tension because there is a special formula for the frictional force. Even when the interviewer drew students' attention to the fact that the other forces (normal force and weight) were the same in both questions and they are both equilibrium problems, only some of the students appeared concerned. Others used convoluted reasoning and asserted that friction has a special formula which should be used whereas tension does not have a formula, and therefore, the free body diagram must be used.

The IPP in questions (5) and (6) below relates to the work done by a person while pushing a box over the same distance at a constant speed parallel to different ramps with or without friction. Although the frictional force in question (6) is irrelevant for the question asked, it was a distracting feature for a majority of students:

(5) You are working at a bookstore. Your first task is to push a box of books 1.5 m along a frictionless ramp at a constant speed. You push parallel to the ramp with a steady force of 500 N. Find the work done on the box **by you**.

(a) 200 J
(b) 300 J
(c) 750 J
(d) 1000 J
(e) Impossible to calculate without knowing the angle of the ramp.

(6) You are working at a bookstore. Your second task is to push a box of books 1.5 m along a rough ramp at a constant speed. You push parallel to the ramp with a steady force of 500 N. A frictional force of 300 N opposes your efforts. Find the work done on the box **by you**.

(a) 200 J
(b) 300 J
(c) 750 J
(d) 1000 J
(e) Impossible to calculate without knowing the angle of the ramp.

Out of the 131 students who answered both questions, 65% and 27% provided the correct responses to questions (5) and (6), respectively. Common incorrect reasonings for question (6) was assuming that friction must play a role in determining the work done by the person and the angle of the ramp was required to calculate this work even though the distance by which the box was moved along the ramp was given. Interviews suggest that many students had difficulty distinguishing between the work done on the box by the person and the total work done. They asserted that the work done by the person cannot be the same in the two problems because friction must make it more difficult for the person to perform the work.

In another IPP related to the equilibrium of a box or a table on a horizontal surface (questions (7) and (8) below), students had to realize that since the net force on the object is zero, the force exerted by you and Arnold on the box must be equal in magnitude in question (7) and the force exerted by you and the frictional force on the table must be equal in magnitude in question (8):

(7) Arnold and you are both pulling on a box of mass M that is at rest on a frictionless surface, as shown below. Arnold is much stronger than you. You pull horizontally as hard as you can, with a force $\vec{f}$, and Arnold keeps the mass from moving by pulling horizontally with a force $\vec{F}$. Which one of the following is a correct statement about the magnitude of Arnold's force $\vec{F}$? g is the magnitude of the acceleration due to gravity.

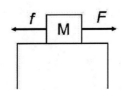

(a) $F = f$ *because the mass is not accelerating.*
(b) $F = f + Mg$ because Arnold must balance out your force and the weight.
(c) $F > f$ because Arnold is stronger and pulls harder.
(d) $F < f$ because Arnold is stronger and need not pull as hard as you.
(e) $F + f = Mg$ to maintain equilibrium.

(8) You are trying to slide a table across a horizontal floor. You push horizontally on the table with a force of 400 N. The table does not move. What is the magnitude of the frictional force the rug exerts on the table? The coefficient of static friction between the table and the rug is 0.60, and the coefficient of kinetic friction is 0.50. The table's weight is 1000 N.
(a) 0 N
(b) 400 N
(c) 500 N
(d) 600 N
(e) 1000 N

We asked 115 students only question (8) while a different 272 students were asked both questions (7) and (8), given in that order. Out of the 115 students, 22% provided a correct response when only question (8) was asked. Out of the 272 students, 69% provided a correct response for question (7) and only 24% for question (8). The most common incorrect response in question (8) was $\mu_s N = 600$ N ($\sim 40\%$) with or without question (7). Similar to the car on the inclined plane problem, the misconceptions about friction were so strong that students who were given both problems did not fully discern their similarity and take advantage of their response to question (7) to analyze the horizontal forces in question (8). To understand the extent to which students have difficulty with the magnitude and direction of the static frictional force, 131 students were also given question (9) below:

(9) A packing crate is at rest on a horizontal surface. It is acted on by three horizontal forces: 600 N to the left, 200 N to the right and friction. The weight of the crate is 400 N. If the 600 N force is removed, the resultant force acting on the crate is
(a) zero
(b) 200 N to the right
(c) 200 N to the left
(d) 400 N to the left
(e) impossible to determine from the information given.

The correct response for question (9) is (a), i.e., there is no resultant force acting on the crate because the 200 N force to the right will be balanced by a 200 N static frictional force in the opposite direction. Only 15% of the students provided the correct response. In question (9), the coefficient of friction was not provided, and similar to

the common misconception in question (4), almost 50% of the students believed that it is impossible to determine the resultant force on the crate without this information.

SUMMARY AND CONCLUSIONS

We explored the effect of misconceptions about friction on problem solving and transfer, and find that they interfere with the ability to reason correctly. The responses to the IPPs involving friction shows that students had difficulty in seeing the deep connection between the IPPs and in transferring their reasoning from the problem not involving friction to the problem with friction. The fact that students did not take advantage of the problem in an IPP not involving friction (which turned out to be easier for them) to answer the corresponding problem involving friction suggests that the misconceptions about friction were quite robust. For example, many students believed that the static friction is always at the maximum value or the kinetic friction is responsible for keeping the car at rest on an incline or the presence of friction must affect the work done by you (even if you apply the same force over the same distance). Even asking students to draw the free body diagrams explicitly in some cases did not help most students.

In such cases where misconceptions about friction prevented transfer, students may benefit from paired problems only after they are provided opportunity to repair their knowledge structure so that there is less room for misconception. Instructional strategies embedded within a coherent curriculum that force students to realize, e.g., that the static frictional force does not have to be at its maximum value or that the work done by a person who is applying a fixed force over a fixed distance cannot depend on friction may be helpful. Eliciting student difficulties by asking them to predict what should happen in concrete situations, helping them realize the discrepancy between their predictions and what actually happens, and then providing guidance and support to enhance their expertise is one such strategy.

ACKNOWLEDGMENTS

We are grateful to the NSF for award DUE-0442087.

REFERENCES

1. Gick and Holyoak, Cognitive Psychology, **15**, 1-38, 1983.
2. Dufresne, Mestre, Thaden-Koch, Gerace and Leonard, 155-215, in **Transfer of learning from a modern multidisciplinary perspective**, J. P. Mestre (Ed.), 393 pages, Greenwich, CT: Information Age Pub., (2005).
3. Lobato, J. Learning Sciences, **15(4)**, 431-439, (2006).
4. Ozimek, Engelhardt, Bennett and Rebello, PERC Proceedings, Eds. Marx, Heron, Franklin, 173-176, (2004).
5. diSessa and Sherin, Int. J. Sci. Ed., **20(10)**, 1155-1191, (1998).

Symbols: Weapons of Math Destruction

Eugene Torigoe and Gary Gladding

Department of Physics, University of Illinois at Urbana-Champaign, Urbana, IL 61801

Abstract.
This paper is part of an ongoing investigation of how students use and understand mathematics in introductory physics. Our previous research [1] revealed that differences in score as large as 50% can be observed between numeric and symbolic versions of the same question. We have expanded our study of numeric and symbolic differences to include 10 pairs of questions on a calculus based introductory physics final exam. We find that not all physics problems exhibit such large differences and that in the cases where a large difference is observed that the largest difference occurs for the poorest students. With these 10 questions we have been able to develop phenomenological categories to characterize the properties of each of the questions. We will discuss what question properties are necessary to observe differences in score on the numeric and symbolic versions. We will also discuss what insights these categories give us about how students think about and use symbols in physics.

Keywords: Physics Education Research, math, symbols, algebra
PACS: 01.40.Fk. 01.30.Cc.

INTRODUCTION

We have been investigating how students use and understand mathematics in introductory physics classes. Our interest in mathematics is related to how students use and understand the meaning of symbols rather than any difficulties they may have with algorithmic mathematical activities like solving two equations and two unknowns. Our previous research [1] revealed that differences in score as large as 50% can be observed between numeric and symbolic versions of the same question. While we were able to measure large effects, the scope of the study was limited because we only studied two questions. The two questions used in the previous study (each with a numeric and a symbolic version) dealt with one dimensional kinematics and the scores for the numeric version of both questions were high. We have expanded our study of numeric and symbolic differences to include 10 pairs of questions that span many different introductory mechanics topics and have a variety of difficulty levels.

METHODOLOGY

The subjects of the study were students who were enrolled in the calculus-based introductory mechanics course, Physics 211, at the University of Illinois, Urbana-Champaign in the spring 2007 semester. Students in Physics 211 take three multiple-choice midterm exams and a cumulative final. While numeric questions are the most common type of question, symbolic questions are not uncommon. There were 765 students who completed one of the two randomly administered versions of the final exam. Each version of the final contained either the numeric or symbolic version of each of the 10 questions.

All but one of the 10 paired questions contained analogous choices for each the numeric and symbolic versions of the question[1]. To discourage cheating many of the choices were rearranged between the versions.

To test the equivalence of the groups we compared the average midterm exam score of the students for each of the final exam versions. The average midterm score and standard error for final 1 was 78.9 ± 0.5, and for final 2 was 78.6 ± 0.5. Based on their average midterm exam scores the two groups are equivalent.

RESULTS AND ANALYSIS

The numeric/symbolic difference by group

Table 1 describes the scores observed on the numeric and symbolic versions of each question. The errors shown represent the standard deviation of the mean. While there exist large differences between numeric and symbolic scores for some questions, there are other questions that show no significant difference. To further study the relationship between the differences between numeric and symbolic questions and overall course performance we divided the class into three subgroups based on the total course points. The groups are the bottom

---

[1] The incorrect choices for the two versions of Question 4 were similar but with i and negative signs omitted from the choices from the numeric version.

CP951, *2007 Physics Education Research Conference*, edited by L. Hsu, C. Henderson, and L. McCullough
© 2007 American Institute of Physics 978-0-7354-0465-6/07/$23.00

TABLE 1. Average and standard error for numeric and symbolic versions of each question in the study

	Q1	Q2	Q3	Q4	Q5
Numeric	91.5 ± 1.4	93.3 ± 1.3	79.6 ± 2.1	90.5 ± 1.5	44.9 ± 2.5
Symbolic	70.4 ± 2.3	56.8 ± 2.5	63.4 ± 2.4	82.3 ± 1.9	31.9 ± 2.3

	Q6	Q7	Q8	Q9	Q10
Numeric	61.2 ± 2.4	76.0 ± 2.1	33.2 ± 2.3	78.6 ± 2.0	48.8 ± 2.3
Symbolic	52.9 ± 2.5	75.6 ± 2.1	29.8 ± 2.2	54.5 ± 2.5	52.8 ± 2.4

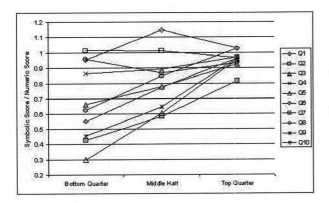

FIGURE 1. The ratio of the symbolic score over the numeric score for each question and for three groups based on total points in Physics 211

quarter, the middle half and the top quarter[2]. For each group and for each question in the study we calculated the ratio of the average score on the symbolic version to the average score on the numeric version. We interpret this ratio to represent the likelihood that the students who could solve the numeric version correctly would also solve the symbolic version correctly. Even though no individual was given both versions of a single question, we believe we are justified in this interpretation because of the equivalence of the groups. The results of this analysis are shown in Figure 1. In the cases where we observed a large difference between the numeric and symbolic versions, the largest difference (smallest ratio) was observed for the bottom quarter of the class. The top quarter of the class, by comparison, rarely showed any distinction between the numeric and symbolic versions. For the top quarter only one question had a ratio lower than 0.93. This correlation suggests that there are differences in the ways top students and bottom students understand symbols in physics.

---

[2] The groups were created in this manner because of our interest in the issue of retention. The groups distinguish students who fail or who are on verge of failing (bottom quarter) with those that pass comfortably (middle half) and those that excel (top quarter).

Phenomenological Categories

We performed an analysis involving a careful inspection of each of the 10 questions including an inspection of the popularity of the correct choices, popularity of the incorrect choices, and an analysis of students written work to create a coding scheme to differentiate the questions. We have identified the following properties based on our analysis.

Difficulty for the top quarter The average for the numeric version of the question.

Multiple equations This code distinguishes whether the problem is commonly solved with one equation or with multiple equations

Coupled The solution is coupled if, for example, there are two equations and two unknowns and both unknowns are present in both equations. This is contrast to a sequential problem where with two equations and two unknowns there is an equation where only one of the unknowns is present.

General equation manipulation This code signifies whether it is possible to obtain one of the incorrect choices by combining general equations or manipulating a single general equation with minimal changes (e.g. replacing x with d).

Specification of a slot variable A slot variable is a symbol found in a general equation. This code signifies that in order to reach the correct symbolic solution the student must replace a slot variable with a compound expression (e.g. replacing the slot variable v with a more specific compound expression $v/2$).

Table 2 shows the codes for each of the 10 questions in this study, as well as the measurement of the ratio of the average score on the symbolic version to that of the numeric version for the bottom quarter.

Separating Physics Difficulties from Math Difficulties

Because we are primarily interested in determining the role of mathematical difficulties we would like to be able to control for conceptual physics difficulties. We

TABLE 2. Coding of question properties for each of the 10 questions in the study

	Difficulty for the top quarter	Multiple equations	Coupled	General equation manipulation	Specification of a slot variable	Ratio of the symbolic score to the numeric score for the bottom quarter
Q.1	0.96	Y	N	Y	N	0.55 ± 0.06
Q.2	0.99	Y	N	Y	Y	0.43 ± 0.06
Q.3	0.97	Y	N	Y	N	0.66 ± 0.10
Q.4	0.98	N	N	N	Y	0.86 ± 0.08
Q.5	0.67	Y	N	N	Y	0.30 ± 0.12
Q.6	0.87	Y	N	N	Y	0.62 ± 0.13
Q.7	0.94	N	N	N	Y	1.01 ± 0.11
Q.8	0.49	Y	N	Y	Y	0.96 ± 0.23
Q.9	0.94	Y	N	Y	Y	0.45 ± 0.08
Q.10	0.82	Y	Y	N	N	0.95 ± 0.19

used the difficulty for the top quarter as a measure of the physics difficulty. Two questions (Questions 5 and 8) appear to be significantly more difficult for the top students. Consequently we removed them from further analysis in this study. This decision is supported by the fact that the average scores for the numeric version for the bottom quarter for each of these questions is 23%, consistent with the random guessing rate of 20%.

No numeric and symbolic difference

From Table 2 one can see that there are three remaining questions (Questions 4, 7, and 10) in which there was virtually no difference between the numeric and symbolic versions for the three groups in the class. Questions 4 and 7 were the only questions in which the solution could be found with a single equation. Further the solution to Question 7 depended mainly on the concept that linear momentum is conserved after a collision that caused rotation.

Students also showed no distinction between the numeric and symbolic versions of Question 10. This question is the only question in our sample involving coupled equations. In order to solve a coupled physics problem one has to setup both equations before one can proceed to the solution. For this reason, we expect that students will be forced to use the same procedure to solve both the numeric and symbolic versions of the question. The results are consistent with out expectation that there would be no difference between the versions.

Large numeric and symbolic difference

For the five remaining questions in our study (Questions 1, 2, 3, 6, and 9) it is apparent that there are many students who can solve a numeric question but who are unlikely to be able to solve the analogous symbolic version. From the data on Table 2 we can see commonalities

between the questions for which a large difference is observed.

All five questions require that multiple equations be used to reach the correct solution. All of the questions can also be solved sequentially which we believe aids in the solution of numeric problems. When working with numbers in a sequential problem, the problem can be broken up into well defined steps during which an equation condenses into a numeric value. Unlike numeric solutions, symbolic solutions do not condense so compactly and often become more complex.

For all but Question 6 an incorrect choice can be obtained by the blind manipulation of general equations. For Questions 1, 2, and 3 the most popular incorrect answer is the choice (or one of the choices in the case of Question 2) that can be obtained by manipulating general equations.

Questions 2, 6, and 9 require that students specify a slot variable as a compound expression. For Question 2 the second most popular incorrect choice corresponded to using v instead of $v/2$ or t instead of $t/2$[3]. For Question 6 the most popular incorrect choice corresponded to a failure to specify the slot variable M. Question 9 (See Figure 2) required that students specify both the slot variable m as $3m$ and the slot variable v as $v/3$. Interestingly very few students failed to specify both slot variables, the most popular incorrect choices corresponded to a failure to specify either one or the other.

The findings of the preceding analysis can be summarized by the following statements: We see no difference in the average scores of symbolic and numeric versions if the correct solution to the problem involves the use of either (i) a single equation or (ii) two coupled equations. We do see significant differences in the average scores of symbolic and numeric versions if (i) the correct solution requires the use of multiple equations, none of which are coupled, or (ii) the correct solution requires a slot vari-

---

[3] There two different common ways of reaching the solution.

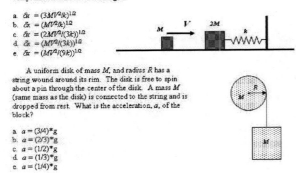

A block of mass M slides on a frictionless surface with a velocity V. It strikes a second block of mass $2M$ that is at rest and is attached to a long, relaxed ideal (massless) spring of spring constant k. Assume that the blocks stick together after colliding and the collision takes place very quickly. Provided that the spring is stiff enough to stop the blocks before striking the wall, determine δx, the maximum amount the spring is compressed from its relaxed length.

a. $\delta x = (3MV^2/k)^{1/2}$
b. $\delta x = (MV^2/k)^{1/2}$
c. $\delta x = (2MV^2/(3k))^{1/2}$
d. $\delta x = (MV^2/(3k))^{1/2}$
e. $\delta x = (MV^2/(9k))^{1/2}$

A uniform disk of mass M, and radius R has a string wound around its rim. The disk is free to spin about a pin through the center of the disk. A mass M (same mass as the disk) is connected to the string and is dropped from rest. What is the acceleration, a, of the block?

a. $a = (3/4)*g$
b. $a = (2/3)*g$
c. $a = (1/2)*g$
d. $a = (1/3)*g$
e. $a = (1/4)*g$

FIGURE 2. The symbolic versions of Question 9 (upper) and Question 10 (lower)

able to be replaced by a compound expression, or (iii) an incorrect choice can be obtained by manipulating general equations with no substantive replacements.

DISCUSSION

Data on 10 different numeric and symbolic pairs of questions that spanned difficulty level and topic have allowed us to develop a set of phenomenological categories to distinguish the questions. We have been able to show correlations between these categories and the differential performance on the numeric and symbolic versions.

A common indicator of difficulty for a symbolic question related to whether or not one of the incorrect choices could be obtained by the blind manipulation of general equations. This error could indicate that some students approach symbolic problems as a manipulation exercise. This may reflect a difference in the way they frame the activity of solving a symbolic problem to the activity of solving a numeric problem[2]. The framing of the exercise determines what resources are activated and what activities (games) are appropriate. This framing argument may also explain errors associated with the failure to specify slot variables, which is a form of blind equation manipulation.

An alternative explanation may be the added cognitive demand placed on the students when they solve symbolic problems[3]. Cognitive load theory posits the existence of the working memory and the long term memory. While long term memory is boundless, the working memory is easily overloaded. As discussed in an earlier section the activities related to sequential problems are very different for numeric and symbolic problems. Numbers allow for the alleviation of cognitive load because

each step condenses into a numeric value, whereas each step in a symbolic problem ends with a potentially complex symbolic expression. Because symbols are carried from step to step, it is more difficult to clearly demarcate different steps.

Notational differences between numeric and symbolic problem solutions may also have a significant effect on cognitive load. There are two distinctions that are explicit in numeric solutions, but which are implicit in symbolic solutions. First, there is no explicit notation when solving symbolic problems to distinguish known from unknown symbols. Second, there is no notational difference between slot variables and specific known or unknown quantities[4]. In numeric problems there is a clear demarcation between a slot variable and a specific known (a number), as well as between an unknown (symbol) and a known (a number). When working on a symbolic problem these factors contribute to the cognitive load. Some students who fail to specify the symbols in the general equation may have an overly general sense of the meaning of symbols. Students may fail to specify slot variables because they may still consider them in some cases to be a general symbol. This could also be the case when students manipulate general equations and get an incorrect answer, they may think that the symbol is so general that even the object association of the symbol is variable.

CONCLUSION

By studying a variety of physics problems with both numeric and symbolic versions we have been able to develop a phenomenological coding to distinguish the cases where differences between numeric and symbolic versions of a physics question exists.

REFERENCES

1. E. Torigoe and G. Gladding, "Same to Us, Different to Them: Numeric Computation versus Symbolic Representation," in *2006 Physics Education Reasearch Conference Proceedings*, edited by L. McCullough. et al., AIP Conference Proceedings 883, AIP, New York, 2007, pp. 153–156.
2. D. Hammer, A. Elby, R. Scherr, and E. Redish, "Resource, Framing, and Transfer," in *Transfer of learning from a modern multidisciplinary perspective*, edited by J. Mestre, Information Age Publishing, Greenwich, CT, 2005, pp. 89–119.
3. J. Sweller, *Learning and Instruction* **4**, 295–312 (1994).

---

[4] Sometimes subscripts can help with this distinction, but our observations find that few students use them

Understanding How Physics Faculty Use Peer Instruction

Chandra Turpen and Noah D. Finkelstein

Department of Physics, University of Colorado, Boulder, Colorado 80309, USA

Abstract. We investigate how the use of physics education research tools is spreading throughout faculty practice and examine efforts to sustain the use of these practices. We specifically focus on analyzing the local use of the innovation Peer Instruction. We present data based on observations of teaching practices of six physics faculty in large enrollment introductory physics courses at our institution. From these observations, we identify three dimensions that describe variations in faculty practices: the purpose of questions, participation with students, and norms of discussion.

Keywords: Educational Change, Peer Instruction, Personal Response Systems, Physics Faculty, Teaching Practices.
PACS: 01.40.-d, 01.40.Di, 01.40.Fk, 01.40.gb

INTRODUCTION

Significant work has been conducted within the Physics Education Research (PER) community to better understand how physics professors come to adopt and adapt reform-based tools and curricula. Prior research has sought to understand both the beliefs and values that professors bring with them, and those beliefs which inform the design of curricular tools [1]. Dancy and Henderson specifically found that although most physics professors' beliefs and values are consistent with most PER-based reforms, professors' self-reported practices do not reflect these PER-consistent values due to systemic constraints surrounding the classroom environment [2]. Our work builds on this prior research by looking specifically at how physics professors make ongoing pedagogical decisions in a working classroom. Through observations of classroom practice in large-enrollment introductory physics classrooms, we seek to characterize similarities and differences among physics educators' actual practices and use of curricular tools. In this paper, we identify three primary dimensions along which we see significant variation in individual professors' implementation concerning one pedagogical tool, Peer Instruction [3]. We begin to describe these differences in practices, so that ultimately we may examine the impacts on student engagement and perception.

METHODS

In this study, six introductory physics courses were observed using ethnographic observation techniques [4] and each of the observed class periods was audio-taped.

For each of the courses, at least twenty percent of all class periods were observed. Daily records of clicker question responses were collected, as well as other course documents such as syllabi, homework assignments, exams, and other web correspondence. In addition, the primary instructors for the courses were interviewed at the end of the course about their broader teaching practices and specifically about their use of clickers or personal response systems. The results presented in this paper primarily focus on the analyses of classroom observation data and course records. The identified dimensions of practice emerged from initially broad observations. Data from these observations and records were successively narrowed through an inductive analysis which led to the identification of characteristics and patterns across course sessions [4].

BACKGROUND

All courses observed in this study were large enrollment introductory undergraduate physics courses with average lecture attendance ranging from 130 to 240 students. Five of the six courses studied used iClicker technology, a radio frequency classroom response system [5]. One of the courses used H-iTT technology, an infrared classroom response system [6]. In this paper, we will refer to both of these technologies as 'clickers.' Five of the six courses studied were courses required for science-based degree programs. The other was an elective course for non-science majors. In two courses where conceptual assessments were given, both courses achieved significant learning gains ranging from two to three times the national norm for non-interactive engagement courses [7].

CP951, *2007 Physics Education Research Conference*, edited by L. Hsu, C. Henderson, and L. McCullough
© 2007 American Institute of Physics 978-0-7354-0465-6/07/$23.00

The lead instructors for these courses varied from tenured professors to temporary instructors. Pseudonyms have been chosen to assure their anonymity. Two of the six instructors observed, Green and White, were novices with respect to the use of Peer Instruction and clickers in large-enrollment courses. Both also happened to be temporary instructors. Three of the instructors observed, Blue, Red, and Purple, were active members of the physics education research group at the University of Colorado at Boulder. Blue, in particular, has been nationally recognized for his excellence in teaching and has pioneered the use of Peer Instruction and clicker systems at our institution. It is also important to mention that Blue mentored Yellow as he learned to use clickers and Peer Instruction in his own large-enrollment introductory physics courses. Although all of these educators used the language of Peer Instruction, based on the work of Mazur [3], to describe their practice, none of them implemented Peer Instruction exactly as described by Mazur. The most notable variation between their practices and Mazur's description is that all faculty observed in this study did not have an explicit "silent" phase of the clicker question where the students came to an answer individually first. Importantly, we observe significant student discussion in all classes. In this way, the use of the term Peer Instruction by physics faculty and our use in this paper should be loosely interpreted.

THREE DIMESIONS OF PRACTICE

We describe three dimensions along which there exist significant variation within our local sample of six physics professors. It is important to note that there also exist significant similarities between these professors, but due to limited space these similarities will not be emphasized here. Each of the professors observed used clickers in the spirit of Peer Instruction as described by Mazur with significant peer discussion and carefully constructed conceptually-based clicker questions. The first dimension of practice identified, the purpose of clicker questions, directly relates to explicit recommendations by Mazur in *Peer Instruction*. We identify two additional dimensions for which no or few explicit recommendations are provided for in Mazur's book. These dimensions include the role of the professor during the clicker question response time and the role of discussion of question solutions.

Purpose of the Clicker Questions

A preliminary, rough measure of the purpose of the clicker questions can be seen from the frequency of clicker questions used by the professor. In the courses

observed, there was great variation in the average number of clicker questions asked per hour of class, ranging from a maximum of 9.2 by White and a minimum of 3.2 by Green. The average number of clicker questions per hour for each professor along with the respective standard error on the mean was Green: 3.2 ± 0.1, Purple: 5.0 ± 0.3, Red: 5.3 ± 0.4, Yellow: 5.9 ± 0.5, Blue: 6.5 ± 0.4, White: 9.2 ± 0.6. It is interesting to note that the two professors at either extreme were the novice clicker users. From these data, there are indications that a clicker question in White's class may look different and have a different purpose compared to a question in Green's class. In addition to the challenging conceptual questions most closely resembling ConcepTests [3], our first pass analysis shows that some professors are occasionally using shorter questions to review or check for understanding. For these clicker questions students are given less time to answer the questions and less time is spent discussing the solution. These types of clicker questions were most frequently used by White, which we can see reflected in this course's unusually high frequency of clicker questions. Further classification of the characteristics of 'check for understanding' type clicker questions is currently being analyzed based on the distribution of students with correct and incorrect answers, the construction of the question itself, the average amount of time given to students to respond to clicker questions, and the amount of time spent discussing clicker question solutions.

Through our analysis and observations we have also identified another type of clicker question used by some of the professors, logistical questions. Logistical questions included questions where the students were asked when they would like the homework due or how difficult the midterm exam was. The purpose of these questions was usually to gather student opinions on how the course was going and possibly inform the structure or direction of the course. Yellow, Green and White almost never (<2% of the time) asked logistical questions, while Blue, Purple, and Red occasionally asked logistical questions (5-15% of the time). We also found that one of the professors observed, Red, was using clicker questions to give graded reading quizzes.

Role of the Professor during the Clicker Question Response Time

During the time interval where students were constructing an answer to a clicker question, what did the professors do? Based on our observations individual physics professors were spending this time differently. One significant difference was the extent to which the professor leaves the stage area of the lecture hall and walks around the classroom amongst the

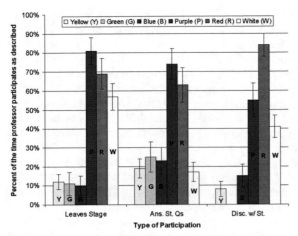

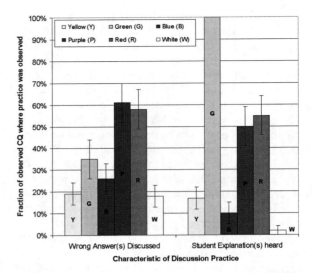

FIGURE 1. The percentage of observed clicker questions where the professor was observed to participate with the students during the response time by leaving the stage (column 1), answering student questions (column 2), or actively discussing with the students (column 3). The error bars shown are the standard error on the proportion.

FIGURE 2. The percentage of observed clicker questions where the wrong answers where discussed and student explanations were heard. The error bars shown are the standard error on the proportion.

students. The first column of data in Figure 1 shows the fraction of observed clicker questions where the professor left the stage area of the classroom. The professors also varied in how they spent their time interacting with the students during this clicker question response time. The fraction of the observed clicker questions where the professor answered students' questions or discussed with the students varied, as shown in columns two and three of Figure 1 respectively.

From Figure 1 we see that Yellow, Green, and Blue are almost never leaving the stage (less than 15% of the time). On the other hand, we see that Purple, Red and White are usually leaving the stage (more than 50% of the time). We also note that the professors who usually left the stage were more likely to answer student questions or actively discuss with the students than the professors who almost never left the stage.

Role of Discussion Concerning the Clicker Question Solution

After the students had finished responding to the clicker question, how was the explanation phase of Peer Instruction conducted? We have identified two characteristics of this discussion that vary across professors: whether incorrect clicker question answers were addressed and whether students participated in the explanation of the clicker question solution. The first column of data in Figure 2 shows the fraction of observed clicker questions where incorrect answers were discussed during the description of the solution. The second column of data in Figure 2 shows the

fraction of the observed clicker questions where student explanation(s) of the clicker question solution were heard during the whole class discussion. From Figure 2 we can see that Purple and Red are usually discussing incorrect clicker question answers and that they are doing so more frequently than the rest of the educators observed.

It is also interesting to note that although Green and Blue are discussing incorrect options about the same fraction of the time, in Green's class these explanations of the incorrect options are coming from the students while in Blue's class the explanations of the incorrect options are coming from the professor's explanation of common student challenges. In Purple and Red's courses, the discussion of incorrect answer options was commonly coming from both the students and the professor. The second column of data in Figure 2 also shows that Green is always using student explanations when explaining the clicker question solution. Purple and Red are usually using student explanations in the construction of the clicker question solution, while Yellow, Blue, and White are rarely (less than 20% of the time) using student explanations.

The number of students explanations usually heard in class for a given question also fluctuated. When students were asked to contribute explanations of the clicker question solution, Yellow on average heard from 2.2 ± 0.2 students, Green: 1.4 ± 0.1, Blue: 1.3 ± 0.3, Purple: 2.3 ± 0.5, and Red: 2.4 ± 0.4. White was not included in this analysis since student explanations were only used once. We see that Blue and Green are primarily hearing from only one student concerning the correct answer, when student explanations are used, while Yellow, Purple and Red are usually hearing from at least two. We can infer that in the classrooms of Yellow, Purple, and Red there is more emphasis on discussing the incorrect answers and associated student reasoning.

DISCUSSION AND CONCLUSIONS

Despite all six instructors implementing the 'same' activity, Peer Instruction, we observe that there are influential decisions that these faculty members make that distinguish their practices. Some of these specific decisions have not been previously discussed in the research literature. We seek to identify categories of practice in order, ultimately, to distinguish which aspects of Peer Instruction are flexible and can be effectively implemented in a multitude of ways and which aspects of Peer Instruction are essential for the effective use of the educational tool. In the present work we have identified three categories which differentiate faculty practice: the purpose or type of questions, participation of faculty in the student response time, and discussion norms surrounding the description of the question solution.

From these categorizations and observations of faculty in class, we found that instructors new to using clickers were more likely to reflect the extremes of the dimensions described. For example, Green always heard student explanations for the questions solutions, while White almost never did. This 'all or nothing' observation of novice faculty practice may be indicative of the fact that novice users do not have highly developed heuristics for deciding when to use certain practices.

We found the similarities between the practices of Blue and Yellow particularly notable, considering Blue was a mentoring instructor for Yellow when Yellow was beginning to learn to use clickers and Peer Instruction. The aspects of practice described here begin to capture some of the similarities in these instructors based on their time spent teaching together. Additionally, we observe very similar practices for two of the PER faculty, Red and Purple. Despite teaching in two different types of environments (one a non-science majors course and one a required course for engineers), their teaching practices surrounding Peer Instruction are almost indistinguishable.

It is also interesting to note that there is diversity of practice even within the subset of physics education research faculty members. While we do find Red and Purple to be quite similar on all dimensions described, Blue's practices vary. This description of practices surrounding Peer Instruction is surely not complete, and is made clear by the fact that Blue's excellent teaching practices do not stand out as unique along the dimensions described. Perhaps more subtle features of practice were not captured by these initial descriptive measures. For instance, we hypothesize that based on years of teaching experience and familiarity with the PER literature, Blue already understands common student difficulties. For this reason, Blue may not necessarily need to solicit student explanations to understand what students are thinking; whereas Yellow, White and Green do not have this background knowledge. In this way their practices would mean something different. Finally, these varying practices may be directed towards achieving different goals, all of which may be realized through Peer Instruction in the generic sense. Unpacking the variation in classroom practices and how these practices shape classroom culture and ultimately student engagement and learning are the subjects of current studies.

ACKNOWLEDGMENTS

This work has been supported by the National Science Foundation (DUE-CCLI 0410744 and REC CAREER 0448176), the AAPT/AIP/APS (Colorado PhysTEC program), and the University of Colorado. We would like to thank the Physics Education Research Group at Colorado for their continuing thoughtful and provocative feedback. We also appreciate the physics department faculty members who participated and actively supported this research. Without their cooperation this project would not be possible.

REFERENCES

1. M. Dancy & C. Henderson, "Framework for articulating instructional practices and conceptions," *Phys. Rev. ST Phys. Educ. Res.* 3, 010103 (2007); E. Yerushalmi, C. Henderson, K. Heller, & V. Kuo, "Physics faculty beliefs and values about the teaching of problem solving," (accepted, to *Phys. Rev. ST Phys. Educ. Res.*)
2. M. Dancy & C. Henderson, "Beyond the Individual Instructor: Systemic Constraints in the Implementation of Research-Informed Practices." *PERC Proceedings*, (2004)
3. E. Mazur, *"Peer Instruction,"* Prentice Hall (1997)
4. J. Creswell, *"Educational Research,"* Pearson Prentice Hall (2005); J. Maxwell, *"Qualitative Research Design,"* Sage Publications (2005); J. Spradley, *"Participant Observation,"* Holt, Rinehart and Winston (1979)
5. iClicker technology information available at: http://www.iclicker.com/
6. H-iTT technology information available at: http://www.h-itt.com/
7. L. Kost, S. Pollock, and N. Finkelstein, "Investigating the Source of the Gender Gap in Introductory Physics," *PERC Proceedings*, (submitted this volume-2007); N. Finkelstein, and S. Pollock, *Phys. Rev. ST Phys. Educ. Res.* 1, 010101 (2005).

Comparing Student Use of Mathematical and Physical Vector Representations

Joel Van Deventer[*] and Michael C. Wittmann[*¶†]

[*]Center for Science and Mathematics Education Research, University of Maine, Orono, ME 04469, USA
[¶]Department of Physics and Astronomy, University of Maine, Orono, ME 04469, USA
[†]College of Education and Human Development, University of Maine, Orono, ME 04469, USA

Abstract. Research has shown that students have difficulties with vectors in college introductory physics courses and high school physics courses; furthermore, students have been shown to perform worse on a vector task with a physical context when compared to the same task in a mathematical context. We have used these results to design isomorphic mathematics and physics free-response vector test questions to evaluate student understanding of vectors in both contexts. To validate our test, we carried out task-based interviews with introductory physics students. We used our results to develop a multiple-choice version of the vector test which was then administered to introductory physics students. We report on our test, giving examples of questions and preliminary findings.

Keywords: Physics Education Research, Vectors
PACS: 01.40.Fk

INTRODUCTION

Students in introductory physics classes have difficulties with vector tasks [1,2]. Some topics of difficulty include: vector magnitude, direction, addition, subtraction, dot product, cross product, and unit vectors. Furthermore, students have difficulty with the vector nature of physics concepts [3,4]. Recently, Shaffer & McDermott [5] observed introductory physics students performing better on a mathematics vector task when compared to performance on an isomorphic physics vector task. Their research has led to our current research focus of comparing student performance on isomorphic math/physics vector tasks. The following are our guiding research questions:

1) How does student performance on the isomorphic vector tasks compare?
2) What tools[1] are students using in math and physics contexts to accomplish vector tasks?
3) Are students transferring scalar and vector tools from mathematics to physics contexts in introductory physics courses?
4) What tools are missing in order to be successful?

Our limited discussion in this paper primarily focuses on questions one and two above.

---

[1]See [6] for specific meaning and background for the term tool.

METHODS

We developed two isomorphic free response vector tests using equivalent questions in different contexts, one math and the other physics. However, two questions on the physics version have added difficulty and break strict isomorphism. Our tests were validated using interviews with early versions of the test. Using interview data and results from other researchers, we developed a multiple-choice version of each test. This tests was administered to an introductory calculus-based physics course.

The interview sample consisted of 10 self-selected students from a 1st semester calculus-based introductory physics course. Interviews were conducted near the end of the semester, when all students had used vectors in many different physics contexts.

The multiple-choice (MC) tests were administered to 28 students in a summer-session introductory calculus-based physics course at the end of the semester. A self-reported survey of the class indicates 83% (24 students) studied vectors in at least a prior high school physics course and half of those (13 students) also studied vectors in at least a high school math course. Both the interview sample and the MC sample were taught by the same instructor and covered the same material.

CP951, 2007 Physics Education Research Conference, edited by L. Hsu, C. Henderson, and L. McCullough
© 2007 American Institute of Physics 978-0-7354-0465-6/07/$23.00

Test Design

Influence from previous research [1-5] led to the development of preliminary free-response questions on graphical and algebraic representations of vectors. For example, we modified a one-dimensional graphical subtraction question from Shaffer & McDermott's research [5], as shown in Figure 1. An example of a MC question on an algebraic expression for a x-component is given in Figure 2. Two test questions are not strictly isomorphic, but contain added difficulty in the physics version. For example, the angle on the x-component algebraic expression question (Fig. 2) is labeled differently between contexts. Question ordering is the same on each version; however, the choices are randomized.

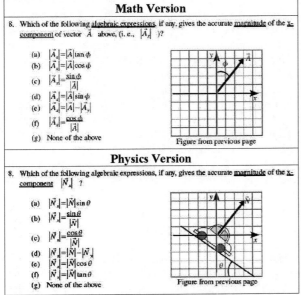

Figure 1. One-dimensional vector subtraction questions.

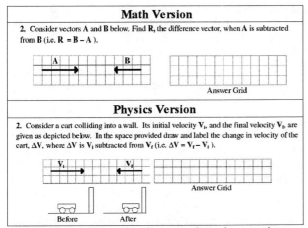

Figure 2. MC x-component algebraic expression questions.

The preliminary interviews were conducted by giving the math version of the vector free-response test first, followed by the physics version. Interviews were videotaped and students were asked to "think out loud" when working. Common incorrect responses given during the interviews and found in the literature [1] were used as distractors in the multiple-choice versions of the tests.

The MC tests were administered in the same manner as the interviews. Students completed and submitted the math version before receiving the physics version.

In the interests of shortening the MC test to a manageable length, while still remaining useful for instructors, we dropped several questions. For example, we dropped questions dealing with graphical magnitude and direction of vectors because students performed very well on these questions during interviews (60-100% correct). Our results are consistent with Nguyen & Meltzer's (2003) [1]. We also dropped the unit vector questions because of high student success rates (80-90% correct). Most of the incorrect responses from questions about unit vectors, magnitude, and direction resulted from poor arithmetic or attentiveness, and did not reflect a lack of student understanding on the topics.

We added to the MC tests a question dealing with the incorrect belief of a dot product having a direction. In interviews, students draw in vectors for the question on dot product magnitude. More on that finding will follow below.

RESULTS AND DISCUSSION

Early analysis of the multiple choice test indicates the mean overall performance on both tests are about the same, at 38% correct on math and 36% correct on physics. Below we only highlight some preliminary observations (see Figure 3) from the interviews and MC tests pertaining to the questions on: one-dimensional vector subtraction (Fig. 1),

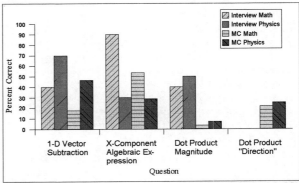

Figure 3. Preliminary interview (N=10) and MC (N=28) data.

x-component algebraic expression (Fig. 2), and dot product.

One-Dimensional Vector Subtraction

We observed a discrepancy between the number of correct answers on the math and physics tests for the one-dimensional subtraction question (see Fig. 3). Interview results show students using added tools when solving the physics problem than were used for the math problem. Many students answering incorrectly on the math test subtracted the magnitudes of the vectors rather than subtracting the vectors themselves. One student's responses to the math and physics questions are given in Figure 4. Her correct response on the physics question is consistent with that of other students who interpreted the minus sign as a marker of direction and implicitly labeled the directions in the physical system with a coordinate system.

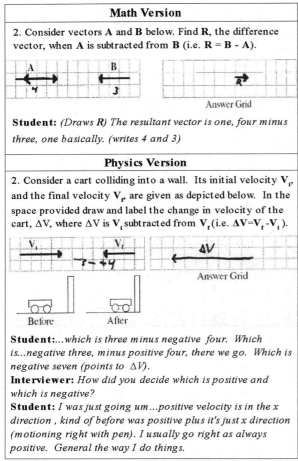

Figure 4. 1-D vector subtraction scanned interview sketches.

Of the seven students who answered correctly in the physics version, six used this method. More interestingly, two of the students who still responded incorrectly in the physics version also used the above tools and derived the correct response. However, these two students noticed the inconsistency between their previous incorrect math answer and their current physics answer. Their response to this inconsistency was to justify their previous math answer and disregard their reasoning on the physics version, resulting in answers like the one shown in the math version of Figure 4.

We also observe such large differences in performance between the math and physics one-dimensional vector subtraction question in the MC data (see Fig. 3). We found overall low performance, with 18% correct on math and 46% correct on physics. The most common incorrect answer on both versions is that shown in the math version in Figure 4. Learning between question versions could account for the increase in correct responses on the physics version since students completed the physics version immediately are the math version.

Algebraic Expression for a Vector Component

In the algebraic component-magnitude question, students are expected to find an algebraic expression for the *x*-component of a vector (Fig. 2). During interviews, nine students answered correctly on the math version and only three on the physics version (see Fig. 3). We often give students inclined-plane or ramp problems similar to this one, and presume students will take that information and appropriately label the known angle with reference to the vector, and coordinate system at hand. However, this subtle difference in versions proves to be a challenge for students. The majority of incorrect answers on the physics version result from inappropriate re-labeling of the angle: students place theta between the *x*-axis and vector **N** instead of the *y*-axis and vector **N**.

In the MC data, we find low performance for these questions, with 54 % correct on math and 29% correct on physics (see Fig. 3). The only incorrect choice selected on the physics version was choice "e" (see physics version answers in Fig. 2), indicating a large portion of students leaving our summer-session calculus-based introductory physics class still have difficulty with their relabeling of angles.

Dot Product

During interviews, students performed poorly on questions about dot product magnitudes on both the math and physics versions (see Fig. 3). Unexpectedly, two students drew in a vector for the dot product.

They were using vector addition tools to evaluate a dot product. Figure 5 shows a student using the parallelogram method of vector addition to find R, which was defined as the dot product of vectors **A** and **B** in the question. Both students consistently indicated direction of a dot product on the isomorphic physics question dealing with work.

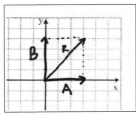

Figure 5. Incorrect interview response to the math dot product question. Students were first asked to draw in the vectors **A** = 3**i** and **B** = 4**j**. Then students were asked to find R, the dot product of vector **A** and **B**. In the MC questions vectors **A** and **B** are given.

We observed students performing extremely poorly on the MC dot product magnitude question, with 4% correct on math and 7% correct on physics. When asked to find the dot product of vectors **A** and **B** above the most common incorrect choice (~50%) was 3**i**+4**j**, indicating student use of vector addition tools to evaluate a dot product. The math version may prime students to use vector addition tools through its use of the non-neutral symbol R, which is typically used to represent a resultant vector. However, the same fraction of students incorrectly answered 3**i**+4**j** in the isomorphic physics version in the context of work, suggesting the math notation did not impact student responses.

The majority of students ended the semester by incorrectly applying "direction" to dot products. The MC tests results show students answering correctly only 21% in math version and 25% in the physics version on the dot product "direction" question (see Fig. 3). The MC math version of the dot product "direction" question is shown in Figure 6. The most common incorrect choice in both versions (~75% math and ~65% physics) was the arrow pointing in the direction of the vector sum of vector **A** and **B** (choice F in Fig. 6).

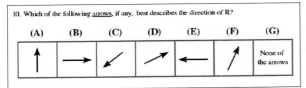

Figure 6. MC math dot product "direction" question. R is defined in the previous dot product magnitude question as being equal to **A·B**, where **A** = 3**i**, and **B** = 4**j**. The correct

answer is *G, none of the arrows,* because a dot product is a scalar and does not have direction.

CONCLUSION

Introductory students' overall performance on our isomorphic math and physics MC vector tests was very similar, 36% and 38% correct respectively. However, closer preliminary analysis of the interviews and MC questions reveals differences. Our introductory physics students often use different tools on the math and physics one-dimension graphical vector subtraction questions. In an interview setting students were observed in the math context to subtract vector magnitudes without regard to direction. When answering the physics question, students used the minus sign and an implicit coordinate system to label direction and appropriately subtract the vectors. MC data indicates a greater portion students subtracting vector magnitudes on the math version than on the physics version. During interviews we found student difficulties with relabeling of angles on an inclined-plane problem, specifically when the presented angle is the angle of the incline; MC data support this finding. Finally, we observed student confusion about dot products, including difficulties with "direction." Specifically, students used vector addition tools to evaluate a dot product in an abstract math context and in the physics context of work.

ACKNOWLEDGMENTS

We thank the instructors and J. R. Thompson for their time and support. We also thank D. E. Meltzer and P. S. Shaffer for their support. This research is supported in part by the National Science Foundation under grants DUE-0441426, DUE-0442388, and REC-0633951.

REFERENCES

1. N.-L. Nguyen and D. E. Meltzer, Am. J. Phys., **71** (6), 630 – 638 (2003).
2. R. D. Knight, Phys. Teach. **33** (2), 74-78 (1995).
3. J. M. Aguirre, Phys. Teach. **26** (4), 212-216 (1988).
4. J. M. Aguirre and G. Erickson, J. Res. Sci. Teach. **21**, 437-457 (1984).
5. P. S. Shaffer and L. C. McDermott, Am. J. Phys. **73** (10), 921-931 (2005).
6. N. S. Rebello, D. A. Zollman, A. R. Allbaugh, P. V. Engelhardt, K. G. Gray, Z. Hrepic, and S. F. Itza-Ortiz, in *Transfer of Learning from a Modern Multidisciplinary Perspective*, edited by J. M. Mestre, Information Age Publishing, Connecticut, 2005, pp. 217-250.

Using Students' Design Tasks to Develop Scientific Abilities

Xueli Zou

California State University, Chico, Chico, CA 95928

Abstract. To help students develop the scientific abilities desired in the 21st century workplace, four different types of student design tasks—observation, verification, application, and investigation experiments—have been developed and implemented in our calculus-based introductory courses. Students working in small groups are engaged in designing and conducting their own experiments to observe some physical phenomena, test a physical principle, build a real-life device, solve a complex problem, or conduct an open-inquiry investigation. A preliminary study has shown that, probed by a performance-based task, the identified scientific abilities are more explicitly demonstrated by design-lab students than non-design lab students. In this paper, detailed examples of the design tasks and assessment results will be reported.

Keywords: Experimental designs, lab skills, scientific abilities, lab instruction, physics education research
PACS: 01.50.Pa, 01.40.gb, 01.40.Fk

INTRODUCTION

This article describes experimental design tasks for students that have been developed, tested, and implemented in introductory physics labs. These design tasks are aimed at developing physics students' scientific abilities [1]. Detailed examples of the design tasks and preliminary assessment results will be reported below. This work is a collaborative effort with Eugenia Etkina and Alan Van Heuvelen at Rutgers University. For more information on scientific abilities and related rubrics, visit http://paer.rutgers.edu/scientificabilities; for more examples of different experimental design tasks, visit http://www.csuchico.edu/~xzou.

DEVELOPMENMT AND IMPLEMENTATION OF STUDENTS' DESIGN TASKS

Four different types of hands-on design tasks—observation, verification, application, and investigation experiments—have been developed for calculus-based introductory physics labs. Students working in small groups are engaged in designing their own experiments either to explore some physical phenomena, verify a physical principle, build a real-life device, solve a complex problem, or conduct an open-inquiry investigation. These design activities, including both qualitative and quantitative tasks, help students not only better understand concepts, but also think creatively and develop important workplace skills in design, complex problem solving, scientific investigation, teamwork, communication, and learning-how-to-learn. Identified scientific abilities in conducting the design tasks, related sub-abilities, and scoring rubrics are reported in detail in Ref. 1.

An *observation experiment* asks students to "play around" using some given equipment so as to explore and observe certain physical phenomena. Fig. 1 describes an example of a qualitative observation experiment used in a calculus-based electricity and magnetism course. These observation experiments are carefully selected and given to students before related concepts are discussed in class. In a traditional instruction setting, these observation experiments are usually conducted by the instructor as lecture demonstrations while associated concepts are introduced. For many students, the experiments are more like magic shows; they have no ideas about how

---

Can you generate currents without a battery?
Using the given materials, including a magnet, a Galvanometer, some wire coils, and wires, conduct some experiments so as to generate a current through the Galvanometer. In your lab notebook, please (1) sketch your experimental setup and describe what you have performed, (2) record in which ways the magnitude of the generated current could be changed, and (3) record in which ways the direction of the generated current could be changed.

FIGURE 1. Example of a qualitative observation experiment

CP951, *2007 Physics Education Research Conference*, edited by L. Hsu, C. Henderson, and L. McCullough
© 2007 American Institute of Physics 978-0-7354-0465-6/07/$23.00

they work. As a result, of course, it is difficult for students to construct the experiments as part of their own knowledge.

While using these experiments as design tasks, students themselves need to come up with some ideas and conduct the experiments. Students not only gain a better understanding of the phenomena, but no longer view them as magic. The experiments also arouse their curiosity to further study the phenomena, which is evident from the many questions students ask when the phenomena are discussed in subsequent classes.

Many observation experiments have been developed covering the important concepts in calculus-based introductory mechanics and electricity and magnetism courses. For some electrostatic phenomena, which are difficult for students themselves to perform successfully due to physical restrictions (e.g., room humidity) in a student lab room, digital videos have been made for students to observe and construct the concepts. Those videos are available at http://www.csuchico.edu/~xzou.

A *verification experiment* asks students to devise an experiment to test a conceptual or mathematical model. Fig. 2 describes a qualitative verification design task given in a calculus-based mechanics laboratory. In contrast to a traditional verification experiment, students are not provided with step-by-step instructions and ready-to-go setups. Instead, they are provided with certain materials and equipment, as well as a procedure outline (see below) [2] and detailed rubrics (see examples in Ref. 1) as scaffolding and real-time feedback.

Basic Procedure of Verification Experiment
Model: *State the model to be tested*
Plan: *Describe some phenomenon that can be analyzed using the model*
Prediction: *Predict the outcome of the situation*
Experiment: *Set up, conduct the experiment, and record, analyze data*
Conclusion: *Decide if experimental evidence supports the model*

Using two carts and the other given equipment on your table, design and perform some situations for which you predict that the force exerted by cart A on cart B is greater than the force exerted by cart B on cart A (or vice versa).

FIGURE 2. Example of a qualitative verification experiment

Students work in a small group of 3 or 4 to conduct their experiments and are encouraged to follow explicitly the basic process that guides practical scientists in conducting their own experiments. It is observed that many introductory physics students have difficulties conducting such verification experiments as a scientist would since they have no similar previous experiences. It is very important at the very beginning of lab to discuss with students what an acceptable scientific design is and what a poor naïve design is using real examples. In particular, a hypothetical-deductive reasoning [3, 4] and the power of prediction in science are addressed and emphasized explicitly with students [5, 6].

Many traditional verification experiments can be readily revised as students' design tasks. In addition, students' common-sense ideas or identified alternative conceptions can be used as "hypotheses" to be tested (see the example in Fig. 2). Ample evidence in physics education research has shown that it is very difficult to address students' "misconceptions" effectively. An experimental design task to test students' own "misconceptions" serves an innovative way to help students reconstruct their knowledge.

An *application experiment* requires students to build a real-life device or invent an experimental approach to measure some physical quantities. Fig. 3 describes a quantitative design task in measuring the maximum coefficient of static friction between a wooden block and a wooden board. To conduct such application designs, students are provided with some materials and equipment and encouraged to follow a basic process of effective problem solving (see below) [2]. A significant component of an application experiment is *evaluation*—students are required to devise an additional approach or experiment to evaluate their own design or result.

Basic Procedure of Application Experiment
Plan: *Define the problem and plan an experiment*
Assumption: *State any assumptions or estimations*
Prediction: *Theoretically predict outcomes of the experiment*
Experiment: *Set up, conduct the experiment, and record, analyze data*
Evaluation: *Perform additional experiment to evaluate the result*

Devise and perform an experiment to determine the maximum coefficient of static friction between a wooden block and a wooden board.

Design and perform an additional experiment to evaluate your result.

FIGURE 3. Example of a qualitative verification experiment

An *investigation experiment* is a more challenging case-study which is combined with all the other three design tasks described above. For example, towards the end of a calculus-based introductory electricity and magnetism lab, students are asked to investigate the magnetic field inside a current-carrying slinky before the concepts are addressed in class. In this lab, students conduct some selected observation experiments, come up with some conceptual ideas to account for the observations, identify possibly related physical quantities, design and conduct their own experiments to explore a quantitative model for the magnetic field inside the slinky, and finally design and conduct an additional experiment to verify or apply their own mathematical model.

ASSESSMENT OF STUDENTS' ACHIVEMENTS

Student experimental design tasks have been implemented and tested in calculus-based introductory physics courses with a class enrollment of about 48 students each semester at California State University, Chico since 2003. Each week the students have three one-hour lectures and are divided into two lab sections to conduct one three-hour lab. Typically, about 60% of the students are majors in engineering, 30% in computer science, and 10% in natural sciences.

To assess the impact of these innovative design tasks on students' development of conceptual understanding and scientific abilities, three types of assessment have been administered. To assess student conceptual understanding, for example, *Conceptual Survey of Electricity and Magnetism* (CSEM) [7], has been given to the second semester calculus-based physics students every seminar since fall 2003. Students' normalized gains [8] are consistently higher than 0.5 [9], which is typical for interactive engagement courses and almost double the national norm from the similar traditional classes [7].

How do students perceive the lab with design components? Do their perceptions mirror the goals of the design tasks? To answer these questions, a Q-sort instrument, *Laboratory Program Variables Inventory* [10], has been given to students since fall 2003. For example, the students from spring 2004 listed the followings as the three "MOST DESCRIPTIVE" features for the design lab: (1) *"Students are asked to design their own experiments." (2) "Lab reports require the interpretation of data." (3) "Laboratory experiments develop skills in the techniques or procedures of physics."* This result [11] shows that the goals of the design tasks had been successfully conveyed to the students.

To assess students' scientific abilities, a performance-based task was given to students in both design and non-design labs at the end of spring semester 2004. Both groups of the students had learned a semester of mechanics in a traditional format, but the students from the design-lab had an additional semester of electricity and magnetism (E&M) with design tasks. Since the content of the task (see below) is in mechanics, it is reasonable to expect

that both student groups would have similar performance; the training with design tasks in E&M would have little impact on how the students conduct the experiment in mechanics.

A short, theoretical, but *fake* "scientific paper" [12] was given to both student groups a week before the lab. The paper theoretically argues that for falling cotton balls the distance as a function of time should be $s = (1/2)kt^2$, where k is a constant but smaller than g = 9.8 m/s². The students were asked to design, conduct, and report an experiment to test the theory in a three-hour lab.

As shown in Table 1, most students from both the design and non-design labs explicitly followed the procedures emphasized in each lab to report their results. Then students' reports were analyzed anonymously and independently by two researchers using the identified scientific abilities (see Table 2) [1] and response taxonomies (see Table 3) developed particularly for this lab based on related rubrics [1]. Discrepancies between scores given to a particular ability in a particular report were carefully discussed between the two researchers and agreements were reached at the end. Percentages of students' scores on the six identified scientific abilities are shown in Table 4 for both the design lab and the non-design lab. One major difference shown in Table 4 is that about 70% of the non-design lab students did not make any predictions in their experiments, while about 90% of the design lab students made reasonable predictions. To compare the students' overall performance from

TABLE 1. Format of students' lab reports

Design Lab (Students = 42, Lab groups = 17)	Non-Design Lab (Students = 25, Lab groups = 10)
81% *explicitly* wrote their reports following the format emphasized in the lab for a testing experiment: **Model** **Plan** **Prediction** **Experiment** **Conclusion**	**96%** *explicitly* wrote their reports following the format emphasized in the lab: **Introduction** **Procedure (or Experiment)** **Conclusion**

TABLE 2. Scientific abilities used to analyze students' reports from the deign and non-design labs

1. Ability to identify the model to be tested
2. Ability to make a reasonable prediction based on the model to be tested
3. Ability to design a reliable experiment to test the prediction
4. Ability to analyze data appropriately
5. Ability to decide if or not to confirm the prediction based on the experiment results
6. Ability to make a reasonable judgment about the model

the two labs, the data was run by a Mann-Whitney U Test (two-tailed). The result is that the total score of the design lab is statistically significantly ($p < 0.002$) higher than the total score of the non-design lab. This preliminary result indicates that the scientific abilities demonstrated by the design-lab students are more explicit than those by the non-design lab students.

In summary, this paper has reported on four different types of hands-on design tasks that have been developed, tested, and implemented as formative assessment activities to advance physics students' scientific abilities. Preliminary assessments have shown that students exposed to the design lab perceived the basic goals of the design tasks, gained a good conceptual understanding, and demonstrated some desired scientific abilities in the given performance-based experiment. Carefully-designed studies will be further conducted to investigate how students' thinking processes are different in performing a design task and a traditional experiment. Moreover, based on the scientific abilities and scoring rubrics developed in this project, a simulated physics experiment is being investigated for potential use as a diagnostic tool to assess students' scientific abilities learned in the introductory physics laboratory.

TABLE 3. Example of response taxonomy for analyzing students' lab reports

Score / Ability	0	1	2	3
2. Ability to make a reasonable prediction based on the model to be tested	No *explicit* prediction.	"My prediction is that the distance formula $s=1/2kt^2$ is not valid because my experience is that $s=1/2gt^2$."	"If our model is correct the cotton balls will take longer to fall than the heavier object. Therefore $k<g$."	"If the model is true, then the slope of s vs. $1/2t^2$ should be constant $k<g$."

TABLE 4. Percentages of students' scores on six identified scientific abilities from the design lab and non-design lab

Score / Ability	Design Lab				Non-Design Lab			
	0	**1**	**2**	**3**	**0**	**1**	**2**	**3**
1. Identifying model	2	17	64	17	12	8	24	56
2. Making prediction	2	7	76	15	72	4	24	0
3. Designing experiment	0	9	17	74	0	28	24	48
4. Analyzing data	0	26	43	31	0	68	0	32
5. Confirming prediction	12	7	17	64	64	16	4	16
6. Making judgment	9	64	17	10	8	92	0	0

ACKNOWLEDGMENTS

Special thanks go to Eugenia Etkina, Alan Van Heuvelen, Sahana Murthy, and other members of the PAER group at Rutgers for their collaboration and support. The NSF funding for this project (DUE #0242845) is also greatly appreciated.

REFERENCES

1. E. Etkina, A. Van Heuvelen, S. White-Brahmia, D. T. Brookes, M. Gentile, S. Murthy, D. Rosengrant, and A. Warren, *Phys. Rev. ST PER* **2**, 020103 (2006).
2. The procedure outline is used as a guide not only for students to design their experiments but also help them explicitly acquire some basic procedural knowledge in conducting scientific experiments. It is observed that the procedure outline is important to lead students conducting their experiments successfully at the beginning of the lab. While their abilities and interests in design tasks grow, the outline is not needed any more.
3. A. E. Lawson, *Am. Bio. Teach.* **62**, 482–495 (2000).
4. A. E. Lawson, "The nature and development of hypothetico-predictive argumentation with implications for science teaching," *Int. J. Sci. Ed.* **25**, 1387–1408 (2003).
5. X. Zou and O. Davies, "Development and assessment of students' skills in designing and conducting introductory physics experiments: Part I," National Conference of American Association of Physics Teachers, Sacramento, CA, August 2004.
6. E. Etkina, S. Murthy, and X. Zou, *Am. J. Phys.* **74**, 979–986 (2006).
7. D. P. Maloney, T. L. O'Kuma, C. J. Hieggelke, and A. Van Heuvelen, *Am. J. Phys.* **69**, S12 (2001).
8. R. R. Hake, *Am. J. Phys.* **66**, 64-74 (1998).
9. X. Zou and S. Murthy, "Development and assessment of students' skills in designing and conducting introductory physics experiments: Part II," National Conference of American Association of Physics Teachers, Sacramento, CA, August 2004.
10. M. Abraham, *Journal of Research in Science Teaching* **19**, 155-165 (1982).
11. X. Zou, Y. Lin, D. Demaree, and G. Aubrecht, "From students' perspectives: A Q-type assessment instrument," National Conference of Physics Education Research 2004: Transfer of Learning, Sacramento, CA, August 2004.
12. T. Erickson, "Using fake scientific papers as a prompt," National Conference of American Association of Physics Teachers, Sacramento, CA, August 2004.

AUTHOR INDEX

A

Adams, B., 184
Aryal, B., 37
Aubrecht, II, G. J., 41, 61

B

Beichner, R., 3
Bernhard, J., 45
Bilak, J., 49
Black, K. E., 53
Bonham, S., 57
Bowman, C., 61
Brewe, E., 77
Brookes, D. T., 65
Bucy, B. R., 152

C

Conlin, L. D., 69
Corpuz, E. G., 73
Cummings, K., 3

D

Dancy, M., 77
Demaree, D., 81
Demel, J. T., 112
Diff, K., 85
Dubson, M., 128

E

Etkina, E., 88, 92, 96, 192
Eylon, B. S., 27, 31, 124

F

Feil, A., 100
Finkelstein, N. D., 128, 132, 136, 164, 204
Freuler, R. J., 112

G

Ganiel, U., 124
Gladding, G., 188, 200
Gray, K. E., 160

Gupta, A., 69, 104

H

Hammer, D., 69, 104
Harlow, D. B., 108
Harper, K. A., 112
Haynicz, J. J., 116
Heckler, A. F., 180
Henderson, C., 77, 120
Hite, Z. D., 112
Hmelo-Silver, C., 92
Hrepic, Z., 144
Hsu, L., 3

J

Jonassen, D., 144
Jordan, R., 96

K

Kapon, S., 124
Karelina, A., 88, 92, 96
Keller, C., 128
Kohl, P. B., 132
Kolodner, J. L., 3
Korkea-aho, J., 176
Kost, L. E., 136
Krakowski, M., 23

L

Lee, V. R., 23
Lising, L. J., 140

M

Mateycik, F., 144
McBride, D. L., 148
McCullough, L., 3
Mestre, J., 65, 100
Mountcastle, D. B., 152, 168
Murthy, S., 156

217

N

Nieminen, P., 176
Nokes, T. J., 7

O

Otero, V. K., 160

P

Perkins, K., 128
Podolefsky, N. S., 164
Pollock, E. B., 168
Pollock, S. J., 128, 136, 172

R

Rebello, N. S., 73, 116, 144
Redish, E. F., 104
Rosengrant, D., 88, 92, 96
Ross, B. H., 7, 11
Ruibal-Villasenor, M., 88, 92, 96

S

Savinainen, A., 176
Scaife, T. M., 180
Scherr, R. E., 69
Schuster, D., 184
Schwartz, D. L., 15
Scott, M. L., 188
Sears, D. A., 15
Shekoyan, V., 192
Sherin, B., 19, 23

Singh

Singh, C., 27, 31, 49, 196
Stelzer, T., 188
Stine-Morrow, E. A. L., 65

T

Tache, N., 85
Talikka, A., 176
Thompson, J. R., 152, 168
Torigoe, E., 200
Turpen, C., 128, 204

U

Undreiu, A., 184

V

Van Deventer, J., 208
Van Heuvelen, A., 88, 92, 96
Viiri, J., 176

W

Wittmann, M. C., 53, 208

Y

Yerushalmi, E., 27, 31

Z

Zollman, D. A., 37, 148
Zou, X., 212